D1202420

MUSHROOMS OF NORTH AMERICA

A CHANTICLEER PRESS EDITION

MUSHROOMS
OF NORTH AMERICA

ORSON K. MILLER, JR.

E. P. DUTTON & CO., INC., NEW YORK

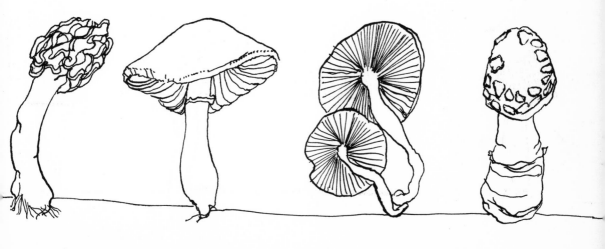

All rights reserved under International and
Pan-American Copyright Conventions

Published in the United States by E. P. Dutton & Co., Inc.
New York.

Fourth Printing

Planned and produced by Chanticleer Press, New York
Manufactured by Dai Nippon Printing Co., Ltd., Tokyo, Japan
Library of Congress Catalog Number LC-72-82162
SBN 0-525-16165-1

Contents

Simplified Picture Key to Major Groups of Fungi[1]

	Fleshy, with gills; with or without stalk	Agarics
	Fleshy, with blunt ridges or wrinkles; stalk present	Chanterelles
	Fleshy, with pores; central stalk; on ground	Boletes
	Tough to woody, with pores; with or without stalk, on wood	Polypores
	Fleshy to tough with cap and stalk and spine-like teeth	Teeth Fungi

[1] If in doubt after using the simplified key, use the Key to Major Groups of Fungi.

	Round to oval spore sacs, with or without stalk Star shaped with spore sac in center Like bird's nest	Puffballs Earthstars Bird's nests
	Green slimy top with dis-agreeable odor; chambered stalk	Stinkhorns
	Like coral, single or many erect fingers, sometimes with a massive base	Coral Fungi
	Jelly like, white, usually bright yellow or even black-ish-brown bodies, most often on wood	Jelly Fungi

	Pine cone-shaped top with ridges and pits; central stalk; on ground only in spring. Flesh brittle	True Morels
	Irregular to saddle shaped, wrinkled top; central stalk; on ground in spring, summer and fall. Flesh brittle	False Morels
	Cup-shaped, urn-shaped, or a flattened disc; stalk short or absent. Flesh brittle	Cup Fungi
	Various shaped; often stalked; usually small; flesh usually leathery to tough	Earth Tongues and Allies

How to Use the Keys

There are 71 keys in this book, including a key to the major groups of fungi and a key to each order or family, genus and each species in the text.

The keys are sequences of alternatives arranged so that you can key out (that is, track down) the group in which a fungus belongs, then the family or genus, and finally the species.

To determine the group, you use the *Key to the Major Groups of Fungi* (you may also get clues from the Simplified Pictorial Key). Start with choice 1 and read both statements designated by that number. At the end of each statement (on the right), will be a boldface number that will lead you to another pair of alternatives. You then choose the description that fits your specimen. If there is a boldface number at the end of the description, you proceed to the pair of alternatives with that number, and repeat the choosing process.

Thus, if your specimen has gills, you can immediately turn to the *Key to the Families of Gilled Mushrooms* (called Agarics). From the various pairs of alternatives in that key, you choose the first one that fits your specimen. If the statement describes the gills as "thick and fleshy, white to buff, waxy . . . .", you will find at the end of the description that your specimen is in the family *Hygrophoraceae*. Turn to the key to the *Hygrophoraceae* and choosing among the alternatives determine which *Hygrophorus* it is and read the full description of the species. If your specimen has features not mentioned in the key or in the full description of a species, it is probably not covered in this book.

A fungus should be identified while it is fresh or, if this is impractical, consult the section on collecting and studying mushrooms for directions on notes to be taken for use in making a later identification. If possible, collect several specimens showing various stages of development.

For unfamiliar terms consult the Glossary in the back pages as well as drawings in the Illustrated Glossary. Figure numbers in parenthesis refer to drawings in the Illustrated Glossary.

Key to Major Groups of Fungi

1 a. Fruiting body (mushroom) a cap with gills (Figs. 83–85), pores (Figs. 88–91), blunt ridges (Figs. 86–87), or teeth (Figs. 92–94); with or without a stalk, **2**

   b. Fruiting body a puffball (Figs. 95–96), earthstar (Fig. 97), "bird's nest" (Figs. 98), "stinkhorn" (Fig. 99), coral-shaped (Figs. 100–103), cup-shaped (Fig. 104), resembling a pine cone (Figs. 105–106), or jelly-like (Figs. 107–108), **6**

2 a. Cap with gills, stalk present or absent (Fig. 83–85), **agarics** (p. 24)

   b. Cap with ridges, pores or teeth, stalk present or absent, **3**

3 a. Cap with blunt, thick, often irregular ridges, stalk present (Figs. 86–87, **chanterelles** (p. 149)

   b. Cap with pores or teeth, stalk present or absent, **4**

4 a. Cap with teeth, stalk present; if stalk is absent fruiting body is on wood (Figs. 92–94), **teeth fungi** (p. 184)

   b. Cap with pores; on the ground or on wood, **5**

5 a. Cap fleshy with pores, with single, usually central stalk; on the ground (Fig. 88), **boletes** (p. 157)

   b. Cap with pores, leathery to woody, if fleshy then stalk eccentric, fused with other stalks or absent; usually on wood (Fig. 89–91), **polypores** (p. 178)

6 a. Round to oval or pear-shaped spore sacs (Figs. 95–96), star-shaped with a spore sac in the center (Fig. 97), or like a miniature bird's nest (Fig. 98), **bird's nests, earthstars,** and **puffballs** (p. 193)

   b. Another shape or form, **7**

7 a. Stalked with a green, slimy top, fragile, usually with a very disagreeable odor, Base surrounded by a fleshy volva (Fig. 99), **stinkhorns** (p. 190)

   b. Another shape or form, **8**

8 a. Resembling coral, single or many erect "fingers", sometimes with a massive base (Figs. 100–103), **coral fungi** (p. 153)

   b. Another shape or form, **9**

9 a. Pine cone-shaped top with ridges and pits (Fig. 105) and a central stalk; or with a wrinkled to smooth, sometimes saddle-shaped top (Fig. 106) and a central stalk, **true morels** (p. 212) and **false morels** (p. 213)

   b. Another shape or form, **10**

10 a. Cup-shaped, urn-shaped or a flattened disc on the ground (Fig. 104) or variously shaped as in Figs. 44–61, **cup fungi** (p. 216) and **earth tongues and allies** (p. 218)

   b. Jelly-like, white, yellow to blackish brown; on wood (Figs. 107–108), **jelly fungi** (p. 208)

Abbreviations Used in the Text

Seasons

Sp. = spring
S. = summer
F. = fall
W. = winter

Distribution

C = Canada
A = Alaska
PNW = Pacific Northwest
PSW = Pacific Southwest
RM = Rocky Mountains
CS = Central States
SO = South
SE = Southeast
NE = Northeast

Preface

For years it has been my conviction that a mushroom book should be published to satisfy the needs of all types of possible users: the casual observer, the ardent amateur mycologist, and the student of biology. Therefore I have included more than 680 species in these pages, with full descriptions of 422 species. A full key to the major groups of fungi is combined with 72 keys to the families, genera, and species. There is a simplified Pictorial Key to the major groups. Unfamiliar terms are explained in a Glossary and in many cases illustrated in an accompanying Illustrated Glossary. More than 280 species are illustrated in color. Every attempt is made to show the fungus in its natural habitat or to illustrate the diagnostic features which one should see in order to be sure of its identification.

The area covered by this book includes the continental United States and Canada. In addition, many of the species are found in adjacent regions of Mexico, in Europe and other parts of the world.

Since scattered areas in North America have weather patterns and climatic variations vastly different from the major continental climatic system, one may expect to encounter many of the species described here at different times from those given here. Thus, in the short rainy season in the Southwest almost all fleshy fungi fruit in one short period.

Each description in the text is numbered consecutively; the same number appears next to the species in the key, and if there is a color illustration, the number appears next to the plate. Figure numbers (e.g., Fig. 15) refer only to the illustrations in the Illustrated Glossary and the picture key. Only positive chemical reactions—odors, tastes, etc.—are given, such as amyloid (blue) spores in Melzer's reagent, or a bitter taste when the flesh of a mushroom is tested.

I especially acknowledge my debt to Dr. Alexander H. Smith who first introduced me to the fascinating field of mycology. His guidance, enthusiasm, high standards, and outstanding ability continue to be a constant source of inspiration to me. The constant companionship of my wife, Hope, her help in my field work combined with her mushroom dishes, tireless typing and refreshing ideas substantially contributed to the completion of this text. Robyn Morgan has produced excellent illustrations for the Illustrated Glossary. Ellen Farr contributed many welcome editorial comments and aided greatly in the final preparation of the manuscript. A number of excellent suggestions were made by Judith B. Tankard of the New York Mycological Club. Finally, the interest and encouragement of countless amateur mushroom hunters throughout North America has added much worthwhile information to this effort.

The majority of the color illustrations are my own. However, many excellent color illustrations were provided by Dr. Howard E. Bigelow, Dr. Kenneth Harrison, Dr. Robert T. Orr, and Mrs. Robert Scates. Their contributions are deeply appreciated.

ORSON K. MILLER, JR.

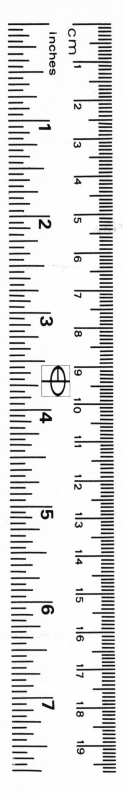

The measurements in this book are given in centimeters. To facilitate conversion of these measurements to inches an acetate ruler with both inch and centimeter markings is inserted. A reproduction of this ruler is also shown here.

Introduction

THE ROLE OF FUNGI

Unlike green plants, fungi lack chlorophyll and must live on organic matter. Many fungi break down or decompose various kinds of dead plants and animals, while others attack living organisms. Still others form a unique association with the roots of higher plants and are known as formers of mycorrhiza. Fungi thus play three basic roles in nature—as saprophytes, parasites, and mycorrhizal associates.

Saprophytes

The fungi included here live on dead organic matter, including leaves, needles, grass, limbs, logs, pine cones, nuts, dung, or even other dead or living fungi. In most cases a fungus will attack only a certain type of organic matter. *Agrocybe pediades,* for example, lives on dead grass, *Pleurotus ostreatus* on dead wood, and *Panaeolus separans* on horse dung. Some species are of course quite versatile: *Coprinus comatus,* the Shaggy Mane, is able to live on buried wood and other organic litter anywhere on earth, but it is always found on bare, hard ground or in grassy areas. Very often a mushroom will fruit each year on the same stump or log or under the same grove of trees. A knowledge of what the mushroom lives on is therefore essential to successful mushroom-hunting.

Parasites

Parasitic fungi invade and sometimes kill living plants and animals. The fungi treated in this book, however, are parasites only of living plants. *Armillariella mellea* will attack and kill the roots of trees, but it is also able to survive quite well as a saprophyte. Some of the woody polypores are parasites on trees. *Fomes annosus* is a destructive root parasite of many American forest trees; controlling it has been the subject of intensive research for many years. A number of other fungi, including several species of Pholiota, are weak parasites and are able to attack and kill only a host previously weakened by other factors. In some unusual cases, one fungus will parasitize another. *Boletus parasiticus* is found only on the puffball *Scleroderma aurantium,* and *Volvariella surrecta* fruits on living specimens of *Clitocybe nebularis.*

Mycorrhizal Associations

Only in recent years has this interesting, mutually beneficial relationship between fungus and higher plants been fully appreciated and understood. The vegetative fungus plant (the mycelium) forms a sheath of fungus tissue around the outside of the tiny rootlets of trees and shrubs. Some of the fungus cells penetrate between the cells of the rootlet itself, but the rootlet is not killed. Through the fungus the host tree or shrub is able to obtain nitrogen, phosphorous, and other nutrients which are otherwise relatively unavailable. The fungus, on the other hand, obtains moisture and protection even under adverse environmental conditions. Trees deprived of their mycorrhizal associates have much reduced growth and are totally incapable of competing with those that have their normal complement of mycorrhizal fungi. The mycorrhizal fungus, in many cases, cannot exist as a saprophyte or parasite and must have the living host. Thus, many of our mushrooms are associated with a wide variety of North American trees. For example, it is quite likely that most Boletes, Amanitas, and Russulas form mycorrhizal associations. *Gyrodon meruioides* is always found under or near species of ash; *Suillus luteus* is found under larch; *Chroogomphus rutilus* under or near five-needle pines, and so forth. Again, it is possible to anticipate where a given mushroom will grow by knowing the higher plants with which it is always associated. If one knows these relationships, he can distinguish a highly necessary and beneficial fungus from one that is a parasite and a potential threat to its host.

COLLECTING AND STUDYING MUSHROOMS

Proper equipment is necessary for successful field trips. Only then can mushrooms be correctly handled, identified, and treated before they are eaten. First it is essential to have a container with stiff sides such as a market basket or pack basket. Each kind of mushroom, called a "collection," should be wrapped separately in waxed paper, paper bags, or coffee cans. This enables the collector to keep edible, poisonous, and unknown species safely separated. Very often fragile mushrooms will fragment if they are all placed loose in a basket and allowed to shift about during collecting; and fragments of undesirable mushrooms might also be mixed up with the edible species. The separate packages also allow a collector to make notes on a small piece of paper, called a "label," indicating what the fungus was growing on, its color, and any other information necessary for identification.

A knife is essential—preferably a sheath knife with a three-to-six-inch blade. With such a knife, mushrooms can easily be removed from wood, hard earth, or any difficult location. It is of great importance to dig under the mushroom in order to get *the entire stipe* (stalk) out of the ground. For example, the base of the stipe, including the volva, must be checked in order to identify a mushroom of the poisonous genus Amanita.

Rounding out the essential gear for an average field trip is a 10X hand lens on a cord long enough to hang around the neck, along with a compass, appropriate maps, and some insect repellent. In order to give full protection to fragile fungi which I may wish to photograph later, I carry a couple of rectangular aluminum tins (2″ x 3″ x 6″) of the type used by backpackers and mountaineers. The specimen is wrapped in waxed paper and placed in the container; it will not shift, and it is protected from other fungi.

Essential in identifying mushrooms is ascertaining the color of the spore print. The stipe must be severed from the cap very close to the gills and placed with the gills or pores resting on a sheet of white paper. Black paper is sometimes recommended, but it tends to mask the delicate buff and pink coloration of the nearly white spores of many species. A strip of paper 1½ x 5 inches long is sufficient to record the spore print. If several fungi are set up for spore prints and wrapped in waxed paper during the collecting trip, each spore print will be visible when the collector arrives home. The specimens should then be laid out on a table and the unidentified ones separated from the rest. The color of the spore print should be noted in natural light and recorded. It is not unusual to leave a cap overnight to obtain a good spore print. Other characteristics should be noted and recorded on the field label or in some other way. Assigning a number to each collection will greatly aid the collector who plans to make a large collection of fungi.

The specimens should be examined systematically, starting, for example, with the cap and working down to the stipe base. One specimen should be cut lengthwise to expose the flesh. Note should be taken of any peculiar odor or taste, change in color when handled or bruised, and the color of the fresh flesh. The cap surface, for example, may be smooth or covered with fibrils (hairs), and it may range from dry or moist to viscid or slimy (glutinous).

To use the fungus descriptions in this book, first use the Keys to determine the

16

genus, and then determine what features identify members of a particular genus. Below is a guide to the details that should be noted on a fresh mushroom. The Illustrated Glossary may be consulted for features mentioned in the outline.

Cap.	Size; shape; color; surface dry, moist, viscid, slimy glutinous; surface with hairs or warts or without hairs (glabrous).
Gills.	Color; whether free, reaching to the stem but not attached (adnexed), attached (adnate), or running down the stem (decurrent); distant, close or crowded; thin or broad; edges hairy or smooth.
Annulus (ring).	Present or absent; single or double; superior or inferior.
Stipe (stalk).	Size;shape; color; single or fused with other stipes; surface with or without hairs or warts.
Flesh.	Color; texture soft, firm, or tough; changing color when bruised; latex color, when present.
Odor.	Mild, pronounced, or reminiscent of some familiar odor.
Taste.	Mild, pleasant, or disagreeable (usually bitter or acrid).
Host.	On the ground, wood, leaves, etc.; under hardwoods or conifers; or in a mixture of both or in lawns and grassy areas, on dung, etc.

MUSHROOM OR FUNGUS NAMES

Each fungus, like other plants, has a scientific name consisting of two parts, the genus and species, and it is always written in Latin. For example, *Clitocybe irina* (genus *Clitocybe* and species *irina*), is the name of a good edible. In the text this name is followed by the names or abbreviated names of individuals, for example: *Clitocybe irina* (Fr.) Bigelow and Smith. This notation means that the species *irina* was first described by Dr. Elias Fries of Sweden (in the 19th century), and that Dr. H. Bigelow and Dr. A. H. Smith, contemporary American mycologists, placed *irina* in the genus *Clitocybe*. *Irina* was previously placed in *Tricholoma* by Dr. Kummer and was called *Tricholoma irinum* (Fr.) Kummer. However, studies of the spores with modern microscopes and, more recently, electron microscopes, have revealed features found only in *Clitocybe* and not in the genus *Tricholoma*. So our notation says, in effect, that this is the fungus that Kummer placed in the genus *Tricholoma*, but that Bigelow and Smith now find it to be a *Clitocybe*. This uniform system enables us to recognize a species in a book in a foreign language or in an older mushroom book in English. In addition, we know at a glance the mycologist who described the species and those who subsequently worked on it. Their books and papers can then be easily referred to for further information.

MICROSCOPIC CHARACTERISTICS

Because an increasing number of ardent amateurs have access to microscopes, the size, shape, and structure of the spores and some information on the location and shape of cystidia (sterile cells) have been included. In addition, the color of spores and other tissues in an iodine solution known as Melzer's solution (see Glossary for the formula) has been given whenever the blue (amyloid) color reaction is of primary importance in distinguishing between closely related genera or species. In the white spored families of gilled mushrooms the amyloid reaction is only mentioned for those species with amyloid spores. The others are not amyloid.

Ascomycetes and Basidiomycetes

The large fleshy fungi are found in two classes of true fungi (Eumycophyta): the

Basidiomycetes and the Ascomycetes. The vast majority of the large fungi are Basidiomycetes. There are many Ascomycetes, but most are rather small, inconspicuous, and of no importance as edible fungi, although they often have great economic importance.

The chief difference between these two classes of fungi is in the way in which the sexual spore is borne. The Ascomycetes produce their spores in a saclike cell (Fig. 77), each sac or *ascus* (Fig. 77) containing a fixed number of spores. The number varies from species to species, but usually four or eight spores are produced in each sac or ascus. The spores carry the name of the class and are called *ascospores.* The location of the spore-bearing surface in each group, for example within the cup of a *Peziza* (Fig. 104) or in the pits of a *Morchella* (Fig. 105), is given in the text. If one blows gently over the surface of a mature *Peziza* cup, a dustlike puff of spores may appear. These are ascospores that have been violently discharged from each ascus. When the empty ascus is viewed with a microscope, a small lid called an *operculum* (Fig. 77) may be seen at the top; when this lid is forced open the spores are released. Ascospores, usually 30μ (microns) long or less, are released in this manner and then wind-blown to a new location. When they land in a favorable environment, they germinate much as a seed does and give rise to a new plant. The fungus plant consists of long threadlike cells (hyphae) collectively called a mycelium.

The Basidiomycetes, on the other hand, produce their spores on spinelike projections, called *basidia,* located on the outside of a club-shaped cell and usually at the top of the cell (Fig. 74). There are usually four, but sometimes one or two spores are born on each club-shaped (clavate) cell (Fig. 74). These spores also carry the name of the class and are called *basidiospores.* Basidiospores are either arranged in a spore-bearing surface or are borne in a spore mass. The spore-bearing surfaces of an agaric are the gills (Figs. 83–85), the pores of the bolete or polypore (Figs. 88–91), the top parts or the "fingers" of the coral fungi (Figs. 100–103), and the "teeth" of the teeth fungi (Figs. 92–94). The spore mass is found inside of the puffball (Fig. 25) or false puffball, or inside each "egg" of the bird's nest fungi (Fig. 35). In the groups which have exposed sporebearing surfaces (not puffballs or bird's nest fungi), the spores are violently discharged into the air and carried away by wind currents. If we place the spore-bearing surface (for example a mushroom cap) on a white sheet of paper, the forcibly discharged spores drop to the surface of the paper. The accumulated millions of these spores provide us with a spore print and reveal the color of the spores of each species. When the spore mass is borne within the fungus, as it is in the puffballs or false puffball, we must examine the spore mass or *gleba* (Fig. 25) at maturity to ascertain the spore color. The basidiospores are not forcibly discharged but merely ripen and fall off the basidium. In most puffballs the immature white gleba changes slowly to olive-brown, yellow-brown, dingy lilac, or some other color. The color change is due mostly to the color of the mature spores, which are often thick-walled and spiny (Fig. 68). Spores are distributed by wind, insects, animals, and so on. The means of dissemination is discussed under each section.

Spores

Spores are as variable in size and shape as are the seeds in green plants. They serve a similar reproductive function, but there the resemblance ends. The seeds of higher plants are complex structures with several tissues and many cells, whereas the spores of the fungi covered in this book are minute and usually consist of one cell. The cell is usually 5 to 10 μ (microns) long and rarely over 30 μ long. There are about 3,000 microns in one eighth of an inch; therefore all fungus spores must be viewed through a microscope. Spores have a number of shapes and are smooth or variously ornamented. The common shapes and the types of ornamentation are illustrated in Figs. 62–72.

We must use a microscope to observe the shape and ornamentation of the spore; but only a spore print is needed to distinguish a spore with an amyloid reaction from a non-amyloid spore. Place a drop of Melzer's solution on a microscope slide set on white paper. Using a razor blade, lift some spore powder from the spore print, and tip it into the drop of Melzer's solution. If the spores turn blue, they are amyloid. The blue reaction results when certain components of starch are present in the wall of the spore. It is helpful to use a magnifying glass or hand lens to observe the reaction. To test the reaction, a spore print from any *Lactarius, Russula,* or *Hericium* can be used because all of their species have spores with a blue reaction in Melzer's solution.

Cystidia

In the Basidiomycetes (but not including the puffballs) which forcibly discharge their spores, the basidia, discussed above, form a spore-bearing surface called a *hymenium;* for example, in the pores of the polypore (Fig. 1) and the gills of a mushroom (Fig. 3). Various types of sterile cells, called *cystidia* (Figs. 75–76, 79), are often found in the hymenium among the basidia, and a knowledge of these cystidia is often essential to determine a species. The presence and sometimes the shape of the cystidia are described in the text only when it is essential for identification. Cystidia may also occur on the cap or stipe and are often used by the mycologist to separate closely related species.

SEASONAL FRUITING OF MUSHROOMS

Many mushroom collectors hunt their favorite edibles during one season of the year. The morel collector, for example, looks for morels only in the spring. As a result, some collectors assume that the season for mushrooms is over when their favorite edible is gone. Yet mushrooms can be found throughout most of the year if one knows what to look for and where to look.

The warming trend in the spring, combined with moist weather, brings out the second largest number of species of mushrooms during the year: morels, false morels, cup fungi, and a number of gilled mushrooms. Most numerous among these are *Entoloma, Agrocybe, Coprinus, Conocybe,* and *Agaricus.* Most boletes, fleshy polypores, hericums, chanterelles, and many mushrooms are not in evidence at this time. In the late spring the number of fungi drops to the lowest point. As summer progresses, a slow build-up of fungi begins with the appearance of *Russula, Lactarius, Amanita, Inocybe, Laccaria,* a few coral fungi, and the first few boletes, to mention only a few. However, the typical spring species now disappear. A cool spell in August, accompanied by showers, will often bring out large fruitings of puffballs like *Calvatia cyathiformis* and *Lycoperdon perlatum. Clitocybe nuda* and *Coprinus comatus* (the "shaggy mane"), along with many boletes and polypores, will appear in great numbers. But hot weather quickly reduces a sizeable fruiting to a few scattered specimens, so the abundance of edibles is short lived at this time.

If the first cooling trend of the fall season, usually in September, is accompanied by fall rains, the largest number of mushrooms of the entire year will appear: for example, the various species of *Fuscoboletinus, Suillus, Leccinum,* and *Boletus* in the boletes; *Polyporus sulphureus, Polyporus frondosus,* and other polypores. The chanterelles, teeth fungi, including *Hericium* and *Dentinum repandum,* many puffballs, coral fungi, and stinkhorns all appear. At this time many of the mushrooms on wood appear, including *Pholiota, Naematoloma,* and *Pleurotus.* This large assemblage of fungi quickly dissipates when cold November nights plunge temperatures below freezing.

Winter is a very poor period for edibles. In areas of North America which experience periods of thaw and warmth during the winter, a few winter mushrooms can be gathered and eaten. The most notable of these are *Flammulina (Collybia) velutipes* and *Pleurotus ostreatus.* On the Pacific Coast winter fruitings of edibles are rather common.

North America is so extensive that it is obvious that local differences in patterns of fruiting will be encountered. This is especially true in mountains and mild southern climates where seasons are markedly shortened or lengthened. Close observation of local patterns of fruiting will enable a collector to find edible fungi throughout the year.

EDIBILITY AND PREPARATION

Because more people each year are discovering the delights of preparing as well as collecting mushrooms, one of the objectives of this guide is to provide information on the principal edible mushrooms in North America.

It should be stressed again that separate packages should be made of each species collected. I use a knife to lift the entire mushroom out of the ground. If it is a familiar edible, the lower stalk and the dirt-covered base may be cut off and fresh, dirt-free specimens wrapped together, lessening the work of cleaning at home. In hot weather the collections should be put in a cool place until they can be prepared for the table.

All fungi should be cleaned and checked, especially for the presence of worm holes or insects. Plunging specimens into cold water for a few minutes will drive out any insects. Each specimen should be cut in half to see if it contains worm holes. It is possible to find worm holes in the stalk but not in the cap of larger specimens. The infested part may be removed and the rest of the fungus eaten.

Usually the cap, upper stalk, and gills or teeth may be used in the agarics and hydnums. The pores of the young bolete are firm and edible; in older bolete specimens the pores become slimy when cooked. In most cases the flavor is not altered, but some people object to the texture. The tubes of boletes can easily be removed by running a finger, knife, or spoon underneath them; the tubes can then usually be cut out in one or two sections. The cap cuticle should be removed in any species which is viscid or glutinous.

In some polypores only a portion of the fungus can be eaten. *Polyporus sulphureus* has a tender margin (outer area) when young. The marginal area is cut off and used, but the rest of the fungus is much too tough to be eaten. The basal portions of some polypores may also be too tough and must be removed.

All of the coral fungus can be eaten unless it is wormy. This is a difficult fungus to clean because it is often partially buried in the soil. A soft bristle brush is useful in cleaning fungi of this type.

The puffball should always be cut in half down through the point of attachment to ascertain that it is a puffball and not a mushroom button. It should also be free from worms or worm holes. Some species have tough, leathery outer walls or skin. These should be peeled before cooking, but others need only be washed.

The morel and false morel should also be cut in half down through the point of attachment. The interior is hollow, and insects, and occasionally slugs, may often be found inside. If there are no worm holes, the entire morel may be eaten after the inside is carefully washed.

There are one or two exceptions to the general rules for the preparations of mushrooms for the table. *Coprinus,* the Inky Caps, *should not* be gathered and stored in the refrigerator. They must be cooked immediately; if not, the enzymes of the still-living fungus cause the gills to dissolve into an inky (or black) liquid, giving this group its name. When this happens, the fungus is neither good to eat nor easy to clean. To stop the enzymatic action it is necessary to simmer them for three or four minutes in a frying pan on medium heat. The high water content of this group will provide sufficient liquid for simmering. The liquid may be used as a stock for soups and gravies or it may be discarded. The fungus should then be refrigerated until needed.

For centuries the mushroom has been looked upon as a staple food by some and a gourmet's delicacy by others. Mushrooms, in fact, may be used for hors d'oeuvres, in soups, raw in salads, in sauces, relishes and gravies, as a garnish for meats and vegetables, as a vegetable, and as the major ingredient in casseroles, omelettes and main dishes. They may also be used as ingredients in bread and cookies, or in pickled or candied form. A selected group of recipes is included in the Appendix.

Dry and powdered mushrooms are used as a seasoning throughout the world. Notable examples of these are the Japanese Shiitake and Matsutake(*Lentinus edodes* and *Armillaria ponderosa*), dried boletes from a wide variety of countries, chanterelles and truffles from Europe, and dried Yung Nge and Muk Nge *(Auricularia)* from China.

By being dried, powdered, canned, or frozen mushrooms may be kept for use during periods of the year when they are not in season. In general, canning seems to be the most satisfactory method, followed by drying and lastly freezing; they are more apt to lose their flavor in the last two methods. Instructions for preserving mushrooms will be found in the recipes section at the back of the book.

MUSHROOM TOXINS

Many people believe that all toadstools are poisonous and that all mushrooms are edible. But there is no broad group of poisonous fungi. Toxins are found in many different families of mushrooms. The only safe way is to learn to recognize which are edible and which are poisonous.

Most toxic species are not fatal to man. Even when eaten in quantity, they produce only nausea, act as a laxative, or induce mild hallucinations. Unfortunately, a small number of mushrooms have toxins that are fatally poisonous while others affect the central nervous system and are very debilitating. They are discussed below, commencing with the most deadly. It should, however, be remembered that very few of the total number of mushrooms have been tested.

Major Groups of Toxins

GROUP I. Phallotoxins and amatoxins. These are complex polypeptide molecules consisting primarily of proteins composed of amino acids. When the polypeptides are first ingested, the taste of the fungus may be quite good. The toxin, however, enters the bloodstream and in several hours the faster acting phallotoxins are converted by liver enzymes into a compound which begins to attack liver cells. Only then, when it would no longer help to pump out the stomach, does the victim suffer extreme pain, profuse vomiting, lethargy, and distorted vision. Some hours later the action of the slower amatoxins is felt and during the interim the victim may even feel somewhat better. The onset of severe pain then continues from four to six days and often culminates in death. Even when death does not result, the illness lasts several weeks and may do permanent damage to the liver.

Known members of this group include *Amanita verna, A. virosa, A. phalloides, Galerina marginata, G. autumnalis,* and *G. venenata.*

GROUP II. Muscarine. The effect of this toxin is mainly to excite the parasympathetic nervous system, which results in the slowing of the heart, dilation of blood vessels, and constricting of the pupils of the eyes. Atropine has long been used to counteract the effect of muscarine in *Amanita muscaria* poisonings. There seems to be no further effect, and the poisonous properties once attributed to muscarine in *A. muscaria* are now largely discounted. However, the increased concentrations in species of *Inocybe* cause rather strong toxic reactions. Known members of this group include *Amanita muscaria, Inocybe patouillardi, I. fastigiata,* and *Clitocybe dealbata.*

GROUP III. Muscimol, ibotenic acid, pantherin, tricholomic acid, and related compounds. These are hallucinogenic substances that act on the central nervous system. They, and not the muscarine (which is also present in small amounts), comprise the chief toxins in *Amanita muscaria.* When a large amount of *A. muscaria* containing these toxins is ingested it often results in severe illness but victims usually recover.

Known members of this group include *Amanita muscaria, A. pantherina, A. solitaria, A. porphyria, A. citrina,* and *Panaeolus campanulatus* and *Panaeolus* sp.

GROUP IV. Psilocybin and psilocin. These well-known compounds are in the LSD (lysergic acid) family of hallucinogenic compounds. They have a strong hallucinogenic effect on the central nervous system, producing visions and smothering sensations, optical distortions; and some victims have reported experiencing religious or mystical experiences. The quantity ingested, physiological condition, personality, and mood of the subject are all apparently involved in the effect. Since as little as 5 mg. of *Psilocybe cubensis* can cause visual hallucinations in 15 minutes, it is obvious that the ingestion of large quantities of some mushrooms in this group could be fatal or cause severe toxic effects. Perhaps the most important factor is the presence of other toxic compounds *in addition to* the hallucinogenic agent in a given species. These may as yet be unknown or they may vary in amount in different strains of the same species.

Known members of this group include *Psilocybe cubensis, P. fimentaria, P. mexicana, P. semilanceata, P. caerulescens, Conocybe cyanescens, C. cyanopus,* and *Panaeolus foenisecii.*

GROUP V. The *Gyromitra* toxins. For many years the presence of helvellic acid was assumed to cause blood poisoning. It was common knowledge that boiling the mushroom would extract the toxin from the tissue. The water was drained off and discarded and some Gyromitras were then reported to be safe to eat. We now know that the helvellic acid is harmless. Instead a toxin called monomethylhydrazine, abbreviated MMH, is responsible. The procedure of boiling the mushroom does, indeed, reduce the levels of this toxin. MMH not only causes blood poisoning but also affects the central nervous system. Symptoms include diarrhea, vomiting and a loss of muscle coordination. Severe headaches and pains may linger for some time. The tolerance of individuals, the amount of MMH present in the fungus, and the method of cooking may combine to cause widely varying toxic reactions. There have been a number of fatalities, both here and in Europe, from ingestion of false morels. In the past we have also been unsure of the number of species of false morels and how to tell them apart, and this has added to the uncertainty of determining which false morels, if any, are safe to eat.

Known members of this group include *Gyromitra (Helvella) esculenta, G. fastigiata, G. brunnea,* and in all probability other closely related Gyromitras; perhaps *Tricholoma sulphureum* also contains toxins in this group.

GROUP VI. Gastrointestinal and other toxins. A number of unknown toxins are rarely fatal but act primarily as purgatives, causing nausea, hallucinations, or alcohol-like intoxication. Some of these could be fatal to young children or, if ingested in quantity, to anyone. This group comprises the largest aggregation of species and requires much additional study. The toxic effects are further discussed under the individual species.

A *partial list* of known members of this group includes *Boletus luridus, B. miniatoolivaceus, Clitocybe dealbata, Collybia dryophila, Coprinus atramentarius, Entoloma lividum, E. strictius, Hebeloma crustuliniforme* and *Hebeloma* sp's., the genus *Inocybe, Lactarius piperatus. L. rufus, L. torminosus, L. uvidus, Chlorophyllum molybdites* (also known as *Lepiota morgani*), *Marasmius urens, Omphalotus olearius, Ramaria formosa, Russula emetica, R. fragilis, R. densifolia,* and *Tricholoma pardinum.*

Basidiomycetes

The Gilled Mushrooms

Key to the Families of Gilled Mushrooms (agarics)

I. Spore print white to yellowish (green in one species)

1 a. Gills free from stalk (Fig. 6), **2**
　　 b. Gills attached to stalk (Fig. 5), **3**
2 a. Gills free; volva present; cup (Fig. 14) or pieces of volva in duff around stalk base; warts (Fig. 9) present or absent on cap; ring (Fig. 9) present or absent, p. 25 *Amanitaceae*
　　 b. Gills free; no volva of any kind; small scales (Fig. 23) present on cap, but no warts; ring present and often breaking free to form a loose ring, p. 34 *Lepiotaceae*
3 a. Gills thick and fleshy, white to buff, waxy when rubbed; cap often brightly colored; always on the ground; ring usually absent; stalk present, p. 38 *Hygrophoraceae*
　　 b. Gills thick or thin but not waxy; on ground or on wood, leaves, needles, dung, etc. stalk present or absent, **4**
4 a. On ground; usually in woods or near trees; no ring; flesh brittle; latex present or absent; stalk always present; spores reticulate (Fig. 64), turns blue in Melzer's solution, p. 46 *Russulaceae*
　　 b. On ground or on wood, leaves, needles, dung, etc.; in many habitats; flesh firm, soft, tough, etc.; stalk present or absent (Fig. 85); latex absent; spores colorless or blue but not amyloid reticulate in Melzer's solution, p. 60 *Tricholomataceae*

II. Spore print salmon to pink

1 a. Gills free from stalk; usually on wood; volva (Fig. 84) present or absent; spores smooth, subglobose to elliptical (Figs. 63–64), p. 105 *Volvariaceae*
　　 b. Gills attached; on the ground; volva never present; spores angular (Fig. 69) to long striate (Fig. 65), p. 107 *Rhodophyllaceae*

III. Spore print black to smoky gray

1 a. Gills decurrent (extend partially down the stalk) (Fig. 5), thick, fleshy, not turning to "ink"; never on dung; always on ground under or near conifers, cap tissue yielding amyloid blue reaction or not; spores subfusiform (almost spindle-shaped) (Fig. 65) without a pore; cuticle of cap (slimy) gelatinous (Fig. 81) to (dry) filamentous (Fig. 82), p. 109 *Gomphidiaceae*
　　 b. Gills usually free (Fig. 6), never decurrent, often becoming inky; often on dung or manured grass; in many habitats; cap tissue never has amyloid reaction; spores elliptical with pore at apex (Fig. 64); cap (never slimy) cuticle cellular (Fig. 80), p. 112 *Coprinaceae*

IV. Spore print purple-brown to chocolate-brown

1 a. Spores purple-brown, nearly round (subglobose) to elliptical with pore at apex (Figs. 63–64); gills attached (Fig. 5) to stalk; white to tan when young; found on wood, dung, or in grassy areas, p. 119 *Strophariaceae*
　　 b. Spores chocolate-brown, subglobose to elliptical, pore at apex (Fig. 64) present or absent; gills free from stalk (Fig. 6), gray to *pink* when young; in grass, on the ground, occasionally on dung, never on wood, p. 126 *Agaricaceae*

V. Spore print clay color; gills decurrent (Fig. 5)

On stumps, logs, and sticks usually of conifers; spores smooth, subfusiform (Fig. 65), p. 130 *Paxillaceae*

VI. Spore print bright yellow-brown, rusty-brown, cinnamon-brown to clay-brown

1 a. Spores bright yellow-brown to clay-brown, with pore at apex (Fig. 64); gills usually free (Fig. 6); often like the *Coprinus* in stature; cap cuticle cellular (Fig. 80), p. 132 *Bolbitiaceae*

   b. Spores rusty-brown to cinnamon-brown, without pore at apex; gills attached (Fig. 5); not like *Coprinus* in stature; cap gelatinous or with threadlike cells (Fig. 81–82), p. 134 *Cortinariaceae*

AMANITACEAE
The Amanitas

There are so many species of Amanitas in North America that including all of them here would be impossible. It is important, then, to know the characteristics common to all of the species. Amanitas all start as round or oval buttons surrounded by a protective tissue layer known as a universal veil. The young mushroom, or button, develops gills, cap, and stalk but these are small and incomplete. At this stage they have sometimes been mistaken for puffballs and eaten with disastrous results. If what seems like a young puffball is cut in half lengthwise and reveals a typical mushroom button, it should be evident at once that it is an *Amanita*.

When the stalk of the *Amanita* button starts to elongate, a very soft universal veil fragments and is carried up on the expanding cap as a series of warts (Fig. 9) or patches of tissue. If it is a tough membrane or universal veil, it is split by the expanding cap and stalk and nothing of the universal veil is left on the cap. Instead, a well-formed volva, or cup (Figs. 14–15), surrounds the base of the stalk. In the closely related genus *Limacella,* the universal veil is a thick, slimy gelatinous tissue, but the development is the same as in the Amanitas. *Volvariella* is the third genus in which one encounters an obvious volva in the gilled fungi. However, all Volvariellas have pink spores while *Amanita* and *Limacella* (in the Amanitaceae) all have white spores. A third important character possessed by *Amanita* and *Limacella* are free gills, so called because they do not touch the stalk (Fig. 6). Therefore, a fungus in the Amanitaceae must have white spores, free gills, and a universal veil which leaves a volva (cup) surrounding the stalk. If the volva is cottony or membranous it is an *Amanita.* A gelatinous universal veil and volva signifies a *Limacella.* If the volva fragments easily we must be careful to dig down under the stalk and lift the entire fungus out of the ground. The fragmented cup can often be seen in the surrounding soil. Two additional characters can be verified with a microscope. All the spores of the Amanitaceae are entire, smooth, and thin-walled (Fig. 62). Figs. 63–64 show spores which are thick-walled, ornamented, or are not entire and possess a pore at the apex. Lastly, if one cuts down through the gills of, preferably, a young fruiting body, the tissue in the center of the gill is divergent, that is, it grows outward from a central strand. This character must also be viewed microscopically.

Edibility. There are, indeed, edible species of *Amanita* in North America, and several of the same species have been eaten in Europe for centuries. But we have many more species in North America, many that are inadequately known, and, in addition, we have several poisonous species that closely resemble the edible ones. I would therefore not recommend collecting or eating any Amanitas. The poisons possessed by Amanitas are described under Mushroom Toxins and include toxins in Groups I, II, and III.

There is little doubt that these large, attractive, showy mushrooms have caused the majority of fatalities attributable to mushrooms and more than all other mushrooms combined. Interestingly enough, I have often encountered teeth marks made by rodents on a wide variety of mushrooms but I have yet to encounter a white Amanita that has been chewed on. The rodents have apparently learned their lessons well. They may shy away from the chlorine-like odor which is often very faint to strong.

Key to the Genera of the Amanitaceae

1 a. Volva and/or universal veil fibrillose (hairy) to membranous, *Amanita*
   b. Volva and/or universal veil glutinous, slimy, *Limacella*

Key to Amanita

1 a. Volva a persistent, saclike cup (Figs. 14–15) surrounding the stalk base, **2**
   b. Volva in fragments over the bulb (often not seen) (Figs. 16–20) or in adjacent soil or tightly appressed to the stalk, **7**
2 a. Fruiting body pure white, no warts on cap, *1. A. virosa*
   b. Fruiting body with colored cap, **3**
3 a. Cap covered with warts (Fig. 9) or a large patch of volval tissue, **4**
   b. Cap free of warts or volval patches, **5**
4 a. Cap slender, dark brown, covered with gray warts; ring absent; widely distributed, *2. A. inaurata*
   b. Cap robust, yellow-brown, covered with a large central white volval patch; ring present; Pacific Coast only, *A. calyptroderma* (see *A. inaurata*)
5 a. Veil absent; cap dark brown to orange-brown or yellowish brown, *3. A. vaginata and varieties*
   b. Veil present (Fig. 9); cap green or bright yellow to orange, **6**
6 a. Cap pale yellowish green to green; veil white; gills white, *4. A. phalloides*
   b. Cap bright yellow to orange; veil yellow; gills yellow, *5. A. caesarea*
7 a. Cap and entire fruiting body white; or with a noticeable chlorine-like smell, **8**
   b. Cap with a definite pigment; no chlorine-like smell, **9**
8 a. Odor of chlorine, usually strong; cap covered with small sticky warts and hairs, *6. A. chlorinosma and related species*
   b. Odorless; cap covered with pyramid-like warts, stocky hairs over margin, *7. A. cokeri*
9 a. Cap brown to reddish brown, **10**
   b. Cap light yellow to yellow, orange, green or red, **13**
10 a. Stalk gray or pinkish with reddish brown stains, **11**
   b. Stalk white, **12**
11 a. Stalk covered with pale gray hairs; cap livid brown with large, gray warts, *8. A. porphyria*
   b. Stalk pinkish and stained reddish brown especially near base; cap light reddish brown with olive-gray warts, *9. A. rubescens.*
12 a. Bulb cleft or split longitudinally (Fig. 19); cap dark brown; spores round, with amyloid blue reaction, *10. A. brunnescens*
   b. Bulb not split; (Figs. 15–17, 20) cap light to dark (sometimes yellow brown) brown; spores elliptical, non-amyloid, *11. A. pantherina*
13 a. Cap green, yellowish-green or very light yellowish, **14**
   b. Cap yellow, orange to red, **15**
14 a. Cap green to light green with *pink-tinted* volval patches. *12. A. citrina*
   b. Cap light yellowish with white volval patches, *13. A. gemmata*

15 a. Cap large (8–24 cm broad), warts white; volva white, **16**

    b. Cap small (3–10 cm broad), warts yellow; volva yellow, **17**

16 a. Stalk base just above bulb with several concentric rings (Fig. 17), *14. A. muscaria*

    b. Stalk base lacking the concentric rings (like Fig. 16), *A. frostiana* (see *14. A. muscaria*)

17 a. Cap bright orange to yellow-orange; veil whitish; flesh and cuticle of stalk near base white, *15. A. flavoconia*

    b. Cap deep, even yellow; veil yellow; flesh and/or cuticle of stalk tinted red, *16. A. flavorubescens*

1 AMANITA VIROSA   Poisonous (deadly)/Common

(Fr.) Quel. ["Destroying Angel"]

Pure white, viscid; ring; saclike volva; no chlorine smell

Cap 3–9.5 cm broad, conic to egg-shaped, convex to flat, viscid when wet to sticky, smooth, *pure white,* margin smooth, not striate or only obscurely so, no volval warts, stains yellow in 3 percent KOH (caustic soda). Flesh firm, white. Gills close to subdistant in age, free, narrow, *white.* Stalk 14–24 cm long, 10–23 mm wide; enlarging gradually to an enlarged base, finely fibrous above ring, hairy below, white. Volva white, membranous, persistent, saclike, and free from the stalk, often 3–4 cm high. Veil membranous, smooth, faintly striate on upper surface, leaving a tattered or skirtlike superior ring, which on occasion is almost missing. Odorless. *Do not taste this fungus.*

Spores 8–10 (12) x 6.5–9.5 μ round, almost round to short elliptical, thin-walled, amyloid blue reaction, *white spore print.*

Solitary or in same groups under hardwoods or in mixed woods in Sp. and S., but mostly in F.; widely distributed in eastern North America, rare but present on the Pacific Coast.

A. verna (Fr.) Quel., a species with round, amyloid-reacting spores, has been described along with a two-spored (that is, the number of spores per basidium; see Fig. 74) species, *A. bisporigera* Atk. They are *all deadly poisonous* and the characteristics described above apply equally well to all three species. I have included a spore range for the group. Usually one collection will yield spores of a consistent shape (that is, round to almost round *vs.* elliptical).

A variant with a very faint chlorine-like odor suggests the possibility of the hybridization or speciation in the SE and the need for further study of the toxins in this group.

2 AMANITA INAURATA Secr.   Nonpoisonous/Infrequent

Cap brownish black, viscid with gray warts; no ring; saclike white volva

Cap 4–12 cm broad, egg-shaped to convex to flat with a low knob in age, viscid, brownish black at first to brownish gray in age, covered with gray warts, margin striate. Flesh thin, soft, white. Gills close, free, with white minute hairy edges. Stalk 3–8 cm long, 10–12 mm thick, enlarging slightly toward base, covered with flattened gray hairs. Volva membranous, persistent, white with orange-brown stains in age, saclike, not attached to stalk. Veil absent.

Spores 10–15 μ round, thin-walled, non-amyloid reaction, white spore print.

Single to scattered, usually under conifers, in S. and F., widely distributed.

This fungus has often been reported as *Amanitopsis strangulata* Fries; the lack of a veil, dark cap with gray warts, and the saclike white volva are the distinctive characters of this stately fungus. *Amanita calyptroderma* (Atk. and Ballen) is a very large, Pacific Coast Amanita with a large rigid cup, conspicuous central volval patch on an orange or orange-brown cap, and a thick veil with elliptical spores (9–11 x 5–6 μ).

3 AMANITA VAGINATA (Fr.) Vitt. ["Grisette"]   **Nonpoisonous**/Common

Cap gray to yellow-brown, viscid, without warts; no ring; saclike white volva

Cap 5–9 cm broad, nearly conic, flat in age often with a small knob, without hairs, viscid, a conspicuously striate margin, warts absent, brown, gray-brown, yellow-brown, orange-yellow to yellowish. Flesh soft, thin, white. Gills close, fairly well separated in age, free, narrow, white. Stalk 10–20 cm long, 10–15 mm wide, white, smooth or with some flattened hairs over the center, enlarging to an oval or club-shaped basal bulb. Volva white, membranous, saclike, persistent, unattached to stalk. Veil *absent.*

Spores 7–14 x 7–13 μ round, thin-walled, non-amyloid reaction, white spore print.

Single to numerous under both hardwood and conifers in Sp., S. and early F.; widely distributed.

A. fulva (Schaeff.) Pers. is based on the orange cap variant of this species. Several varieties (for example, var. *alba* and var. *livida*) have also been described, using cap color as the criterion. The spores vary in size. *Amanitopsis* is an old genus name for Amanitas without a ring. I have found the gray-brown cap most commonly under western conifers and the orange or yellow-brown cap in eastern forests. Mr. Andrew Norman of New York and some of his friends eat this Amanita and report that it has excellent flavor. I would not recommend eating Amanitas for any amateur mushroom collector.

4 AMANITA PHALLOIDES Fries ["Death Cap"]   **Poisonous (deadly)**/Rare

Cap pale green, viscid, warts absent; ring; saclike volva; no chlorine smell

Cap 7–15 cm broad, convex, viscid, smooth, *pale yellowish green to greenish,* flattened radiating hairs but margin not striate, *no warts.* Flesh firm, white to light green just below cap cuticle. Gills close, free, broad, white. Stalk 8–14 cm long, 10–20 mm thick enlarging somewhat toward the base, white, smooth. Volva membranous, *persistent, saclike, not flattened to stalk, white.* Veil membranous, free, remains as superior ring hanging skirtlike (Fig. 9) on stalk. Odor slightly disagreeable.

Spores 8–11 x 7–9 μ almost round, amyloid blue reaction, thin-walled, white spore print.

On ground, most likely under European trees or plantations established in America in S. and F.

The green viscid cap without warts, white saclike volva, and almost round spores giving amyloid reaction are the important field characteristics. Dr. Harold Burdsall collected it on one occasion under Norway spruce in Laurel, Maryland. In America, *A. brunnescens* has often been mistaken for *A. phalloides* (Coker, 1917).

5 AMANITA CAESAREA (Fr.) Schw.   **Nonpoisonous**/Numerous
["Caesar's Mushroom"]

Cap yellow-orange, viscid, no warts; gills yellow; ring yellow; white saclike volva

Cap 5–15 cm broad (recorded up to 30 cm), conical at first, convex, sometimes with a low knob, viscid, smooth with long striations toward the margin, margin bright yellow shading to orange or orange-red toward the center, no warts. Flesh firm, white, yellowish just under cap cuticle. Gills close, free, broad, *yellow.* Stalk 7–20 mm thick, nearly equal, hollow, dry, smooth, without hairs (glabrous) or thinly cottony (floccose), yellow. Volva membranous, persistent, *white* cup enclosing the base at maturity. Veil membranous, soft, yellow to orange, a persistent, superior (above the middle of the stalk) skirtlike ring is formed.

Spores 7.5–14.5 x 5–8.5 µ oval, thin-walled, non-amyloid reaction, white spore print.

Scattered in groups, sometimes in fairy rings under hardwood forests in S. in PSW, SO, SE, NE, CS and adjacent C.

A. flavoconia Atk. looks similar but the volva breaks up into yellow pieces. The gills are white with a nearly white to buff stalk. *A. caesarea* is one of the truly beautiful mushrooms encountered in eastern woodlands and only recorded recently from the Southwest. Often recorded as edible but there are too many similar North American Amanitas for me to recommend it for the table.

6 AMANITA CHLORINOSMA (Austin) Lloyd   Poisonous/Common

Cap white, powdery; volva has powdery patches; odor of chlorine

Cap 7–30 cm broad, convex to nearly flat in age, *covered with a thick, sticky, powdery layer,* adorned with small warts, white to dull white. Flesh firm and white throughout. Gills crowded, free, broad, white to cream color with minutely hairy edges. Stalk 10–30 cm long, 10–25 mm thick, enlarging toward base which forms an egg-shaped bulb, and a rootlike base set in the soil; surface covered with a thick, powdery layer. Volva remains only as powdery patches on the bulb. Veil mealy and sticky, ring superior but not always persistent at maturity. *Odor of chlorine,* usually strong.

Spores 8–11 x 5–6 µ ellipsoid, thin-walled, amyloid reaction, white spore print.

Under hardwood forests in lower NE to SE and SO during S. and F.

This robust, stately *Amanita* is distinguished by the powdery white cap and stalk and the strong, disagreeable odor of chlorine. No white *Amanita* should be eaten or even tasted. A dark gray to smoke-gray cap species in eastern North America, *A. onusta* (Howe) Sacc., also has a strong chlorine-like smell and amyloid-reacting spores 8–11 x 5–8 µ. It has often been called *A. cinereoconia* Atk., which has a *powdery,* finely cotton-like surface on the gray cap, narrow, amyloid-reacting spores (8.5–12.5 x 4.5 6.5 µ), a similar chlorine smell, and is found in the same range.

7 AMANITA COKERI (Gilb. and Kuehn.) Gilb.   Edibility unknown/Numerous

Cap ivory-white, brown warts, viscid; volva in patches; no chlorine smell

Cap 6–15 cm broad, large, convex, flat in age or even slightly depressed, viscid, shiny ivory-white, covered with pyramid-like or brownish volval warts which grade into dense, sticky hairs over the margin and hang from it when young. Flesh firm, white. Gills close together to fairly well separated in age, narrowly free or attached, white. Stalk 6–14 cm long, 15–30 mm thick, white, nearly smooth above ring, loose sticky hairs below, widening to a basal bulb which has recurved scales arranged in several circles just above it (Fig. 20); the bulb and lower stalk often have pinkish brown stains and the bulb often has shallow cracks and splits. Volva adheres to the bulb, splitting into irregular patches. Veil white, fibrous, separating unevenly and leaving long strands attached in taffy-like fashion to a wide area of the stalk, ring superior, hanging skirtlike. Odorless.

Spores 10–13.5 x 7–9 µ ellipsoid, thin-walled, amyloid reaction, white spore print.

Single, several or scattered under Virginia pine-hardwood stands in SE in F.

An often used incorrect name for this species is *Amanita solitaria* (Bull. ex Fr.) Merat. Following hard fall rains, this beautiful, stately fungus can become quite smooth and have only scattered warts. I have found it most commonly during mid-September in Maryland and Virginia. Like the other white Amanitas it should not be considered edible. *A. silvicola* Kauf. of the PNW under conifers with a pure white, fluccose, viscid cap and amyloid, elliptical spores 9–12 x 5–6 µ is a similar species.

8 AMANITA PORPHYRIA   Nonpoisonous/Infrequent

(Alb. and Schw. ex Fr.) Secr.

Cap drab brown, viscid, large flattened warts; ring ashy gray; volva in patches

Cap 4–9 cm broad, convex to broadly convex, viscid, fine radial hairs, *deep drab brown to livid brown* with a few *large, gray to pinkish gray, flattened patches,* margin incurved at first, straight in age. Flesh firm, white, unchanging when bruised. Gills close, free but narrowly so, and white. Stalk 6–10 cm long, 8–20 mm thick, enlarging slightly toward base, terminating in a round to flattened bulb that has a distinct collar, evenly gray to drab brown above and below ring from bands of flattened hairs; volva fragile, forming a collar several mm high and adhering in scattered patches to the bulb. Veil *ashy gray* remaining as a hanging, superior ring.

Spores 7–9 μ round, thin-walled, amyloid reaction, white spore print.

Scattered under mixed hardwood-conifers and conifer stands in S. and F.; widely distributed. The livid brown cap, ashy gray large volval patches, ashy-colored stalk and round bulb distinguish this from other Amanitas. *Amanita spreta* Pk, is also gray, lacks a basal swelling (Fig. 18) but has a thick white volva and spores 11–12 x 6–7 μ. It is found in eastern North America.

9 AMANITA RUBESCENS   Nonpoisonous/Common

(Pers. ex Fr.) S. F. Gray ["Blusher"]

Cap red-brown, pinkish warts; flesh bruises red; stalk at base flushed red; volva in reddish patches

Cap 5–15 cm broad, convex or sometimes knobbed (umbonate), viscid to sticky, dull reddish brown shaded with buff or gray, covered with olive-gray to whitish or even pinkish irregular warts often with some of them clustered along the obscurely striate margin. Flesh soft, white *staining reddish when bruised.* Gills close, white, nearly free, soon stained pinkish to red. Stalk 8–20 cm long, 8–20 mm thick, enlarging downward to a swollen oval basal bulb, tinged dingy pinkish above ring and usually with more *reddish stains below.* Volva not persistent, consisting of reddish pieces, some adhering to the bulb and some in the surrounding dirt. Veil membranous and pallid, staining reddish, fragile, leaving an irregular or shredded superior ring.

Spores 7.5–10 x 5–7 μ elliptical, thin-walled, dark amyloid reaction, white spore print. Sterile cells on gill edge abundant and balloon shaped. Grows alone or in groups, under hardwoods in eastern North America mainly in S. but also in Sp. and F.

The red stains, convex knobbed shape of the cap, and the gray warts are the distinguishing characteristics. Although it is rated excellent in many European guides, too many Amanitas are closely related to this one to permit me to recommend it as edible.

10 AMANITA BRUNNESCENS Atk.   Poisonous/Common

Cap brown, viscid, with cottony warts; stalk base split, white

Cap 3–15 cm broad, convex, flat, viscid, dark brown, margin covered with white, cottony warts and patches, faintly striate. Flesh thin, white, staining reddish brown. Gills close, free, broad, white. Stalk 4–13 cm long, 8–20 mm thick, enlarged downward with a large oval basal bulb which is conspicuously split or cleft longitudinally (Fig. 19) white, smooth above ring, hairy below. Volva adheres in irregular patches to the basal bulb and sometimes an obscure collar is formed around top. Veil membranous, white, breaking irregularly to hang in shreads or patches as a superior ring or sometimes pulling free from the stalk altogether.

Spores 7–9.5 μ round, thin-walled, amyloid reaction, white spore print.

Scattered and common under hardwood and mixed woods throughout eastern North America as far north as the boreal forest in S. and F.

The white cleft basal bulb and viscid brown cap with white warts are distinctive characteristics. Less common is a pallid variety with a very light tan to whitish cap.

11 AMANITA PANTHERINA   Poisonous (deadly)/Common

(D. C. ex Fr.) Schumm.

Cap brown, viscid, warts pointed, white; basal bulb with single ring; volva in white patches

Cap 5–12 cm broad, convex to nearly flat in age, viscid, soft white often pyramid-like warts over a light to rich brown surface that is usually buff to yellowish just at the striate margin which may have veil patches adhering to it. Flesh firm, white, buff just beneath cap cuticle. Gills close to crowded, free, white, edges finely scalloped. Stalk 6–12 cm long, 1–2.5 cm thick, white, nearly smooth above and hairy below the ring, gradually enlarging to the egg-shaped basal bulb (Fig. 16.) which has a roll or collar where stalk and bulb meet; it occasionally has a single ring above the bulb. Volval tissue includes a cottony roll and a thin cottony layer on the bulb, which is inconspicuous and may rub off in the dirt. Veil adheres closely to the cap and forms a series of regular patches on the margin of the cap; in age it forms a membranous, superior, hanging ring.

Spores 9–12 x 6.5–8 μ elliptical, thin-walled, ny amyloid reaction, white spore print.

Several to many under conifers or mixed woods, in Sp., S. and F., widely distributed.

This is the common deadly poisonous *Amanita* of the RM, PNW region. In the PNW it hybridizes with the nonpoisonous *A. gemmata* (Fr.) Gillet (same as *A. junquillea* Quél.), producing a series of confusing forms with caps ranging from rich brown to light yellow and containing widely variable amounts of toxin. This example is a strong argument for avoiding all Amanitas in North America.

12 AMANITA CITRINA (Schaeff.) S. F. Gray   Poisonous/Common

Cap green, viscid, with pink-tinted flat warts; volva flattened to round bulb

Cap 4–10 cm broad, convex to convex with a broad knob, viscid, *greenish yellow* to light green with a whitish margin, covered irregularly with flattened white to pallid volval patches which are *pink tinted* but may be washed off in wet weather, margin incurved at first, not striate. Flesh soft, white. Gills close, free, white, and with minutely hairy edges. Stalk 8-12 cm long, 8–15 mm thick, enlarging downward with a nearly round basal bulb, sometimes with a noticeable collar, white, smooth above and cottony below ring. Volva soft, flattened to bulb, sometimes free at the top, forming a collar. Veil fragile, membranous, white to buff, superior, persistent, hanging about or flattened down to the stalk.

Spores 7–10 μ round, thin-walled amyloid reaction, white spore print.

Single to scattered under conifers, hardwoods, or mixed woods in eastern North America in late S. through late F. in SE and S. Very abundant under eastern white pine or pure stands of Virginia pine or SE pine hardwood forests. I have seen hundreds of specimens in a day in cold wet late F. weather in Maryland.

13 AMANITA GEMMATA (Fr.) Gillet   Poisonous to Nonpoisonous/Common

Cap yellow, viscid, warts white; stalk with small round bulb and collar

Cap 3–8 (up to 12) cm broad, convex to flat or even slightly depressed in age, viscid to glutinous, smooth, dull yellow often with a pinkish cast, covered with whitish volval warts, sometimes tinted cinnamon in age, but readily washed off in wet weather; margin striate, often covered with sand, clay, and debris accumulated during expansion. Flesh soft, thin, white. Gills close, free, narrow, white. Stalk 5–14 cm long, 5–20 mm thick, nearly equal, white, smooth above ring, sparse hairs below; small round, white, basal bulb with a narrow collar. Volva flattened down to bulb, or pieces left in the soil, forming a narrow collar. Veil fragile, membranous, forming a ragged, white, superior ring or absent at maturity.

Spores 7.5–11 x 6–9 μ short elliptical to almost round, thin-walled, no amyloid reaction, white spore print.

Single but more often in groups under mixed hardwood-conifer forests in Sp., S. and F., widely distributed. I have found it under pure stands of Virginia pine in Maryland and lodgepole pine in Idaho.

This often tall but fragile *Amanita* hybridizes with *A. pantherina,* producing a series of forms between the two with varying amounts of toxin as well as intermediate cap coloration. I therefore consider this entire group poisonous and recommend again that *no Amanitas in North America be considered edible. Amanita russuloides* Peck is closely related but has a basal bulb like the one in *A. muscaria* and narrow spores (8–12 x 5.5–7.5 μ). *A. cothurnata* Atk. is nearly white, closely related to *A. gemmata,* and has similar spores.

14 AMANITA MUSCARIA   Poisonous (hallucinogenic)/Common
(Fr.) S. F. Gray ["Fly Agaric"]

Cap yellow, orange to red, viscid, warts white; stalk with several concentric rings above bulb

Cap 8–24 cm broad, convex to flat in age, viscid, adorned with whitish warts or small patches, and at first hanging marginal veil remains (all part of the universal veil; straw-yellow, orange to bright blood-red, usually darkest at center. Flesh firm, white throughout. Gills crowded, free, broad, white, minutely hairy edges. Stalk 8–15 cm long, 20–30 mm thick, enlarging toward base and becoming bulbous at the base, white, covered with silky hairs. The only remains of the volva are 2 or 3 concentric rings (Fig. 17.) above the bulb. Veil membranous, superior, several cm long, white, persistent.

Spores 8–11 x 6–8 μ ellipsoid, thin-walled, no amyloid reaction, white spore print.

Scattered or abundant under hardwoods and conifers from Sp., S. to F.; widely distributed.

A. frostiana (Pk.) Sacc. is a SE species that has a thinner stalk, lacks the concentric rings above the bulb, and has no amyloid reaction, almost round, spores 7.5–9.5 x 7–9 μ. I know mushroom hunters who have eaten large portions of *A. frostiana* without ill effect. In addition, certain areas have races of *A. muscaria* that seem to have less or different proportions of toxins than other areas. However, this is a dangerous fungus and should be avoided. *Amanita parcivolvata* (Pk.) Gilbert is closely related to *A. muscaria* but it has a powdery veil, leaves no ring, and the yellowish volval patches fall off readily. It is also smaller in size. It has spores of the same size and no amyloid reaction, its toxicity is unknown, and it is reported only from northeastern North America. *A. velatipes* Atk. has no orange-red cap coloration and has a well-developed, "booted" volva, but I have not seen it.

I have collected the white form of *A. muscaria* under jack pine in Michigan and Virginia pine in Maryland. The orange to orange-red cap color is common in New York and New England. A deep red cap color is often found along the west coast. A faded yellow-orange cap color is typical of fruitings under Rocky Mountain conifers.

15 AMANITA FLAVOCONIA Atk.   Poisonous (?)/Numerous

Cap orange, viscid, warts yellow; gills white; ring white; volva in yellow patches

Cap 3–8 cm broad, convex to nearly flat, viscid, very bright orange to yellow but whole cap evenly colored with scattered yellow, often large, warts; margin faintly striate and at first often clothed in volval remains. Flesh thin, firm, cream to buff near cap, with the rest white. Gills close, free, white, with minutely hairy edges. Stalk 5–12 cm long, 5–15 mm thick, enlarging toward base to an almost oval basal bulb, dry, finely powdered (pruinose), and pale yellow to white. Volva bright yellow, not persistent, breaking up around the basal bulb, and often left in the soil. Veil membranous, whitish, leaving a persistent, superior, ring with a buff margin.

Spores 7–9.5 x 4.5–6 μ ellipsoid, thin-walled, amyloid reaction, white spore print

Single or several under hardwoods or occasionally conifers in CS, SE, NE and adjacent C. in S. and F.

A closely related species is *A. flavorubescens* Atk., which has a striking yellow cap and partial veil. Sometimes the light reddish coloration is present at the base of the stalk but the stalk should be cut open to see whether there is red-stained flesh at the base.

16 AMANITA FLAVORUBESCENS Atk.   Poisonous/Infrequent

Cap yellow, viscid, warts yellow; gills white; ring white; volva in yellow patches; stalk base with red flesh

Cap 6–12 cm broad, convex to flat in age, fibrous, slightly viscid, striate just at margin, deep even yellow with numerous large yellow warts which may soon fall off. Flesh firm, white except in stalk base, which is light red. Gills close to fairly well separated, free, white with pinkish stains in age. Stalk 5–18 cm long, 10–22 mm thick, enlarging toward base, which is bulbous, white to light yellowish with scanty hairs and sometimes with light pinkish tissue near the base. Volva deep yellow, quickly disintegrating in the soil around base. Veil membranous, same color as cap, but white above, leaving a superior, ragged, skirtlike ring.

Spores 7.5–10 x 5–6.5 μ elliptical to almost round, thin-walled, amyloid reaction, white spore print.

Single or several under hardwoods or occasionally conifers in CS, SE, NE and adjacent C, in S. and F. Pomerleau (1966) reports it from Quebec and I have seen it in Maryland in May.

A. frostiana Pk. is very similar but has nearly round, non-amyloid reacting spores 7–9.5 x 7–9 μ, and the cap and warts are not intense yellow but whitish to buff.

17 LIMACELLA GLISCHRA (Morg.) Murr.   Edibility unknown/Rare

Cap brown, slimy; stalk slimy from a glutinous volva

Cap 20–42 mm broad, convex, broadly convex with a low knob in age, slimy *glutinous,* smooth, yellow-brown to reddish brown, margin often with hanging pieces of glutin. Flesh soft, white, light reddish brown near cap cuticle. Gills close, free, broad, white to very pale cinnamon. Stalk 4–7 cm long, 8–12 mm thick, nearly equal, glutinous, pallid to dull whitish or color similar to that of cap. Volva glutinous leaving slime over cap and stalk. Veil glutinous with a hairy veil beneath, pale pinkish cinnamon, leaving a glutinous, hairy ring and pieces hanging from the cap margin.

Spores 3–5.5 μ round to almost round, thin-walled, white spore print.

Alone or in small numbers under hardwoods and conifers in S. and F. following wet weather; widely distributed.

Limacella illinita (Fr.) Earle has a smooth, hairless (glabrous), white, glutinous cap; a white stalk and hairy veil. *Limacella glioderma* (Fr.) Earle has a glutinous, dark brown cap and a dry, hairy veil.

LEPIOTACEAE
The Lepiotas

The *Lepiotas,* or parasol mushrooms as they are sometimes called, are placed close to the Amanitas because of the superficial resemblance between some of the Lepiotas and certain Amanitas. The Lepiotas also have free gills and white spores but they do not have a volva. The squamules (small scales) on the cap are part of the cap tissue and not the superficial remains of the universal veil so commonly encountered in the Amanitas. The spores of the larger Lepiotas are thick-walled and often have a pore at the apex. Lepiotas with this characteristic are placed in the genus *Leucoagaricus.* Green spores are found in the genus *Chlorophyllum* and the species with entire, thin-walled spores are placed in *Lepiota.* The larger Lepiotas, which resemble Amanitas, also have a characteristic superior ring, which often becomes detached from the stalk and can be moved up and down like a napkin ring. In addition, a cross section of the gills observed through the microscope shows interwoven gill tissue and not divergent gill tissue typically found in the Amanitaceae.

Edibility. This family contains both edible and poisonous species. The green-spored *Chlorophyllum* causes an unidentified poison discussed under Mushroom Toxins, Group VI. Many of the very small species of *Lepiota* have not been tested as edibles and, in fact, are too small to bother with. The larger Lepiotas are prize edibles. The cap is most often cut into long strips and sautéed in butter, but it is also used in stuffing for poultry.

Key to Lepiota

1 a. Fruiting body large; cap 4–30 cm broad, stalk 8 mm or thicker, **2**
   b. Fruiting body small; cap 0.8–5 cm broad, stalk 1–6 mm thick, **6**
2 a. Spore print and mature gills green, form fairy rings in grass,
     18. Chlorophyllum molybdites
   b. Spore print and mature gills white; in various habitats, **3**
3 a. Cap white (sometimes tinted gray) and smooth, margin smooth,
     19. Leucoagaricus naucina
   b. Cap colored, with scales, **4**
4 a. Cap with pointed, pyramid-like brown warts, *20. Lepiota acutesquamosa*
   b. Cap with flat or raised scales, **5**
5 a. Stalk very long (15–40 cm); ring easily loosened and movable on stalk; spores 12–18 μ long, *21. Lepiota procera*
   b. Stalk 8–15 cm long; ring not easily loosened; spores 6–10 μ long,
     22. Lepiota rachodes, L. brunnea, L. americana
6 a. Cap white, powdery, margin striate, *23. Lepiota cepaestipes*
   b. Cap colored, at least over the center, margin smooth, **7**

7 a. Cap minute (8–18 mm broad), nearly smooth; stalk 1–1.5 mm thick, *24. Lepiota seminuda*
   b. Cap larger (25–50) mm broad), scaly; stalk 3 mm thick or more, *25. Lepiota clypeolaria* and related species

18 CHLOROPHYLLUM MOLYBDITES Mass.   **Poisonous**/Common

Cap large and scaly; gills and spore print green; found in fairy rings

Cap 7–30 cm broad, egg-shaped, convex-knobbed to nearly flat, dry, basically white with numerous scales tinged buff to cinnamon. Flesh firm, white. Gills close, free, broad, white, coloring greenish in age, bruising yellow to brownish. Stalk 10–25 cm long, 10–25 mm thick, enlarging toward base, without hairs, white with brownish stains over base; thick, white, superior ring with fringed edge and discoloring brownish beneath.

Spores 8–13 x 6.5–8 μ ellipsoid, thickened wall, small pore at apex, *green* spore print.

Scattered or in partial or complete fairy rings in lawns or grassland, during wet rainy periods in Sp., S., and F.; widely distributed.

This Lepiota-like fungus with the green spores in fairy rings in grass is a unique species. It may also be found under *Lepiota morgani* Peck or *L. molybdites* (Fr.) Sacc. This species and its distinctive characteristics should be known before other scaly-cap Lepiotas are eaten; see Group VI toxins.

19 LEUCOAGARICUS NAUCINA (Fr.) Sing.   **Edible***/Common

Cap, gills, and stalk white; in grass

Cap 4–12 cm broad, egg-shaped, convex in age with a low knob, dry, unpolished, white to dull white but sometimes with minute, flattened, gray small scales (squamules). Flesh thick, white. Gills close, free, white. Stalk 6–12 cm long, 10–20 mm thick, nearly equal with an enlarged base, without hairs above and silky below the ring. Veil membranous, cottony, white, leaving a collar-like superior ring which has a double-fringed edge.

Spores 7–9 x 5–6 μ almost egg-shaped, thickened wall, minute pore at apex, turns red-brown in Melzer's solution, *white* spore print.

Scattered to abundant in lawns, pastures, grassy areas, or, less frequently, under hardwoods and conifers, during wet, cool weather in late S. and F.; widely distributed. I have encountered large fruitings in central Michigan in September. It is also common in lawns throughout New England and New York at that time.

20 LEPIOTA ACUTESQUAMOSA   **Edible**/Numerous
(Weinm.) Kummer

Cap covered with brown, pointed warts; gills and stalk white

Cap 4–12 cm broad, nearly conic to convex, or convex with an obscure knob, dry, covered with erect, often pointed, dark brown pyramid-like warts on a pale brown or cinnamon-brown surface. Flesh soft, white. Gills crowded, free, white, minutely hairy edges. Stalk 5–11 cm long, 8–15 mm thick, nearly equal with an obscure basal bulb, dry, hairy, dingy white. Veil cottony, fragile, membranous, leaving a superior, skirtlike ring which is *not* movable.

Spores 7–11 x 2–3.5 μ long-ellipsoid, thin-walled, pale yellow in Melzer's solution, white spore print.

Scattered on sawdust or rich humus under hardwoods or in mixed woods in S. and F.; widely distributed in eastern North America.

---

*This species closely resembles the white Amanitas and should therefore be avoided. *Leucoagaricus naucina* does not have a cup (volva), is egg-shaped with a distinctly different veil which leaves a collar-like ring, and the spores are distinctive. In addition, none of the white Amanitas develop the numerous, fine, gray flattened hairs that very frequently appear on the cap of *L. naucina.* Both *A. virosa* and *A. verna* have a skirtlike ring, saclike volva, and remain white.

The unusual pyramid-like warts are distinctive. *Lepiota friesii* of some European authors is apparently the same species. *Lepiota asperula* Atk. is similar but has small, dextrinoid (red in Melzer's solution) spores (3–5 x 1.5–2.5 μ) and a small cap (1–4 cm broad).

21 LEPIOTA PROCERA   Edible, choice/Numerous

(Fr.) Kummer ["Parasol Mushroom"]

Cap covered with reddish brown scales; stalk tall, covered with brown scales

Cap 6–24 cm broad, egg-shaped, convex to broadly convex in age with a low umbo (knob), dry, white under a dense layer of flat or raised, reddish brown scales which appear to be in concentric rings, less dense near the margin. Flesh soft, white, not changing when bruised. Gills close, clearly free, broad, even, white with minutely hairy edges. Stalk 15–40 cm long, 8–15 mm thick, enlarging gradually toward the base which forms a small bulb, with fine, silky, pinkish tan hairs above the ring but with brown cottony scales below, separating in ringlike fashion to reveal the white flesh beneath; basal bulb is covered with fine, white mycelium. Veil thick, soft, white, leaving a collar-like superior ring which is easily detached and moveable with an edge composed of two rows of hairs.

Spores 12–18 x 8–12 μ egg-shaped, thick-walled, minute pore at apex, purple-brown in Melzer's solution, white spore print.

Several or scattered in weeds, lawns, brush, or under hardwoods and conifers in S. and F., known from CS, So, SE, NE, and adjacent C.

The tall slender aspect of the parasol mushroom along with its white gills, white spores, brown hairy stalk, distinctive ring, and lack of color change when bruised are the chief distinguishing characteristics. Note that the scales are part of the cap and not superficial volval warts as in *Amanita,* and that the spores and gills are never green as in *Chlorophyllum.*

My wife and I have long rated the parasol mushroom as one of the outstanding edible species. The cap is the best part, and we have found specimens fruiting in New Hampshire with caps so large that the fruiting bodies had toppled over.

22 LEPIOTA RACHODES (Vitt.) Quel.   Edible/Rare

Cap reddish gray squamules (small scales), dry; flesh yellow-orange when bruised

Cap 6–12 (up to 20) cm broad, conic, rounded or blunt to convex in age, dry grayish to reddish gray, breaking up on expansion to form a ringed pattern of flat or slightly raised cinnamon-brown small scales with white flesh between. Flesh soft, white, pinkish beneath cuticle, *stains orange-yellow on cap and lower stalk, fading to red-brown when bruised;* upper stalk has hollow area in center. Gills close, free, broad, white to buff with dark brown stains. Stalk 9–15 cm long, 15–30 mm thick, club- to spindle-shaped, white above ring to white below, with scattered brown hairs, stained red-brown near the base. Veil soft, membranous, white, leaving a persistent ring in a wide belt with a flaring ring which has a red-brown, fringed margin.

Spores 6–9.5 x 6–7 μ short elliptical, thickened wall, red-brown in Melzer's solution, white spore print.

In humus, compost piles, wood chip mulch, and such, in late S. and F.; widely distributed.

Lepiota brunnea Farlow and Burt has a darker brown cap, smoky brown hairs in stalk, and flesh staining brown when bruised. I have actually obtained blackish brown stains on the stalk. *L. americana* Peck reddens where bruised or handled and has a reddish brown cap but seems to differ little from *L. brunnea* in other respects. Both species have spores which are similar to those of *L. rachodes.*

23 LEPIOTA CEPAESTIPES   **Edible**/Infrequent

(Sow. ex Fr.) Kummer

Cap white, appearing mealy, dry; gills crowded; stalk white

Cap 25–35 (up to 70) mm broad, conic to bell-shaped, powdery with small or large scales, margin conspicuously striate, at first white but then light pinkish. Flesh white, soft, sometimes bruises straw-yellow near cuticle. Gills white, crowded, free, with even edges. Stalk 4–12 cm long, 3–6 mm thick, nearly equal with a small basal bulb, covered with fine short hairs (pubescent) to smooth and hairless, white may be flesh-colored or straw-yellow where handled; arises from a dense mass of white mycelium and rhizomorphs. Veil hairy, leaving a well-developed, persistent, superior ring which easily becomes detached.

Spores 6.5–10 x 6–8 μ broadly elliptical, thick-walled with a pore at the apex, yellow-brown in Melzer's solution, white spore print.

In large clusters on sawdust, wood chips, compost, rotten logs, and other rich humus, in Sp., S., and F, in eastern North America.

The striking white, mealy caps with the striate margin in dense clusters make this a distinctive *Lepiota.*

24 LEPIOTA SEMINUDA (Lasch) Kummer   **Edibility unknown** /Infrequent

Cap minute, appears without hairs, pinkish; stalk thin; like a Mycena

Cap 8–18 mm broad, conic, convex to plane or upturned at margin with a low broad knob, dry, minutely granular but appears smooth, without hairs, dull white with a pinkish to cinnamon disc. Flesh very thin, white. Gills close, free, broad, white. Stalk 25–40 mm long, 1–1.5 mm thick, enlarging slightly toward base, dry, white above to light reddish brown below, clothed overall with minute white hairs. Veil white, fibrous, remains as hairs on margin of young cap and as an obscure annular zone which soon disappears.

Spores 4–5 x 2.5–3 μ ellipsoid, thin-walled, pale brown in Melzer's solution, white spore print.

Grows in groups on needle duff under conifers in S. and F.; apparently widely distributed; known from PNW by the author.

This small distinctive *Lepiota* appears to be a *Mycena* at first glance, but on close inspection the typical free gills are observed. There are several closely related species according to Dr. Helen Smith (1954) who has studied *Lepiotas* in North America. It is certainly too small to be considered an edible.

25 LEPIOTA CLYPEOLARIA   **Poisonous**/Numerous

(Bull. ex Fr.) Kummer

Cap convex-knobbed, with brownish scales; stalk covered with wooly hairs

Cap 2.5–7 cm broad, egg-shaped, convex, convex-knobbed, dry, continuous yellowish to brownish tissue breaking up as it expands into numerous scales (ragged appearing) and hairy patches even over the center, which remains intact the longest. Flesh soft, white. Gills close, free, broad, white. Stalk 3–10 cm long, 3–4 mm thick, nearly equal, covered with whitish to buff, woolly hairs. Veil of cottony strands with pieces left on cap margin, leaves ring of fine hairs which disappears at maturity.

Spores 10–18 x 4–6 μ spindle-shaped (fusiform), thin-walled, yellow-brown in Melzer's solution; white spore print.

Single to numerous, most often under conifers in late S. and F.; widely distributed.

There are a number of small Lepiotas with scaly caps and stalks. Some are known to be poisonous; since they are hard to distinguish, none of them should be eaten. *Lepiota cristata* (Alg. and Schw. ex Fr.) Kummer is usually somewhat smaller, not as "ragged" appearing, solid reddish brown at center with reddish brown cap scales, and a nearly smooth, hairless stalk. *L. lutea* (Bolt.) Quel. is a small species with soft powdery yellow hairs and is most often found in greenhouses or in potted plants.

HYGROPHORACEAE

Perhaps no other group of gilled fungi has so many colorful members. Reds, oranges, and yellows are most frequently encountered, but a number of species also have drab gray or brown caps. The shape of the cap ranges from sharply conic to flat and the surface may be slimy, viscid, or dry. However, there are several characters which bind this group together. The most important are the gills which are *thick, waxy,* and attached to the stalk. The gills are usually adnate to short and decurrent (extending partially down the stalk). If one holds a cap upside down and rubs the gills hard enough to crush them partially, *the waxy feeling will be quite easily detected.* A waxy layer will also adhere to the fingers. The spores are always smooth, white, and entire. Lastly, all the species grow on duff or on the ground and not on logs, limbs, stumps, dung, or any other specific host. They also do not produce a latex when cut or bruised. If one keeps these characters in mind, a *Hygrophorus* is not hard to identify.

Most of the species fruit in the summer and early fall. In the western mountains several species fruit near or under melting snowbanks in the spring.

Edibility. I have not encountered any reports of poisonings caused by species of *Hygrophorus.* I have found several of them quite tasty. I suspect that the vast majority are bland and are therefore rejected by most mushroom eaters. At least a few have a disagreeable taste which would preclude their use. Some of the good edibles are also very fleshy and have a texture that would be suitable for many culinary purposes. Almost any place where one would use a commercial mushroom would be suitable for the fleshy *Hygrophorus* species.

In species with a slimy cap or very waxy gills, it would be well to remove them. In most cases the entire mushroom may be eaten.

Key to Hygrophorus

1 a. Fruiting body pure white or flushed faintly pink to buff at maturity, **2**

   b. Fruiting body colored, **11**

2 a. Cap margin and top of stalk covered with yellow hairs, *26. H. chrysodon*

   b. Cap margin and stalk pure white to faintly pinkish but no yellow hairs present, **3**

3 a. Cap and gills very pale pinkish at maturity, aromatic odor (see 27. *H. eburneus*)
   H. cossus

   b. Cap and gills pure white, no odor, **4**

4 a. Cap and stalk glutinous, **5**

   b. Cap dry, viscid to glutinous, but stalk dry to moist, **6**

5 a. Slender, stalk thin (2–8 mm broad), *27. H. eburneus*
   b. Robust, stalk thick (15–35 mm broad) (see 28. *H. subalpinus*), *H. ponderatus*
6 a. Cap viscid to glutinous, **7**
   b. Cap dry, **10**
7 a. Gills pinkish buff, often found under spruce (see 29. *H. borealis*), *H. piceae*
   b. Gills white, under conifers or hardwoods, **8**
8 a. Fruiting body robust, stalk 10–20 mm thick, **9**
   b. Fruiting body thin, stalk 2–6 mm thick; widely distributed, S. F. and W.
      (see *H. borealis*), *H. niveus*
9 a. In spring in western North America, *28. H. subalpinus*
   b. In late fall in eastern North America (see 28. *H. subalpinus*), *H. sordidus*
10 a. Cap large (10–45 mm broad), *29. H. borealis*
   b. Cap small (3–7 mm broad) (see 29. *H. borealis*), *H. niveicolor*
11 a. Cap bright yellow, green, orange, pink, or red, **12**
   b. Cap brown, olive-brown, gray-brown to pinkish brown or red-brown, **23**
12 a. Cap green with some buff color in age, viscid, *30. H. psittacinus*
   b. Cap some other color, **13**
13 a. Cap pointed to narrowly conic, viscid, **14**
   b. Cap ovoid (egg-shaped) to convex, if conic then not pointed, **15**
14 a. Cap scarlet-orange to red, bruising black, especially on stalk *31. H. conicus*
   b. Cap yellow to yellow-orange, does not bruise black (see 31. *H. conicus*),
      H. acutoconicus
15 a. Cap bright orange, viscid, **16**
   b. Cap not orange, or if orange then infused with yellow or red, **18**
16 a. Distinct and disagreeable odor (fishy or skunklike), *32. H. laetus*
   b. No odor, **17**
17 a. Gills decurrent (extending down stalk), cap soon flat, *33. H. nitidus*
   b. Gills adnexed (notched), cap convex, *34. H. flavescens*
18 a. Cap orange-red over center to yellowish orange over margin; veil present and
      glutinous, *35. H. speciosus*
   b. Cap yellow or red, veil absent, **19**
19 a. Cap yellow, yellow-orange sometimes tinted olive, *36. H. marginatus*
   b. Cap bright red or pink, **20**
20 a. Cap pink, odor, when detected, fragant, *37. H. pudorinus*
   b. Cap bright red, **21**
21 a. Gills long decurrent (extending down stalk), distant; cap soon flat,
      38. H. cantharellus
   b. Gills attached, close; cap convex, **22**
22 a. Cap dry, broadly convex, scarlet, *39. H. miniatus*
   b. Cap moist to tacky, conic to convex-knobbed, blood-red, *40. H. coccineus*
23 a. Odor fragrant, strong, *41. H. agathosmus*
   b. Odor not distinctive, **24**
24 a. Cap pinkish brown streaked with reddish purple, often bruising yellow near
      margin, **25**
   b. Cap brownish gray to drab gray, **27**
25 a. Hairy veil present (see 42. *H. russula*), *H. purpurascens*
   b. Veil absent, **26**
26 a. Gills distant (see 42. *H. russula*), *H. erubescens*
   b. Gills close to crowded, *42. H. russula*
27 a. Stalk with conspicuous gray-brown hairs over the white lower half, **28**
   b. Stalk without hairs as described above, **30**

28 a. Cap and stalk dry, *18. H. inocybiformis*
b. Cap and stalk viscid to glutinous, **29**
29 a. Stalk covered near its top with small dark gray tufts of hairs; spores small (7–9 x 4–5 μ) (see *H. inocybiformis*), *H. pustulatus*
b. Top of stalk not as described above; spores large (9–12 x 5–8 μ) (see *H. inocybiformis*), *H. olivaceoalbus*
30 a. Cap drab gray; gills white with an ashy tint, *44. H. camarophyllus*
b. Cap olive-brown; gills pale pinkish (see *44. H. camarophyllus*), *44. H. calophyllus*

26 HYGROPHORUS CHRYSODON (Fr.) Fries   Nonpoisonous/Common

Cap with minute yellow hairs, dense over the margin; zone of yellow hairs over apex of stalk

Cap 3–8 cm broad, convex to convex with a low knob in age, dry to viscid in wet weather, white but *flushed yellowish from yellow, minute hairs,* which are especially dense over the margin. Flesh thick, soft, white. Gills extend down stalk, well-separated, white, waxy. Stalk 3–8 cm long, 6–18 mm thick, equal, viscid, white except at apex which is yellow from a thick *zone of yellow hairs.*
Spores 7–10 x 3.5–4.5 μ elliptical, smooth, white spore print.
Scattered to numerous under conifers, in S., F., and W.; widely distributed. The distinctive yellow hairs on the stalk and cap margin, which may be joined in groups to form granules, distinguish this species from all the other essentially white members of this genus.

27 HYGROPHORUS EBURNEUS (Fr.) Fries   Edibility unknown/Common

Cap and stalk white and viscid

Cap 2–10 cm broad, convex to nearly flat, often with a low knob in age, viscid to glutinous, minute, short hairs (pubescent) to smooth, pure white. Flesh thick, white. Gills short and extending down stalk, fairly well separated, pure white, waxy. Stalk 4.5–18 cm long, 2–8 (up to 15) mm thick, nearly equal or smaller toward the base, glutinous, minute white scales at top, pure white, hollow in age. Odor not distinctive.
Spores 6–9 x 3.5–5 μ elliptical, smooth, white spore print.
Scattered to numerous, usually under conifers but also noted under various hardwoods, S., F., and W.; widely distributed.
The *viscid cap and stalk* distinguish this species from the other common pure white species. *H. cossus* Fries (*H. eburneus* var. *cossus* of some authors) differs only in its pale pinkish buff cap and gills and a distinctive aromatic odor.

28 HYGROPHORUS SUBALPINUS A. H. Smith   Edible/Numerous

Cap large, white, viscid; stalk robust, dry; near snowbanks in spring

Cap 5–25 cm broad, convex, soon flat, viscid, smooth without hairs, white. Flesh firm, white, thick ($\pm$ 1 cm). Gills extending down stalk, close, white, *very waxy.* Stalk 3–4 cm long, 1–2 cm thick, dry, white, sometimes with a bulbous base. Veil hairy, leaving a thin ring in center of stalk.
Spores 8–10 x 4.5–6 μ elliptical, smooth, white spore print.
Single to numerous under conifers, usually in Sp. near or under melting snowbanks, rarely in late F., PNW and RM.
Only the waxy gills distinguish this species in the field from an *Armillaria. Hygro-*

phorus ponderatus Britz is a southern winter species that grows under conifers and has a viscid stalk but is otherwise very similar. *Hygrophorus sordidus* Peck, a white species in eastern North America under hardwoods, has smaller spores than *H. subalpinus* and has a dry stalk. It is also a large robust fungus unlike *H. borealis* and those described after it. Ranger Hank Shank, an expert collector, has eaten *H. subalpinus* and we agree with him that it is somewhat bland and certainly not as good as many Western edibles, but not to be overlooked. It has good texture and would be good in a variety of casserole dishes and as a substitute for water chestnuts in Chinese dishes.

29 HYGROPHORUS BOREALIS Pk. Edibility unknown/Common

Cap white, moist not viscid; stalk thin, white, dry

Cap 10–45 mm broad, convex, nearly flat in age, most not viscid, smooth, white, margin slightly striate in age. Gills extending down stalk, fairly well separated, white, waxy. Stalk 2–9 cm long, 2–8 cm thick, or narrowed toward base, dry, smooth, without hairs, white. Flesh soft, white.

Spores 7–12 x 4.5–6.5 µ elliptical, smooth, white spore print.

Scattered to numerous under hardwoods and conifers in S. and F.; widely distributed.

H. niveus Fries is very similar and is also pure white but has a viscid cap. *Hygrophorus niveicolor* (Murr.) Smith and Hessler is white but minute (3–7 mm broad), dry with large spores (10–14 x 7–9 µ). If a large, robust, species is found near or under the edge of snowbanks in the PNW and RM, see *H. subalpinus* A. H. Smith. *Hygrophorus piceae* Kuehn and Romagn. has a white, viscid cap but attached, pale pinkish buff gills, a dry stalk, and is often found under spruce.

30 HYGROPHORUS PSITTACINUS (Fr.) Fries Unknown/Common

Cap rich grass-green, viscid; stalk green, slimy viscid

Cap 1–3 cm broad, conic but soon convex, bell-shaped in age, *glutinous to viscid, rich grass green,* infused with buff or pink in age. Flesh thin, same color as cap but more pale. Gills adnate (attached), fairly well separated, light green to buff. Stalk 3–7 cm long, 2–5 mm thick, equal, *slimy viscid,* green. Veil absent.

Spores 6.5–10 x 4–6 µ elliptical, smooth, thin-walled, white spore print.

Scattered to numerous on the ground under hardwoods and conifers, in Sp., S., and F.; widely distributed.

The collection in Plate 30 was found on the Flathead Indian Reservation in northwestern Montana following heavy summer thundershowers. This is a unique and beautiful green fungus.

31 HYGROPHORUS CONICUS (Fr.) Fries Nonpoisonous/Common

Cap conic, orange-red, viscid; flesh blackening when bruised

Cap 2–9 cm broad, *conic to a narrow peak,* tacky to viscid, without hairs, scarlet-orange to red and usually more orange near margin, often tinted olive. Flesh thin, fragile, same color as cap, *bruises black.* Gills nearly free, close, white tinted yellowish to olive-yellow, bruises black, waxy. Stalk 4–12 cm long, 3–10 mm thick, equal, moist not viscid, appears twisted, hollow, base white, the rest is colored like the cap, but the surface *blackens where bruised when mature.*

Spores 8–14 x 5–7 µ elliptical, smooth, white spore print.

Single, scattered to numerous, in Sp., S., F., and W.; widely distributed. In New Hampshire, Montana and Idaho, and in Alberta, Canada, I have frequently encountered the species under conifers following heavy rainshowers in late July and August.

This delicate, fragile species is listed as edible in Europe. *Hygrophorus acutoconicus* (Clem.) A. H. Smith looks very similar to it but *does not bruise black* anywhere, although the base may be colored black in age and tends to be more yellow to orange-yellow. The cap also has more yellow in it and is often more narrow and pointed.

32 HYGROPHORUS LAETUS (Fr.) Fries   Edibility unknown/Numerous

Cap flat, orange, slimy viscid; stalk viscid; odor fishy or skunklike

Cap 1–3.5 cm broad, convex to flat or slightly depressed, *slimy viscid,* orange to olive-orange, margin striate. Flesh thin, light orange. Gills adnate or extending down stalk (decurrent), subdistant, pinkish to violet-gray. Stalk 3–12 cm long, 2–6 mm thick, equal, smooth, without hairs, color of cap, *slimy viscid.* Veil absent. Odor *fishy or skunklike.*
    Spores 5–8 x 3–5 μ elliptical, smooth, white spore print.
    Scattered to numerous in moss and on the ground during wet weather, in Sp., S., F., and early W.; widely distributed.
    A. H. Smith and I collected these in October in northern Idaho. They would freeze at night, but when they thawed out in the laboratory, spore deposits were readily obtained. The slimy viscid orange stalk and cap, combined with the strong fishy to skunk smell, sets this fungus apart from all others.

33 HYGROPHORUS NITIDUS Berk. and Curt.   Edibility unknown/Numerous

Cap bright orange, convex, viscid; stalk dry, orange-yellow; no odor

Cap 1–4 cm broad, convex, viscid, smooth, without hairs, flat in age or slightly depressed, margin incurved, even bright apricot-yellow to orange. Flesh soft, yellowish. Gills extending down stalk, fairly well separated, pale yellow, waxy. Stalk 3–8 cm long, 2–5 mm thick, equal or slightly larger at top, dry, smooth, without hair, same color as cap. Veil absent. *No odor.*
    Spores 6.5–9 x 4–6 μ elliptical, smooth, white spore print.
    Scattered to numerous in low, wet mossy areas under hardwoods in S. and F.; eastern North America.
    Few mushrooms can be mistaken for this bright orange-yellow species. I have found it in great numbers in Catoctin Mountain Park in the Maryland mountains. *Hygrophorus laetus* (Fr.) Fries is somewhat similar but has a glutinous stalk and a fishy to skunk odor.

34 HYGROPHORUS FLAVESCENS   Nonpoisonous/Common
(Kauf.) Smith and Hesler

Cap nearly flat in age, orange-yellow, viscid; stalk dry, mostly white; no odor

Cap 2.5–7 cm broad, broadly convex, nearly flat in age, viscid then soon dry, bright orange-yellow fading at margin to light orange-yellow. Flesh thin, yellowish. Gills adnexed (notched), close, yellow, waxy. Stalk 4–7 cm long, 8–12 mm thick, equal, compressed or fluted, moist to dry (but not viscid), white, flushed with orange in center. Veil absent. No odor.
    Spores 7–9 x 4–5 μ elliptical, smooth, white spore print.
    Scattered to numerous under hardwoods and conifers, in Sp., S., and F., but often in W. in PSW; widely distributed.

The cap is never really red and the stalk is moist or waxy but not truly viscid, and there is no fishy or skunklike odor.

35 HYGROPHORUS SPECIOSUS Pk.   **Unknown**/Common

Cap convex, orange-red, glutinous; stalk glutinous, white with orange stains

Cap 2–5 cm broad, convex, convex with a low knob in some; smooth, glutinous, orange-red to orange over the center and sometimes fading to yellowish orange over the margin. Flesh soft, white to yellowish. Gills extending down stalk, fairly well separated, white to yellowish. Stalk 4–10 cm long, 4–10 mm thick, enlarging toward base, glutinous, white or whitish with dull orange stains. Veil thin, glutinous.
    Spores 8–10 x 4.5–6 μ elliptical, smooth, white spore print.
    Scattered, single to somewhat clustered, in needle duff and moist areas under or near conifers, in late S. and F.; widely distributed. A frequent habitat is larch and I have found it in abundance under subalpine fir at 6,000 to 8,000 feet in eastern Washington State.

36 HYGROPHORUS MARGINATUS Pk.   **Edibility unknown**/Numerous

Cap conic, yellow, moist; gills orange-yellow; does not bruise black

Cap 1–5 cm broad, egg-shaped conic, convex in age with a low knob, moist, smooth, without hair, but occasionally with loose squamules or small scales, yellow to yellow-orange with an olive tint, margin faintly striate. Flesh thin, same color as cap, does not bruise black. Gills adnate to adnexed (notched), fairly well separated, orange to yellow or yellow-orange, edges sometimes darker, veined, waxy. Stalk 4–10 cm long, 3–6 mm thick, equal or enlarging toward the base, dry, smooth, without hairs. Veil absent.
    Spores 7–10 x 4–6 μ elliptical, smooth, white spore print.
    Single to numerous on soil under hardwoods and conifers, in S. and F.; widely distributed.
    This *Hygrophorus* does not have the tall conic cap of *H. conicus* and does not blacken when bruised. The gills retain their color when other parts have faded. It is common in Pennsylvania, New York and New England.

37 HYGROPHORUS PUDORINUS (Fr.) Fries   **Nonpoisonous**/Common

Cap convex-knobbed, pale pink, viscid; white tufts of hairs over apex of stalk

Cap 5–12 cm broad, egg-shaped, convex to broadly convex with a broad knob, without hairs, smooth, viscid, pinkish, pinkish buff, pale flesh color, margin inrolled at first, slowly expanding in age. Flesh thick, firm, white or slightly pinkish. Gills attached to stalk at first to short and extending partially down the stalk in age, distant, flushed white with light pinkish buff or light salmon, many veins, waxy. Stalk 4–9 cm long, 8–20 mm thick, equal or slightly smaller at base, dry, conspicuous minute white tufts of hairs at top, the rest has flattened hairs, tinted pinkish, areas near base staining yellow. Veil absent. Odor faintly fragant, sometimes not detectable, usually strongest when first unwrapped after collection. Taste unpleasant.
    Spores 6.5–9.5 x 4–5.5 μ elliptical, smooth, white spore print.
    Scattered or in troops under or near conifers in F. and early W.; widely distributed.
    Smith and Hesler (1963) have described several forms and varieties of this variable species. The robust pastel pink-colored viscid cap, combined with the curious white tufts of hairs on the upper stalk, are the best field characteristics. The taste is unpleasant, but it is apparently not poisonous.

43

38 HYGROPHORUS CANTHARELLUS Nonpoisonous/Numerous

(Schw.) Fries

Cap flat in age, deep red, dry; gills extend down stalk, yellow-orange; stalk, thin, dry

Cap 1–3.5 cm broad, convex to flat with a knob or depressed in center with a knob, margin incurved and often wavy, dry, smooth, without hairs, *deep red to orange-red.* Flesh thin, reddish orange to yellow. Gills extend down stalk, distant, yellow to orange *but lighter than cap,* waxy. Stalk 4–10 cm long, 1.5–4 mm thick, equal, dry, smooth, without hairs, same color as cap or somewhat lighter. Veil absent.
Spores 7–12 x 4–6 μ elliptical, smooth, white spore print.
Scattered to numerous on debris, humus and well-decayed logs in S. and F., eastern North America. The tall slender form, red cap, and the stalk with yellowish gills extending down stalk distinguish this species from *H. miniatus* as well as other *Hygrophorii.*

39 HYGROPHORUS MINIATUS (Fr.) Fries Nonpoisonous/Common

Cap convex, brilliant scarlet, dry; gills and stalk red to orange-red; in moss

Cap 2–4 cm broad, broadly convex, dry, smooth, without hairs, brilliant scarlet, fading only in age, margin incurved at first and striate when moist. Flesh thin, scarlet to orange. Gills adnate to adnexed (notched), sometimes short and extending down stalk, close to fairly well separated, near color of cap to orange or yellow, waxy. Stalk 3–5 cm long, 3–4 mm thick, equal, dry, same color as cap. Veil absent.
Spores 6–10 x 4–6 μ elliptical, thin-walled, smooth, white spore print.
Grows in groups, usually in moss or wood debris in hardwoods or mixed woods, in S. and F.; widely distributed. I have often found it almost buried in moss in various localities from New England and New York to Alaska.

40 HYGROPHORUS COCCINEUS (Fr.) Fries Edible/Numerous

Cap conic to convex, blood-red, moist; stalk flattened to fluted

Cap 2–5 cm broad, conic with a blunt top, convex with a low knob in age, *moist to tacky but not truly viscid,* bright blood-red, sometimes with an orange hue, sometimes fading to whitish. Flesh soft, easily breaking, reddish to orange. Gills broadly attached, close. yellowish orange to orange-red, separated by veins. waxy. Stalk 3–7 cm long, 3–8 mm thick, equal, often flattened and fluted, moist, smooth, without hairs, red to orange-red, yellowish just at base. Veil absent.
Spores 7–11 x 4–5 μ elliptical, smooth, white spore print.
Scattered to numerous on soil under both hardwoods and conifers, in late S. and F.; widely distributed.
H. puniceus (Fr.) Fries looks similar and is found in similar habitats, but has a *viscid cap* and a stalk that is seldom fluted and is more often white at the base. It is also somewhat larger, with a thicker stalk (10–15 mm thick) and cap up to 7 cm broad. I have collected and studied both within three weeks in Maryland and find them rather difficult to distinguish except for the cap cuticle.

41 HYGROPHORUS AGATHOSMUS Fr. Edibility unknown/Infrequent

Cap convex, ash-gray, viscid; odor fragrant and of bitter almonds

Cap 4–8 cm broad, convex, flat in age, viscid to glutinous, ash-gray to drab, smooth, hairless, margin inrolled at first. Flesh soft, white to grayish. Gills adnate to broadly adnate, close, white to dingy gray in age, waxy. Stalk 4–8 (up to 16) cm long, 6–14 (up to 25) mm thick, equal, dry, smooth, without hairs, whitish to pale gray in age. Veil absent. Odor *fragrant and of bitter almonds.*

Spores 7–10 x 4.5–5.5 μ elliptical, smooth; white spore print.

Scattered to numerous under spruce, pine, and mixed conifers in S. and F.; most frequent in PNW and RM, occasionally found elsewhere in North America.

The odor combined with cap color and other characteristics make this a most distinctive species. One fall I encountered a large cluster on the lawn at the Priest River Experimental Forest in northern Idaho. It is listed as edible in France, but I have no information about it from North America.

42 HYGROPHORUS RUSSULA (Fr.) Quel   **Edible**/Common

Cap pink, streaked with reddish purple hairs, viscid in wet weather

Cap 5–12 cm broad, broadly convex, nearly flat in age, smooth, dry (viscid at first or in wet weather), pink to dingy pinkish brown streaked with reddish purple hairs, sometimes bruises yellow. Flesh thick, firm, white to pinkish. Gills adnate (attached to stalk) or extending partially down stalk, close to crowded, pale pink, dingy purplish red, stains in age, waxy. Stalk 2–7 cm long, 15–35 mm thick, nearly equal, dry, white at first but streaked with pink or even reddish brown in age. Veil absent.

Spores 6–8 x 3–5 μ elliptical, smooth, white spore print.

Scattered or in groups under oaks or hardwood-conifer forests in late S. and F.; widely distributed but most frequently seen in the east. I have collected it and seen it in quantity in western Massachusetts.

H. erubescens Fr. is very similar but has distant gills and *H. erubescens* var. gracilis has a long stalk. *Hygrophorus purpurascens* (Fr.) Fries, frequently found in early S. in the West nearly buried in conifer duff, has a partial veil but is otherwise very similar. The latter has a very good flavor and our family has eaten it with pleasure. I have not eaten *H. russula* but it is reported to have a good flavor.

43 HYGROPHORUS INOCYBIFORMIS   **Edibility unknown**/Infrequent
A. H. Smith

Cap brownish gray, hairy, dry; stalk dry with gray-brown hairs

Cap 3–6 cm broad, conic, convex with a low knob, *dry,* hairy, dark gray, drab to brownish gray. Flesh thin, soft, white, gray just under cap cuticle. Gills short, extending partially down stalk, fairly well separated, pallid to olive-buff, waxy. Stalk 3–6 cm long, 5–12 mm thick, equal, dry, white and silky at apex but with scattered gray-brown hairs below. Veil of thin gray-brown hairs.

Spores 9–14 x 5–8 μ elliptical, smooth, white spore print.

Scattered under western conifers, especially Englemann spruce and white fir in S. and F.; in PNW, RM, and PSW.

This resembles an *Inocybe* as well as a *Tricholoma.* The brown spore print of an *Inocybe* and the non-waxy gills of a *Tricholoma* settle any doubt. *Hygrophorus olivaceo-albus* (Fr.) Fries looks very similar to it but has a thick glutinous cap and stalk and is more widely distributed (in CS, NE, and adjacent C). *Hygrophorus pustulatus* (Fr.) Fries has granular scales, a slimy veil, smaller spores (7–9 x 4–5 μ) and is also occasionally found in eastern North America.

(Fr.) Dumee

Cap drab gray, sticky when wet, not hairy; gills extend down stalk, tinted ash-gray

Cap 4–13 cm broad, convex to nearly flat, perhaps slightly depressed, moist to sticky when wet, drab gray, overall smoky drab with fine radially arranged dark lines. Flesh firm, white. Gill adnate (attached) extending partially down stalk (short decurrent), small veins, close, white with an ashy hue, very waxy. Stalk 3–13 cm long, 19–25 mm thick, equal, silky with minutely soft, fine hairs to smooth, without hairs near base, white at apex, pale gray to smoky drab color of cap below.

Spores 7–9 x 4–5 μ elliptical, smooth, thin-walled, white spore print.

In clusters or groups under pine and spruce in Sp., S., and F.; widely distributed.

Usually found in Sp. in the PNW along melting snowbanks but elsewhere in S. and F. Also robust and similar but with pale pinkish gills and reported only from PNW and RM is *H. calophyllus* Karst.

RUSSULACEAE
The Milk Mushrooms

There are two genera, *Russula* and *Lactarius,* in this family. Both have white, buff to yellow spores with warts and ridges that have an amyloid (blue) reaction in Melzer's solution. The tissue in the pileus (cap) is composed of large globose (round) cells intermixed with cylindrical cells. There are no veils present in the family and many of the species are typified by white gills and stipe (stalk) and red, orange, yellow, or green caps. The cap cuticle peels off rather easily and the flesh is typically brittle and dry. Many species have a bitter or acrid taste which should be tested in fresh specimens.

All of the species in both genera grow on soil and probably all are mycorrhizal. Many are associated with specific conifers and hardwoods and will be found only where a specific tree or higher plant is distributed in North America. A large number of species in both genera fruit in the summer and make up a conspicuous component of the summer mushroom flora. The rest are found during the fall along with many other mushrooms. There are only a few species that fruit in either the spring or late fall.

The genus *Lactarius* contains a latex which can be seen by cutting the gills or upper stipe. Different species have latex of different colors and in some cases the latex will be white at first and then change color on exposure to air. The latex color must be noted in order to identify the species. *Russula* has no latex, but various color changes result when the tissue of some of them is cut or bruised. This reaction is also used to distinguish one species of *Russula* from another.

A bright orange parasite 422. *Hypomyces lactifluorum* (Schw.) Tul. grows over the surface of the gills and cap of several species in this family. On close examination small orange pustules can be seen on the nearly obliterated surface where the gills should be. The saclike asci (Fig. 77) are located in the pustules of this odd Ascomycete. It is often listed as edible but there is no way of determining what gilled mushroom served as the host.

Edibility. There are many good edibles in both *Russula* and *Lactarius* but there are also a number of poisonous species of the gastrointestinal (Group VI) toxins. In general, the bitter-tasting and brown- or red-staining Russulas should be avoided. The *Lactarius* species which have white latex that turns yellow or lilac can cause acute gastrointestinal upset and should also be avoided. There are numerous recipes for preparing and serving the edible members of both genera. They are used in casseroles, components of stuffing, sauteed with onions in butter, and so forth. Many of the good edibles, such as *Lactarius deliciosus,* fruit in great numbers, providing the cook with enough mushrooms to prepare some of the major dishes of the casserole type.

Key to Lactarius

1 a. Fruiting body white, large (4–12 cm broad); latex white, unchanging, **2**
   b. Fruiting body colored; latex white or variously colored, **4**
2 a. Gills crowded, forked; cap glabrous (smooth, without hairs) *45. L. piperatus*
   b. Gills distant, not forked; cap finely tomentose (woolly), **3**
3 a. Margin of young caps without a cottony roll *46. L. vellereus*
   b. Margin of young caps with a conspicuous cottony roll (see *46. L. vellereus*), *L. deceptivus*
4 a. Latex white, unchanging, sometimes staining bruised areas rose, brown, or lilac, or changing quickly to yellow, **5**
   b. Latex carrot color, blood-red to deep blue; staining bruised areas green, **22**
5 a. Latex white, not staining the gills or flesh, **6**
   b. Latex white but soon staining gills or flesh rose, brown, or lilac, or changing quickly to yellow on contact with air, **15**
6 a. Cap olive-brown; stalk olive-brown and spotted, *47. L. necator* and related species
   b. Cap and stalk variously colored but not olive or green, **7**
7 a. Cap viscid, sticky in dry weather, or matted hairy, **8**
   b. Cap dry to moist, never matted hairy, **10**
8 a. Cap zonate (banded) to azonate (lacking bands), yellowish buff to orange, **9**
   b. Cap azonate, brown, never matted hairy, *48. L. affinis*
9 a. Cap glabrous, always zonate, orange, with a smooth margin, *49. L. zonarius*
   b. Cap glabrous (hairless) only in center, yellowish buff, with densely matted hairs at margin, *50. L. torminosus*
10 a. Cap small (9–30 mm), gray to pinkish gray, *51. L. griseus*
   b. Cap larger, different colored from above, **11**
11 a. Gills *distant* (well separated); yellow coloration on cap and stalk, *52. L. hygrophoroides*
   b. Gills close; cinnamon to red-brown cap and stipe, **12**
12 a. Odor aromatic, and even stronger when dried, **13**
   b. Odor lacking, **14**
13 a. Cap 2–12 cm, cinnamon; taste mild becoming acrid, *53. L. helvus*
   b. Cap small (1.4–4 cm), red-brown; taste mild, *54. L. camphoratus*
14 a. Cap large (2.5–10 cm), red-brown; under conifers, *55. L. rufus*
   b. Cap smaller (1.5–5 cm), maroon-red; under alder, hardwoods or mixed conifer-hardwoods, *56. L. subdulcis*
15 a. Gills and flesh staining lilac where bruised, **16**
   b. Gills and/or flesh staining rose, brown, greenish gray or yellow, or latex changing to yellow, **17**
16 a. Cap glabrous, reddish brown, margin smooth, *57. L. uvidus*
   b. Cap hairy, orange-buff, margin hairy, *58. L. representaneus*

17 a. Cap gray-brown; latex white staining bruised gills bluish or greenish gray; under conifers, *59. L. mucidus*
   b. Cap not gray-brown; not staining as above, **18**
18 a. Gills and/or flesh bruise brown, latex unchanging, **19**
   b. Gills and flesh bruise yellow; latex changing to yellow, **21**
19 a. Cap blackish brown, velvety; gills or flesh bruise rose to pink, *60. L. lignyotus*
   b. Cap orange-brown to red-brown, **20**
20 a. Cap red-brown, wrinkled to corrugated, *61. L. corrugis*
   b. Cap orange-brown to dark brown, smooth, *62. L. volemus*
21 a. Cap matted hairy, margin hairy, *63. L. scrobiculatus*
   b. Cap nearly hairless, margin smooth, *64. L. chrysorheus*
22 a. Latex blue or flesh bruises blue, *65. L. indigo* and related species
   b. Latex orange or red, **23**
23 a. Latex orange, leaving green stains on the gills, *66. L. deliciosus*
   b. Latex red, leaving green stains on gills, *67. L. sanguifluus* and related species

45 LACTARIUS PIPERATUS (L. ex Fr.) S. F. Gray   Poisonous?/Numerous

Cap dull white, dry, smooth, without hairs; gills crowded, forked; white latex unchanging

Cap 4–12 cm broad, convex-depressed to deeply depressed in age, dull white, *dry*, smooth, without hair, lacking bands of color (azonate) margin inrolled at first but up-turned in age. Flesh thick, firm, white. Latex copious, white, unchanging. Gills short, partially extending down stalk, *crowded, forked,* white to cream in age. Stalk 2–6 cm long, 1–2 cm thick, equal or narrowed downward, dry, smooth, without hairs, white. Taste very acrid.
   Spores 6–8.5 x 6–6.5 µ almost round, amyloid reacting warts and ridges (Fig. 64), white spore print.
   Scattered to numerous under hardwood forests in S. and F. throughout eastern North America but infrequent in eastern C.
   Its intensely acrid taste when fresh is said to disappear when it is cooked, and the flavor is apparently good. One must be careful to compare this species with *L. vellereus,* which has a *woolly* cap, *distant gills* which do not fork, and a white latex which *stains the bruised or cut surface brown.* Poisonings have been reported in very closely related species but the toxic principles are not known (see Group VI). I do not recommend eating this species.

46 LACTARIUS VELLEREUS (Fr.) Fries   Poisonous/Common

Cap dull white, dry, woolly; gills distant, white; latex white stains tissue brown

Cap 5–12 cm broad, convex, convex-depressed, often with a central, naval-like depression (umbilicate), dry, minutely woolly (tomentose), white but sometimes yellow-ish, margin inrolled at first without cottony roll. Flesh firm, white, thick, white latex staining cinnamon to brown in time. Gills very short, extending down the stalk, distant, not forked, white, never with a green tint; when cut latex creates brown stain. Stalk 1.5–6.5 cm long, 12–30 mm thick, equal, dry, minutely velvety, dull white. Taste very bitter!
   Spores 6.5–9.5 x 5–6.5 µ elliptical, covered with fine low amyloid-reacting warts, white spore print.
   Single or scattered under hardwoods or hardwood-conifer forests in S. and F. in eastern North America. *Lactarius deceptivus* Peck looks very similar but has a distinc-tive cottony roll on the cap margin as well as larger and broader spores (9–14 x 7–10 µ). If no latex is found, especially in dry weather, or the gills show a green tint, one should

compare it with *Russula brevipes* Peck (often incorrectly called *R. delica* Fries by North American authors).

47 LACTARIUS NECATOR (Pers. ex Fr.) Lundell Edibility unknown/Infrequent

Cap olive-brown, lacking color bands; latex white unchanging; stalk olive-brown, spotted

Cap 5–14 cm broad, broadly convex, umbilicate (with a central, naval-like depression) to convex-depressed in age, sticky to viscid, lacking color bands, few hairs at center to hairy over the margin, which is inrolled at first, *olive-brown,* darker at center. Flesh firm, dull white. Latex white, unchanging. Gills short and extend partially down stalk, crowded, dull whitish to buff, bruising black or blackish brown. Stalk 2.5–6 cm long, 12–25 mm wide, equal, dry to sticky or viscid when moist, olive–buff at apex, the rest is olive-brown, *usually with spots which are darker* olive-brown, sometimes hollow. Taste very acrid.
Spores 7–9 x 5.5–7 μ almost round, amyloid reaction in warts and ridges (reticulate), buff spore print.
Single or in small groups under conifer forest or sometimes mixed conifer-hardwood forest in late S. and F.; widely distributed at least in northern North America. I have found it under uncut stands of old western white pine and under eastern white pine along the Massachusetts-Vermont border.
Also called *L. turpis* (Weinm.) Fr., the spotted olive-brown stalk, olive-brown cap, and unchanging white latex are the noteworthy characteristics. *Lactarius atroviridis* Peck is a closely related eastern species, distinguished primarily by the rough, scabrous surface of the cap. The taste raw would be disagreeable, but I can find no reports on its flavor.

48 LACTARIUS AFFINIS Pk. Nonpoisonous/Numerous

Cap butterscotch-brown, viscid, smooth, without hairs; stalk viscid; latex white, unchanging

Cap 5–15 cm broad, convex-umbilicate (naval-shaped), smooth, without hairs, viscid, lacking color bands, rich yellow-brown, almost butterscotch color. Flesh firm to soft, white to cream, yellowish brown just under cap cuticle. Latex white, copious, unchanging. Gills short, extend partially down stalk (decurrent) or nearly adnate, fairly well separated, buff, staining color of cap. Stalk 2.5–8 cm long, 14–24 mm thick, equal, smooth, without hairs, viscid, buff often with stains the color of the cap. Taste strongly acrid.
Spores 8.5–11 x 7–8.5 μ almost round to short elliptical, warts with amyloid reaction and ridges, white spore print.
Scattered or in groups, sometimes abundant, under conifers (especially spruce), or mixed forests in S. and F.; widely distributed.
This large *Lactarius* is distinguished by the viscid, butterscotch-brown cap and the unchanging white latex. *Russula foetens* (see 73.) has a somewhat similar cap but exudes no latex.

49 LACTARIUS ZONARIUS (Secr.) Fries Edibility unknown/Infrequent

Cap orange, with bands of color (zoned), viscid; latex white, unchanging or sometimes faint buff

Cap 3–7 (up to 12) cm broad, convex with a depressed center and inrolled margin, sticky to viscid, zoned with color bands, orange alternating with reddish orange. Flesh firm, white. Latex white, sometimes with a buff tint on exposure to air. Gills attached to stalk or extending partially down stalk, close, light buff. Stalk 2–6 cm long, 10–25 mm thick, equal to narrow just at base, smooth, without hairs, light orange to yellowish, hollow pithy core in age. Taste acrid.

Spores 9–12 x 7.5–8.5 μ elliptical, amyloid reaction in warts and ridges, cinnamon-buff spore print.

Scattered or in groups under conifers in Sp. and F. in the RM and PNW.

The characteristic zones combined with the white latex are typical of this fungus. I have found it in central Idaho up to and above 7,000 feet, often in wet seepage areas under spruce-lodgepole pine stands.

50 LACTARIUS TORMINOSUS
Poisonous/Common

(Schaeff. ex Fr.) S. F. Gray

Cap yellow tinted rose, with bands of color, viscid, margin densely hairy; latex white, unchanging

Cap 4–10 cm broad, broadly convex, flat with the center depressed in age, sticky to viscid, zoned with bands of color, yellowish to buff to rose color, smooth, without hairs at center but margin covered with dense whitish or light pinkish soft hairs. Flesh white to pale pink, unchanging. Latex *white, unchanging.* Gills close, extending down stalk, narrow, white to yellowish. Stalk 3–8 cm long, 12–22 mm thick, cylindric or narrowed at base, dry, smooth, without hairs or faintly pruniose (with a powdery cover), light pink to yellowish or yellow spotted, hollow in age. Taste very acrid.

Spores 10–17 x 5–8 μ broadly elliptical, amyloid reaction in warts and ridges, white spore print.

Scattered to numerous, especially common under birch in S. and F.; widely distributed. I have seen quantities of this species under paper birch near Northway, Alaska.

L. scrobiculatus, when immature, is somewhat similar but the white latex instantly changes to sulphur yellow. The poisonous toxins are unknown (see Group VI).

51 LACTARIUS GRISEUS Pk.
Nonpoisonous/Numerous

Cap grayish pink, dry; stalk light pink; latex white, unchanging

Cap 9–30 mm broad, broadly convex to convex-depressed in age, dry, grayish brown, grayish pink, pale gray to wine-red. Flesh firm, white. Latex white, sparge, unchanging. Gills extending partially down stalk, close, buff to cream. Stalk 13–25 mm long, 2–6 mm thick, equal, dry, minutely roughened to hairless, light pinkish to light cinnamon. Taste slightly acrid.

Spores 6–9 x 5–7 μ broadly elliptical to almost round, amyloid reaction in warts and ridges, white spore print.

Scattered or in groups in moss or on well-decayed wood in wet areas under conifers in F. in eastern North America.

This is the smallest member of the genus, but a small specimen of *L. cinereus* Peck may sometimes be found. However, the latter has a viscid cap, white gills, and a dry, ashy-colored stalk.

52 LACTARIUS HYGROPHOROIDES Berk. and Curt.
Edible/Numerous

Cap yellow tinted brown, dry, velvety; gills white, set apart; latex white, plentiful, unchanging

Cap 2.5–10 cm broad, convex, convex-depressed in age, dry, appearing velvety, lacking color bands, yellowish to yellow-brown. Flesh brittle, pallid, unchanging when bruised. Latex white, copious, unchanging in air. Gills partially extending down stalk, well separated, white to buff or cream color. Stalk 2–5 cm long, 8–20 mm thick, equal, without hairs, same color as cap, often bright yellow near base.

Spores 7–10 x 6–7 μ broadly elliptical to almost round, amyloid reaction in ridges, white spore print.

Single to scattered under hardwood forests in S. and F. in eastern America.

The wide gills extending down the stalk remind one of a *Hygrophorus* but the white latex that at times drips rather rapidly from the cut gills will clear up this error.

53 LACTARIUS HELVUS (Fr.) Fries   Edibility unknown/Infrequent

Cap pale cinnamon, dry; latex watery white; odor of camphor; taste acrid

Cap (2) 4–12 cm broad, convex, flat with depressed center with a small knob, dry, nearly hairless, *pale tan to cinnamon or light reddish gray.* Flesh white, soft. Latex white, watery. Gills extending partially down stalk, close, white to yellowish with a pink tinge. Stalk 4–8 cm long, 5–15 mm thick, minute hairs (pubescence), dry, color of cap. Taste mild but then very slowly becomes acrid. Odor of camphor even when dried.

Spores 6–8.5 x 5–6 μ almost round to short elliptical, warts and ridges have amyloid reaction, white spore print.

Scattered or in groups in moss in wet areas in a wide variety of habitats in S. and F.; widely distributed.

The larger size, pale cinnamon cap, mild to slightly acrid taste, and sweet, distinctive smell distinguish it from *L. camphoratus.*

54 LACTARIUS CAMPHORATUS (Bull. ex Fr.) Fries   Edible/Numerous

Cap brownish red, dry; latex white; odor aromatic; taste mild

Cap 14–40 mm broad, convex, convex-knobbed in age, lacking zones of color, dry, smooth, without hairs, dark brownish red, margin inrolled at first. Flesh pale or same color as cap. Latex white, unchanging. Gills extend down stalk a short distance, close, white to yellowish or red-brown in age. Stalk 1–5 cm long, 4–8 mm thick, nearly equal, dry, smooth, without hairs or with slight powdery cover, *same color as cap.* Taste *mild.* Odor aromatic (similar to sweet clover), very distinctive, not really like camphor.

Spores 6–8.5 x 6–7.5 μ almost round, coarse warts have amyloid reaction, white spore print.

Scattered or in groups on the ground or on very rotten wood often near or under conifers, in S. and F.; widely distributed.

The mild white latex, distinctive sweet clover-like smell whether fresh or dried, and a dark red-brown cap are the chief characteristics. *Lactarius rimosellus* Peck is similar in every way except that the cap is very cracked (rimose) and the dried specimens are very aromatic.

55 LACTARIUS RUFUS (Scop. ex Fr.) Fries   Poisonous/Common

Cap red-brown, dry; latex white; odor mild; taste very acrid

Cap 2.5–10 cm broad, convex-knobbed (Fig. 5) to convex-depressed or flat with a depressed center, smooth, hairless, moist to dry, lacking bands of color, even red-brown to dark reddish chestnut. Flesh white with a pinkish tinge, hollow in stalk in age. Latex white, unchanging. Gills extending partially down stalk, close, light orange to pinkish salmon. Stalk 3–9 cm long, 8–15 mm thick, equal or slightly narrowed downward, dry, smooth, hairless or sometimes with slight powdery surface, salmon color to light red-brown. Taste very acrid. Odor not distinctive.

Spores 7–9 x 5–7 μ broadly elliptical, amyloid (blue) reaction in netlike ridges (reticulum), white spore print.

Scattered or in groups in wet areas under conifers, in S. and F.; widely distributed.

The dark red-brown cap, comparatively larger size, and typical association with conifers distinguish this from the smaller reddish tan *L. subdulcis,* which is usually associated with hardwoods. This has Group VI toxins.

56 LACTARIUS SUBDULCIS   Edible, good/Numerous

(Bull. ex Fr.) S. F. Gray

Cap maroon-red, waxy; latex white; odor mild; taste faintly acrid

Cap 1.5–5 cm broad, convex to flat (Fig. 5), and depressed in center in age, with a small knob, waxy and moist (may seem viscid), smooth, hairless, lacking bands of color, maroon-red to reddish tan. Flesh white, does not bruise. Latex white, unchanging. Gills extending partially down stalk, close, buff to salmon or light flesh color. Stalk 12–40 mm long, 3–10 mm thick, equal, smooth, hairless, nearly white near apex to orange-brown, often hollow in age. Taste mild, acrid, or faintly acrid to bitter with time. Odor not distinctive.

Spores 7–10 x 6–8 μ broadly elliptical, amyloid reaction in the low warts, white spore print.

Scattered or in groups under hardwoods or conifers, in Sp., S., and F.; widely distributed. Usually along the edges of wet areas or sometimes in bogs in conifer forests.

I have made a number of collections in one area on the same day in August near McCall, Idaho, and the intensity of the acrid to bitter taste varies a good deal. In California this fungus is called the candy mushroom and is used to make persimmon and candy mushroom pudding.

57 LACTARIUS UVIDUS (Fr.) Fries   Poisonous/Numerous

Cap brownish gray, smooth, without hair, margin smooth; latex white, stains lilac

Cap 2.5–9 cm broad, convex, convex-depressed sometimes with a slight knob, viscid, without hairs, sometimes faintly zoned, brownish gray to wine-colored brown with a tint of lilac, margin inrolled at first and *smooth.* Flesh white, staining lilac. Latex white, bruised or cut areas stain lilac. Gills extending partially down stalk, close, white bruising lilac. Stalk 2.5–6 cm long, 10–18 mm thick, slightly larger at base, sticky to viscid, white near apex with the rest light wine-colored gray. Taste mild at first to acrid.

Spores 7–12 x 6–8 μ broadly elliptical, high warts and ridges have amyloid reaction, white spore print.

Scattered or in groups under northern conifers in S. and F. in PNW, RM, A, CS, NE, and C.

This smooth, hairless, viscid *Lactarius* has white latex that stains lilac and is not recommended as an edible. I have no information on the toxins.

58 LACTARIUS REPRESENTANEUS Britz.   Poisonous*/Common

Cap orange-buff, hairy, margin hairy; latex white staining lilac

Cap 7–15 cm broad, convex, broadly convex-depressed, pruinose (with powdery

---

*The lilac staining members of *Lactarius* should not be eaten. *Lactarius speciosus* Bull. is a similar southern species, differing only in the banded arrangement of the cap hairs.

cover) to nearly hairless in depressed center, the rest densely woolly with long hairs that are fringed along the inrolled margin, viscid, center nearly solid orange, the rest buff with orange-buff hairs, margin nearly flat in age. Flesh firm, white staining lilac. Latex white, *staining cut portion lilac.* Gills extend down stalk, close, light buff with conspicuous lilac stains when cut or bruised. Stalk 4–8 cm long, 20–30 mm thick, slightly larger at base, dry, light orange-buff, *covered with slightly depressed shiny orange spots,* hollow or with a soft pith. Taste acrid.

Spores 9–11 x 7.5–9 μ short elliptical, amyloid (blue) warts and ridges, white spore print.

Scattered or in groups under northern conifers in late S. and F.; in PNW, RM, A, CS, NE, and C. I have seen this species in quantity near Junean, Alaska, under Sitka spruce and western hemlock, and near Fairbanks, Alaska, under aspen and spruce.

59 LACTARIUS MUCIDUS Burl. Nonpoisonous/Numerous

Cap gray-brown, viscid, lacking color bands; latex white at first but stains tissue bluish or greenish gray

Cap 2.5–8 cm broad, convex to broadly convex with a low knob, often depressed in age, viscid, smooth, hairless, lacking color bands, *gray-brown* in center to *gray* over the margin, which is inrolled at first. Flesh soft, very thin stalk sometimes hollow. Latex white but stains wounded areas bluish to greenish gray. Gills adnate (attached), close, narrow, shiny white, also stains bluish gray when bruised. Stalk 2–6 cm long, 8–10 mm thick, nearly equal, sticky to viscid, smooth, hairless, sometimes wrinkled, *same color as cap.* Taste very acrid.

Spores 7.5–10 x 6–8 μ short elliptical to almost round, amyloid reaction in ridges, white spore print.

Scattered or in groups in needle duff under conifers in S. and F.; widely distributed. I have found it frequently in northern Idaho.

The acrid taste would make it undesirable as an edible. A similar species, *L. varius* Peck, has a moist to sometimes sticky but not viscid cap. Another species, *L. trivialis* (Fr.) Fries, has a gray cap which is infused with purple or drab gray, and a yellow spore print.

60 LACTARIUS LIGNYOTUS Fr. Nonpoisonous/Numerous

Cap velvety, striking blackish brown; latex white staining cut tissue pinkish

Cap 2–10 cm broad, convex to broadly convex-knobbed, dry, sometimes wrinkled, velvety, *striking blackish brown.* Flesh *brittle,* white, staining rose to pink where bruised. Latex white, thin and watery, staining flesh pinkish. Gills attached or short decurrent (extending partially down stalk), close, white, bruising red, brown edges near stalk. Stalk 4–10 cm long, 4–15 mm thick, equal, dry, furrowed near apex, smooth, with a powdery-velvety cover below, *same color as cap.* Taste slightly acrid or mild.

Spores 9–11 x 8–10 μ almost round, amyloid reaction in ridges, deep buff spore print.

Several or in groups, often in moss, sometimes in bogs, under conifers in S. and F.; widely distributed. I have seen this beautiful species under eastern white pine in New Hampshire in late summer.

Section *Plinthogali* of *Lactarius* was intensively studied by A. H. Smith and L. R. Hesler; they found more than 25 species in North America. *L. gerardii* Peck has well-separated gills and a white spore print. *L. fuliginosus* Fries has an ashy brown to sooty brown cap and round spores 8–9.5 x 8–9 μ with an amyloid (blue) reaction of the net-like ridges.

61 LACTARIUS CORRUGIS Pk.   **Edible**/Infrequent

Cap wrinkled, dark red-brown, dry; latex white slowly stains brown

Cap 6–12 cm broad, convex, convex-depressed in age, dry, lacking color bands, velvety, wrinkled to corrugate overall but especially pronounced over the margin, dark red-brown. Flesh firm, white with weak brownish stains. Latex white, copious, gradually staining brown. Gills extending partially down stalk, close, cinnamon-yellow, cut or bruised areas staining brown. Stalk 6–7 cm long, 15–25 mm thick, equal, dry, felty, same color as cap or somewhat lighter. Odor not distinctive.

Spores 9–12 μ round, amyloid spines, white spore print.

Solitary or several together under hardwoods in eastern America in late S. and early F.

62 LACTARIUS VOLEMUS Fr.   **Edible, good**/Numerous

Cap hazel-brown, smooth, without hairs, dry; latex white, plentiful, slowly stains dark brown

Cap 3–6 (up to 12) cm broad, convex, soon convex-depressed, smooth, without hairs, sometimes with minute cracks, dry, varying in color from hazel-brown to orange-brown, or light orange-buff. Flesh firm, pallid, *stains brown* in time when cut or bruised. Latex white, plentiful, often drips readily when cut, in air it changes slowly to brown. Gills adnate to extending partially down stalk, *close,* white to pinkish buff. Stalk 2.5–6 (up to 10) cm long, 10–20 mm thick, nearly equal, dry, pinkish buff to light orange often streaked or stained light reddish brown, sometimes hollow in age. Odor rather disagreeable, becoming stronger after collecting.

Spores 7–10 μ almost round, amyloid (blue) reaction of the netlike ridges; white spore print.

Scattered to groups in hardwood or mixed conifer-hardwood forests in S. in eastern North America.

L. hygrophoroides has well-separated (distant) gills, latex which does not stain the gills brown, a yellowish to yellow-brown cap, and elliptical-shaped spores. *Lactarius corrugis* has a dark red-brown corrugated or wrinkled cap with a velvety surface.

63 LACTARIUS SCROBICULATUS   **Poisonous**/Common
(Scop. ex Fr.) Fries

Cap light yellowish orange, viscid, sometimes zoned with bands, hairy margin; latex white changing to yellow

Cap 5–15 cm broad, broadly convex, shallow to deeply depressed in age, sticky to viscid, generally lacking bands, smooth, hairless in center to matted hairs extending beyond inrolled margin in young caps, pale yellow to ochre-yellow or even tawny-brown over the center. Flesh white, staining yellow. Latex scanty, white changing quickly to sulphur-yellow. Gills adnate to extending partially down stalk, crowded, whitish to light yellow, staining dark yellow. Stalk 3–6 cm long, 10–30 mm thick, equal, hairless, dry, same color as cap with polished, brighter spots, hollow in age. Taste mild to slightly acrid.

Spores 7–9 x 6–7.5 μ broadly elliptical to almost round, (blue) amyloid reaction in warts and ridges; white spore print.

Scattered or in groups under conifers in late S. and F.; widely distributed but seen very commonly in PNW.

L. torminosus has white unchanging latex and the stalk is not as consistently covered with shiny spots. *L. resimus* Fries resembles *L. scrobiculatus* but the stalk does not have the depressed shiny spots and the cap is nearly white at first. All species with white latex that change to yellow should be avoided (see Group VI toxins).

64 LACTARIUS CHRYSORHEUS Fr. Poisonous/Common

Cap buff to cinnamon, sticky; latex white changing to yellow; taste acrid

Cap 4–10 cm broad, convex, flat to slightly depressed in age with an elevated margin, subviscid to sticky in wet weather, faintly zoned with bands near margin to totally lacking bands, pinkish cinnamon to cinnamon-buff, sometimes darkening to reddish cinnamon, wavy margin in some. Flesh firm, white, changing rapidly to yellow. Latex copious, white changing to *deep yellow*. Gills adnate to extending partially down stalk, close, white to pale pinkish cinnamon, often with reddish brown stains in age. Stalk 4–7 cm long, 10–25 mm thick, equal, dry, with powdery coating, off-white to pale pink, often spotted or stained brownish purple. Taste *acrid*, slowly increasing with time.

Spores 6–9 x 5–7 μ short elliptical, amyloid (blue) reaction of the netlike ridges, light buff spore print.

Grows in groups under hardwoods and conifers in late S. and F.; widely distributed.

I have seen large fruitings, season after season, under a Norway spruce plantation in Laurel, Maryland, but it was also found in less abundance scattered throughout the surrounding native oak and Virginia pine woods. Section *Crocei*, in Smith and Hesler's study (1960), in which this fungus is placed consists of a number of closely related species all with white latex which turns yellow on exposure to air. *Lactarius theiogalus* Fries is closely related but has a tawny-red cap.

65 LACTARIUS INDIGO (Schw.) Fries Edible/Rare

Cap blue, dry, sometimes with zones of blue; latex deep indigo blue; stalk blue

Cap 5–15 cm broad, convex, convex-depressed in age, margin inrolled, smooth, hairless, viscid, zoned blue to deep blue, indigo to gray in age and staining green. Flesh firm, indigo staining green. Latex *deep indigo*, slowly dark green on exposure to air. Gills extending a short distance down stalk, close, indigo, staining green when bruised. Stalk 2–8 cm long, 1–2.5 cm thick, equal or narrowed at base, indigo, dry, hollow. Taste slightly bitter.

Spores 8–10 x 5.5–7.5 μ elliptical, amyloid reaction in warts and ridges; yellow to orange-yellow spore print.

Scattered or in groups under conifer and hardwood stands in S. and F., found commonly only in SE, rare northward. Periods of heavy rainfall precede its appearance.

The deep indigo fruiting body and latex leave no doubt about the identity of *L. indigo*. *Lactarius paradoxus* Beard and Burl. has a grayish indigo cap, pale orange gills, flesh of cap and stalk bluish, and red latex when gills and stalk are cut. Found in eastern North America.

66 LACTARIUS DELICIOSUS (Fr.) S. F. Gray Edible, good/Common

Cap carrot-color mixed with green, viscid when wet; latex orange staining green

Cap 5–15 cm broad, convex to broadly convex, margin incurved, *viscid to sticky* in wet weather, sometimes with several faint color zones, carrot-color mixed in various proportions with dingy green, usually more green with age. Flesh pallid, very light orange to stained with green in age. Latex carrot-colored, leaving green stains. Gills extending partially down stalk, close, bright orange, staining green in age or after injury. Stalk 2–5 cm long, 1.5–3 cm thick, equal, narrow just at base, dry, nearly hairless, lighter orange turning green when handled or in age. Taste slowly and slightly acrid.

Spores 8–11 x 7–9 μ almost round to short elliptical, amyloid reaction in the warts and ridges, buff spore print.

Single to scattered under conifer forests in late S. and F., widely distributed. In Alaska and the Yukon, I have found small specimens almost buried in moss but large fruiting bodies, often in abundance, are encountered elsewhere in North America.

The carrot-color combined with the irregular green staining make this and its close relatives a very distinctive group. *L. deliciosus* var. *areolatus* A. H. Smith has cracks in the cap surface, and the young caps are dark red, and is known from the Payette Lakes region of central Idaho. *L. thyinos* A. H. Smith is separated by its viscid stalk, gills fairly well separated, mild taste, and location in cedar bogs in CS, NE, and adjacent C. *L. chelidonius* Peck is a rare, widely distributed, gray-capped species, staining bluish and then greenish, with yellow to pale carrot-colored latex.

67 LACTARIUS SANGUIFLUUS Fr.   Edible/Infrequent

Cap carrot-color mixed with green, bands of color; latex blood-red, scanty

Cap 6–12 cm broad, convex, depressed in center with inrolled margin, carrot-colored zones alternating with pale zones, green stained in age. Flesh brittle, pale buff to orange. Latex scanty, blood-red to purple-red. Gills adnate (attached) to extending partially down stalk, close, slight reddish tinged with purple, green stained in age.

Stalk 2–5 cm long, 1–3 cm thick, narrow just at base, dry, pale orange color, hollow. Taste mild.

Spores 7.5–10 x 6.5–8 μ elliptical, with amyloid reaction of the ridges, pale yellow spore print.

Scattered or in groups under conifers in the PNW, RM, and PSW fruiting in S. and F. Often seen in early S. near McCall, Idaho.

In western North America the blood red latex is the distinguishing characteristic; however, in eastern North America, *L. subpurpureus* Peck has a scanty, muddy red latex and is closely associated with eastern hemlock. In Maryland and Virginia it appears in the F. and is often locally abundant. The cap has clear zones of color, and pink with dingy green stains developing in age. It is edible.

Key to Russula

1 a. Fruiting body remaining white or white at first, but soon becomes smoky gray to blackish, **2**

  b. Fruiting body colored at first, **5**

2 a. Fruiting body white at first, soon smoky gray to blackish; stalk flesh bruising red fading to black or bruising black, *68. R. densifolia* and related species

  b. Fruiting body white, sometimes with yellowish or brown stains but never becoming blackish, **3**

3 a. Gills decurrent (extend down stalk), crowded; cap dry, large (9–20 cm broad), *69. R. brevipes*

  b. Gills adnate, close to fairly well separated; cap viscid or dry, smaller, **4**

4 a. Cap viscid, *70. R. albidula*

  b. Cap dry (see *70. R. albidula*), *R. albella*

5 a. Cap green to gray-gree, smooth or broken up into small patches, **6**
   b. Cap with olive tints mixed with other colors or cap not green, **7**
6 a. Cap green, smooth (see 71. *R. virescens*), *R. aeruginea*
   b. Cap gray-green, with small irregular patches, *71. R. virescens*
7 a. Cap with a dark purplish center and light pink margin, *72. R. fallax*
   b. Cap color pattern not as above, **8**
8 a. Cap ochre to yellow-brown, very viscid, *73. R. foetens*
   b. Cap not brown, but viscid, **9**
9 a. Cap yellow to yellow-orange, **10**
   b. Cap red, orange-red to purplish red, **11**
10 a. Cap small (2.5–7 cm broad) clear yellow; flesh does not change color when bruised; under birch, *74. R. lutea*
   b. Cap large (9–12 cm broad) yellow-orange; flesh bruises gray; found under conifers, *75. R. ochroleuca*
11 a. Cap bright red; gills white; very acrid taste; usually in deep moss *76. R. emetica* and related species
   b. Cap red but infused with shades of orange-purple or even tinted olive; gills buff to yellow, **12**
12 a. Cap purplish red, sometimes tinted olive; odor of crab or lobster; stalk white flushed pinkish, *77. R. xerampelina*
   b. Cap red to orange-red, often tinted yellow or purple; stalk white, *78. R. paludosa*

68 RUSSULA DENSIFOLIA (Secr.) Gillet   Poisonous/Common

Cap white to smoky gray, viscid when wet; flesh bruises red and then turns black

Cap 8–20 cm broad, convex, convex-depressed, dry or slightly sticky to viscid in wet weather, dingy white and soon smoky to gray or blackish, margin striate. Flesh whitish *turning red* when bruised and then at length *to black*. Gills adnate (attached to stalk), fairly well separated, thick, broad, whitish bruising dingy red to black. Stalk 2–7 cm long, 1–4 cm thick, smooth, hairless, whitish streaked with brown to reddish brown stains. Taste acrid.
   Spores 7–10 x 6–8 μ almost round, with amyloid reaction in ridges, white spore print.
   Single or in groups under hardwood, conifer, or mixed forests in S. and F.; widely distributed. I have found it in quantity in New England, New York and Maryland under eastern white pine and Virginia pine.
   R. albonigra (Krombh.) Fries is very similar but stains directly to black when bruised. *Russula nigricans* (Bull.) Fries is a European species with distant gills and a thin cuticle. It is found infrequently in North America and has often been confused with *R. densifolia*. The toxins are unknown (see Group VI toxins).

69 RUSSULA BREVIPES Pk.   Edible/Common

Cap large, white staining brown, depressed, dry; gills fairly well separated, not forked near stalk

Cap 9–20 cm broad, broadly convex, depressed in center, dry, minutely woolly (tomentose), white stained dull yellow or brown, margin inrolled at first, not striate. Flesh firm, white. Gills extend down stalk, close to crowded and sometimes forked, with veins between, white stained cinnamon to brown. Stalk 3–8 cm long, 2.5–4 cm thick, equal, dry, smooth, hairless, dull white with brown stains. Taste slightly acrid. Odor somewhat disagreeable.

Spores 8–11 x 6.5–8.5 μ, elliptical with amyloid reaction in warts and ridges, cream-colored spore print.

Scattered under conifers or mixed conifer-hardwoods in S. and F.; widely distributed.

R. delica Fries has nearly distant gills. It has not been found in North America, and early reports of it must be referred to *R. brevipes* or one of the closely related species in the Subsection *Lactarioideae* of *Russula* according to Shaffer (1964). *R. vesicatoria* Burl. is similar but the gills frequently fork near the stalk. I have seen this species under Virginia pine in the F. in Maryland and Virginia.

70 RUSSULA ALBIDULA Pk.   Nonpoisonous/Common

Cap medium white, viscid, convex, taste acrid

Cap 2.5–7 cm broad, broadly convex, smooth, hairless, *viscid,* white, margin faintly striate. Flesh fragile, white. Gills adnate (attached to stalk), close to fairly well separated, sometimes forking, white. Stalk 2.5–7 cm long, 10–23 mm thick, equal, dry, smooth, hairless, white, with a fine white down at the base. Taste acrid.

Spores 7–10 μ almost round, with amyloid reaction in the ridges, white spore print.

In oak woods and mixed hardwoods and conifers in S., F., and early W. in the NE, CS, and SE.

R. albidula is often found partially buried in sandy or clay soils in the SE. A very closely related white species, *R. albella* Pk., is slightly larger with a dry cap. It is not buried in sand or clay.

71 RUSSULA VIRESCENS Fr.   Nonpoisonous/Infrequent

Cap green to gray-green, broken up into small patches

Cap 5–12 cm broad, nearly round first, convex, then flat or somewhat depressed in age, dry, *surface broken up into many small irregular patches,* green to gray-green, margin not striate or very slightly so. Flesh firm, white, solid throughout. Gills nearly free, close, even, white, sometimes forked. Stalk 3–7 cm long, 1–2 cm thick, equal, dry, smooth, hairless, white.

Spores 6–8 μ almost round, amyloid reaction in warts and ridges, white spore print.

Solitary to scattered under hardwoods or mixed conifer hardwoods in S.; widely distributed. In Montana it seems to be associated with paper birch. *Russula aeruginea* Lindbl. is darker green, smooth, and nearly viscid in wet weather. I would not recommend these Russulas for the table because the study of them in North America is not complete.

72 RUSSULA FALLAX Cke.   Edibility unknown/Common

Cap with purplish center and pink margin

Cap 2.5–6 cm broad, broadly convex to slightly depressed with a depressed center, viscid, dark purplish center with an olive zone fading to a pink margin. Flesh soft, white. Gills adnexed (notched), fairly well separated, white. Stalk 2–4 cm long, 5–10 mm thick, equal, dry, with longitudinal wrinkles, white, often hollow. Taste acrid.

Spores 6–8 x 5–7 μ almost round, amyloid reaction in warts, white spore print.

Solitary or in groups on moist wet ground in mixed woods, often among moss in sphagnum bogs in S. and early F.

73 RUSSULA FOETENS Fr.   Nonpoisonous/Common

Cap yellow-brown, viscid; young gills have water drops on them; stalk white staining brown

Cap 5–14 cm broad, broadly convex to convex-depressed, viscid to sticky in wet weather, smooth, hairless, margin deeply and widely striate, dark ochre to yellow-brown. Flesh thin, dingy white. Gills adnexed (notched), close, broad, drops of water often seen on them when young, whitish, with dingy bruises in age, with veins between, and sometimes forked. Stalk 3–6 cm long, 10–25 mm thick, equal, smooth, hairless, dull white, staining brownish to yellowish in age. Odor faint, becoming disagreeable as it dries. Taste acrid.
Spores 7–9 x 6–7.5 μ almost round, amyloid reaction in warts, white spore print.
Scattered or in groups under conifers and hardwoods in S.; widely distributed.

74 RUSSULA LUTEA (Huds. ex Fr.) S. F. Gray   Edible/Common

Cap yellow, viscid, margin striate; flesh white, unchanging when cut

Cap 2.5–7 cm broad, convex to flat or slightly depressed in center, smooth, hair-less, viscid, clear bright yellow with a striate margin in age. Flesh fragile, white. Gills free, fairly well separated, narrow, yellow to ochre, often with veins between. Stalk 3–5 cm long, 5–20 mm thick, equal, dry, smooth, hairless, white.
Spores 7–10 x 6.5–8 μ broadly elliptical, amyloid reaction in warts, white spore print. print.
Scattered or in groups under hardwoods, especially paper birch, in S. and early F., in PNW, A, CS, Ne, and C. Paper birch is present wherever I have found this *Russula* in the PNW and A. Unlike many other Russulas with yellow color in the cap, *R. lutea* is a clear unchanging yellow.

75 RUSSULA OCHROLEUCA (Pers.) Fries   Edibility unknown/Infrequent

Cap yellow-orange, viscid, margin striate; flesh white turning gray when cut open

Cap 9–12 cm broad, convex, convex-depressed, smooth, hairless, viscid, striate just at margin, yellow-orange to orange. Flesh soft, white but when cut turns gray immedi-ately. Gills attached, crowded, light yellow. Stalk 7.5–10 cm long, 27–30 mm thick, equal, white to grayish white in age, dry, faint netlike markings.
Spores 8–10 x 6–8 μ almost round, amyloid reaction in ridges and warts, yellow spore print.
Scattered under western conifers in S., and noted under lodgepole pine. Known from the PNW and Europe, but it is probably more widely distributed in North America. The distinctive *immediate change of the white flesh to ash-gray* and the yellow-orange, viscid cap are the important characteristics. *R. decolorans* Fries has an orange-red cap but is quite similar in other ways.

76 RUSSULA EMETICA (Fr.) Pers.   Poisonous?/Common

Cap bright red, viscid, margin striate; gills and stalk white; taste strongly acrid

Cap 4–12 cm broad, broadly convex to convex-depressed in age, viscid, smooth, hairless, *bright red,* fading in age, margin deeply striate. Flesh very soft, white, pink just under cap cuticle. Gills adnate, close, sometimes forked, white. Stalk 4–12 cm long, 10–25 mm thick, equal, dull white, hollow in age. Taste instantly strongly acrid.

Spores 7–10 x 6.5–8 μ broadly elliptical, amyloid reaction in warts, white spore print.

Scattered or in groups on the ground or more commonly in deep moss, in S. and F.; widely distributed. I have often seen the cap level with tips of the moss in the SE.

It has been reported as edible, which must mean that the acrid taste disappears in cooking. Since, however, we have a number of red Russulas in North America and they are poorly known, I cannot recommend any of them. *Russula fragilis* (Pers. ex Fr.) Fries is smaller, and the flesh is not pink under the cuticle. It is also poisonous but, as with *R. emetica,* the poisons are unknown (see Group VI toxins).

77 RUSSULA XERAMPELINA Fr.   Edibility unknown/Infrequent

Cap purplish red, viscid; odor of crab or lobster

Cap 3–16 cm broad, broadly convex, convex-depressed, viscid at first to dry in age, purplish red, lighter or even olive at margin which is striate. Flesh firm, white. Gills notched, close, thick, yellow to orange-yellow. Stalk 5–8 cm long, 1.5–5 cm thick, flaring or even bulbous at the base, white flushed pinkish, wrinkled, grooved or even faintly reticulated, hollow in age. Odor of *crab or lobster usually quite strong* but stronger when dried.

Spores 8–11 x 6–8.5 μ round, amyloid reaction in warts, yellow spore print.

Scattered or in groups under conifer forests in S. widely distributed.

The purplish red cap and distinctive crab or lobster smell are the chief characteristics. In addition, the gills turn gray when dried. The flesh turns deep green in ferric sulphate ($FeSO_4$). It is reported as edible in Europe but I have no information on its edibility in North America. A small variety is found in tundra above the Arctic Circle.

78 RUSSULA PALUDOSA Britz.   Edible/Numerous

Cap orange-red, viscid; stalk ridged; odor mild

Cap 6–14 cm broad, convex, convex-depressed to nearly flat in age, viscid, smooth, hairless, bright red to orange-red, flushed with purple or infused sometimes with yellow, margin weakly striate. Flesh soft, white. Gills deeply adnexed (notched), fairly well separated, warm buff. Stalk 3.5–7 cm long, 20–25 mm thick, equal, dull white, dry, smooth, hairless, with low longitudinal ridges. Taste mildly acrid. Odor mild or slightly fishy.

Spores 9–12 x 8–10 μ almost round, amyloid reaction in warts and spines, yellow spore print.

Scattered, sometimes gregarious under mixed hardwoods and conifers. Common throughout Northern North America. In northwestern Montana found under paper birch and western hemlock.

I would not recommend eating red Russulas because of the number of unknown species. *Russula flava* Romaq. is closely related but the taste is reported to be mild and the change to gray is only evident after drying.

TRICHOLOMATACEAE

This is a large and diverse assemblage of gilled fungi. They are found in many habitats and on various kinds of organic matter. However, it is not too difficult to identify the family if you eliminate other white-spored families as you examine a specific fungus.

There are only five families of white-spored agarics (gilled fungi). These are the Amanitaceae, Lepiotaceae, Hygrophoraceae, Russulaceae, and Tricholomataceae. The first two families have free gills, the Hygrophoraceae have waxy gills, and the Russulaceae have brittle flesh, no annulus (ring), and an amyloid reaction of the spore ornamentation. One genus, *Lactarius,* has a milky latex. All four of these families are almost always found growing on the ground. Therefore, any white-spored fungus that grows on wood or does not meet the criteria described above is a member of the Tricholomataceae.

Within the Tricholomataceae, it is somewhat more difficult to distinguish the various genera. I have arbitrarily divided them into those exclusively on wood and those either primarily on the ground or rarely found on wood. It is noteworthy that those always on wood have, for the most part, eccentric (off-center) stipes or lack a stipe altogether.

Edibility. There are many edible fungi in this family and very few species which are poisonous. Probably the most toxic species known in this family is *Clitocybe dealbata* (Group VI toxins), although *Tricholoma* has a rather sinister reputation. Many genera, including *Tricholoma,* have not been thoroughly studied in North America. Some species of *Collybia* are known to cause severe gastrointestinal disturbances (see *Collybia dryophila*). Many other species are too small, tough or bitter to be suited for the table.

On the edible side we find *Armillariella mellea* and *A. tabescens,* which are both choice and often found in quantity. *Pleurotus ostreatus* and *P. sapidus* have an excellent flavor. Young caps of *Lentinus lepideus* and *L. ponderosus* are also quite good, and *Tricholoma flavovirens* is rated highly by some people. *Marasmius oreades* is small but usually found in lawns and grassy areas in sufficient quantity to be collected for the table. It is rated as choice by many people here and in Europe. It takes about twenty minutes to cook; simmer it slowly in water in which butter and perhaps a little lemon juice have been added. After cooling, it may be used like any other species and is especially good in gravies and as a component of "Marasmius Cookies" (see Recipes).

Key to the Genera of the Tricholomataceae
On wood; stalk central, eccentric (off center), or lacking, **1**
On ground, leaves, needles, humus, dung, *rare on wood,* etc; stalk usually central, **10**

1 On wood; stalk central, off center or lacking

    1 a. Stalk central or nearly so, **2**
       b. Stalk eccentric, very short or absent, **7**
    2 a. Fleshy and easily broken, **3**
       b. Flesh tough to pliant and difficult to break, **6**
    3 a. In cespitose clusters, stalk with a ring (Fig. 3), *Armillariella mellea*
          (see also *Cystoderma*)
       b. Solitary or cespitose; stalk without a ring, **4**
    4 a. Cespitose; stalk without a ring, *Armillariella tabescens*
       b. Single, not cespitose, **5**
    5 a. Cap usually small (4–20 mm broad); gills short or long decurrent (extend down stalk), *Omphalina*
       b. Cap usually larger; gills adnate to adnexed (notched), *Tricholomopsis*

6 a. Gill edges serrate (saw-toothlike), *Lentinus*
   b. Gill edges entire, *Xeromphalina* (if cap covered with stiff hairs see *Crinipellis*)

7 a. Saw-toothed or hairy gill edges, **8**
   b. Smooth gill edges, **9**

8 a. Always sessile (stalk absent); small; very hairy gill edges,
      86. *Schizophyllum commune*
   b. Sessile or with an eccentric stalk; saw-toothed gill edges; amyloid (blue) reaction in spores, *Lentinellus*

9 a. Cap pink with characteristic ridges and pits; spores strongly warted (Fig. 63), Rhodotus
   b. Cap smooth or without ridges and pits; spores never tuberculate; stalk short or absent; flesh soft to tough, *Pleurotus* and allies

10 Found on ground; central stalk; on soil, leaves, needles, humus, dung, etc.

10 a. Stalk with ring, **11**
   b. Stalk without a ring, **12**

11 a. Ring double or single; veil and cap not appearing mealy, *Armillarias* and allies
   b. Ring always single; veil and cap appearing mealy to powdery, *Cystoderma*

12 a. Fruiting body revives in water; stalk thin (1–4 mm), long, pliant, **13**
   b. Fruiting body does not revive in water; or stalk thicker, fleshy and/or brittle, **15**

13 a. Cap covered with long stiff hairs, *Crinipellis*
   b. Cap smooth or felty, **14**

14 a. Stalk blackish brown; cap cuticle not cellular; spores blue in Melzer's solution, *Xeromphalina*
   b. Stalk black to yellow-brown; cap cuticle cellular (Fig. 80); spores colorless in Melzer's solution, *Marasmius*

15 a. Parasitic on decaying mushrooms, 118. *Asterophora* (see *Collybia tuberosa*)
   b. Not parasitic on decaying mushrooms, **16**

16 a. Gills decurrent (Fig. 5), *Clitocybe* and allies
   b. Gills adnate (attached) to adnexed (notched) (Fig. 5), **17**

17 a. Cap margin inrolled at first; stalk brittle or pliant, *Collybia*
   b. Cap margin straight, if inrolled stalk fleshy, **18**

18 a. Stalk very thin (1–4 mm thick), brittle; gills adnate to decurrent, *Mycena*
   b. Stalk thicker and fleshy to pliant, **19**

19 a. Gills fleshy, pink to violet in color; spores spiny (Fig. 64), *Laccaria*
   b. Gills not fleshy and usually not pink to violet; spores smooth to tuberculate (strongly warted), **20**

20 a. Gills staining blackish brown; stalk water-soaked, often tan, *Lyophyllum*
   b. Gills not staining, **21**

21 a. Cap large (6–45 cm broad); surface of gills and stalk appear chalky white, *Leucopaxillus*
   b. Caps generally smaller; flesh not chalky white, **22**

22 a. Stalk tall and thin for the size of the cap; caps characteristically convex-umbonate (knobbed); gills adnexed (notched) (Fig. 5), 152. *Melanoleuca*
   b. Stalk usually thick, fleshy, robust; cap convex, convex-depressed; gills adnexed to adnate, *Tricholoma*

TRICHOLOMOPSIS

The central, fleshy stipe (stalk) and convex caps remind one of a *Tricholoma*. But unlike *Tricholoma* all species of *Tricholomopsis* grow on wood and have abundant cystidia (sterile cells) on the gill edge (Figs. 75 and 76). There are 17 species in North America (Smith, 1960) but many of them are encountered infrequently. The spores are elliptical, smooth, and have a white spore print.

Edibility *Tricholomopsis rutilans* is the only edible species, but I have rarely encountered a sufficient quantity to collect it for the table. None of the other species is reported to be toxic.

Key to Tricholomopsis

1. a. Cap large (5–20 cm broad), blackish brown to gray; gills distant (well separated) very broad *79. T. platyphylla*
   b. Cap smaller, yellowish, yellow-orange to brick-red, **2**
2. a. Cap yellowish beneath a thick layer of brick-red hairs, 80. *T. rutilans*
   b. Cap with gray to brown hairs over a yellowish tissue, **3**
3. a. Cap with many gray to gray-brown hairs (see 80 T. rutilans) *T. decora*
   b. Cap with sparse brown hairs; never gray (see 80. T. rutilans) *T. sulfureoides*

79 TRICHOLOMOPSIS PLATYPHYLLA   Nonpoisonous/Common

(Fr.) Sing.

Cap gray to blackish brown, dry, inrolled; gills well separated, very broad

Cap 5–20 cm broad, convex, flat or slightly depressed in age, dry to moist, without hairs or with scattered hairs, from blackish brown to gray or whitish gray, margin inrolled at first, with short faint striations when wet. Flesh thin, pliant, white to gray. Gills adnate to adnexed (notched), distant, very broad, veined, white to gray. Stalk 6–12 cm long, 1–3 cm thick, equal or enlarging toward base, smooth, without hairs, or with flat-lying hairs, dry or moist, white to grayish white, hollow in age. Veil absent.
Spores 7–10 x 4.5–6 μ elliptical, smooth, white spore print. Cystidia (sterile cells) abundant on gill edge.
Single to scattered on well-decayed logs, stumps, or boards of hardwoods and conifers, fruiting following wet weather in Sp., S., and F.; widely distributed. I have found this species most commonly in New England, New York, and generally in the eastern hardwood region, but I encounter it almost every season somewhere in the PNW.
The broad, fairly well separated gills, large size, and the gray-brown coloration of the cap seem to be the most prominent field characteristics. Also known as *Collybia platyphylla* (Fr.) Kummer.

80 TRICHOLOMOPSIS RUTILANS (Fr.) Sing.   Edible/Common

Cap yellow beneath dense red hairs; flesh pale yellow

Cap 2–15 cm broad, convex to flat with a low knob, covered with brick-red to purplish red hairs, yellow underneath, which shows through between the hairs. Flesh thick, firm, pale yellow. Gills adnate to adnexed (notched) in age, crowded, edges roughened, yellow. Stalk 4–10 cm long, 8–20 mm thick, equal or enlarged toward the base, dry to moist, covered with red hairs over a pale yellowish ground color, staining yellow when bruised. Veil absent. Taste faintly radish-like.

Spores 5–6 x 3.5–4.5 μ elliptical, smooth, white spore print. Cystidia (sterile cells) abundant, club-shaped, on gill edge.

Single but usually several on *conifer* logs and stumps in S. and F.; widely distributed. The largest specimens I have seen were on conifer wood under Sitka spruce near Juneau, Alaska. It is very common on conifer logs and stumps, especially Norway spruce, in New Hampshire. Known also as *Tricholoma rutilans* (Schaeff. ex Fr.) Kummer.

Tricholomopsis decora (Fr.) Singer is quite similar, also on conifer wood, but the small scales and hairs on the caps and stalk are gray to gray-brown over a yellow-orange ground color, and the spores are longer (6–7.5 μ). *T. sulfureoides* (Pk.) Singer has a pale yellow cap with sparse hairs on both cap and stalk darkening to brown but never gray, and a hairy veil which soon disappears. I have found it only on eastern hemlock.

LENTINUS

This is a group of centrally stalked species with tough flesh; they appear on wood. They all have serrate (sawtooth) gill edges although some species develop this characteristic in a pronounced fashion only in age. The spores are smooth, elliptical (Fig. 64), entire and not amyloid (blue) in Melzer's solution. *Lentinellus* also has serrate gill edges but the stipe is eccentric or lacking and the subglobose (nearly round) spores have an amyloid reaction and minute spines. Species of *Panus,* which have smooth gill edges, can easily be confused with *Lentinus.* If you are in doubt, the mushroom should be keyed out in both keys.

Edibility. Because these mushrooms have tough flesh few people eat them. When young they are not so tough, and both *L. lepideus* and *L. ponderosus* have a good flavor. They can be used in most dishes by parboiling. When a dish needs a firm, fleshy mushroom, these species can be used but, at best, they are chewy.

Key to Lentinus

1 a. Ring present, 81. *L. lepideus*
   b. Ring absent, **2**
2 a. Gills irregular, often covered completely by a membranous tissue (see *L. Tigrinus*), *L. tigrinus* var. *squamosus*
   b. Gills normal, **3**
3 a. Cap large (up to 33cm broad); gills adnexed (notched); fragrant odor
     (see 81. *L. lepideus*), *L. ponderosus*
   b. Cap smaller (2–6 cm broad); gills decurrent (extend down stalk); resinous odor, *82. L. tigrinus*

81 LENTINUS LEPIDEUS Fr.   **Edible**/Common

Cap buff with reddish brown small scales (squamules), ring on stalk; gill edges toothed

Cap 5–12 cm broad, convex, nearly flat in age, viscid when young but soon dry, white to buff at first, soon breaking up into small scales which are cinnamon to wood-brown and raised in age, margin incurved at first, straight in age. Flesh tough, white except for stalk base which is yellow to rusty brown. Gills adnexed (notched), close, edges toothed, white to buff in age, rusty brown stains when bruised. Stalk 3–10 cm long, 10–15 mm thick, equal, white and minutely hairy above ring, recurved small scales develop below, white to reddish brown in age. Veil membranous, buff leaving a usually persistent, superior ring, but in age it may weather away. Taste somewhat disagreeable. Odor fragrant, anise-like.

Spores 7–15 x 4–6.5 μ long-elliptical, smooth, thin-walled, pure white to buff spore print.

Single to several on logs and stumps of conifers, occasionally hardwoods, in Sp., and F., following but not during cool wet weather; widely distributed.

L. ponderosus O. K. Miller is similar but lacks a ring, is larger (up to 33 cm broad), the cap has more pinkish buff coloration, and it is known only from the PNW. Both species are tough, become woody and remain in place for a long time. They also require considerable cooking to soften them sufficiently before they can be eaten. Hank Shank, a forest ranger and biologist with the U.S. Forest Service, from McCall, Idaho, rates both as good but he prefers the young caps. I concur and recommend it.

82 LENTINUS TIGRINUS (Bull.) Fries   Nonpoisonous/Infrequent

Cap buff with many brown-tipped hairs; no ring; gill edges toothed; often in cespitose clusters

Cap 2–6 cm broad (up to 14 cm), convex to convex-knobbed, sometimes a depressed knob in age, dry, buff to light cinnamon with many small, flat, brown-tipped hairs, very dense at the center but less so near the incurved margin. Flesh tough, white. Gills extend down stalk, close, edges toothed to finely serrate, buff to pinkish buff. Stalk 2–8 cm long, 4–8 mm thick, equal, dry, buff with brown small scales which become a dense, short, almost complete hairy covering over the base, often with a long rootlike base that extends deep into the wood or ground. Veil absent. Odor like resin.

Spores 6.5–12 x 2.5–3.5 μ narrowly elliptical, smooth, white spore print. Cystidia (sterile cells) on gill edge narrowly club-shaped. Cystidia on cap thick-walled.

Single, but most often in densely cespitose clusters from a rooting base, occasionally in large clusters on hardwood logs, stumps or dead roots, in Sp., S., and early F.; widely distributed in eastern North America. I have seen very large fruiting bodies collected in Ohio. Most caps are small and measure 2–6 cm broad.

Also known as *Panus tigrinus* (Bull. ex Fr.) Singer. A variety with abnormal gills covered with a dense tissue, *L. tigrinus* var. *squamosus* (also called *Lentodium squamulosum* Morgan) is frequently encountered.

XEROMPHALINA

A group of rather small and fragile looking species numbering eleven in North America (Miller, 1968). They are usually found in "troops" on well-decayed wood, needles, and humus. If the entire dried fruiting body is placed in water, it will revive and appear fresh. When revived the stipe is as pliant as it is in fresh specimens. This capacity to revive when moistened is also shared by *Marasmius* and *Crinipellis*. The stipe in *Xeromphalina* is thin (2–5 mm wide), pliant, usually dark brown over the lower two-thirds, and covered with a velvety pubescence (fine, short hairs). The cap appears smooth and the white spores turn blue (amyloid) in Melzer's solution. *Crinipellis* is frequently encountered and has stiff hairs on the cap but the species of *Marasmius* do not. Both have white spores but they are non-amyloid in Melzer's solution.

Edibility. These are small, often bitter tasting species which have no value as edibles. There is no record of toxins in this genus.

Key to Xeromphalina

1 a. Gills attached directly to the stalk (adnate), **2**
    b. Gills extend down stalk (decurrent), **3**
2 a. On conifer wood in western North America, *83. X. fulvipes*
    b. On hardwoods in eastern North America (see 83. *X. fulvipes*), *X. tenuipes*
3 a. On seeds, needles, and twigs of conifers and sometimes aspen leaves,
       84. X. cauticinalis
    b. On logs and stumps, **4**
4 a. On hardwood logs and stumps (see 85. *X. campanella*), *X. kauffmanii*
    b. On conifer logs and stumps, **5**
5 a. Only on redwood (see 85. *X. campanella*), *X. orikiana*
    b. On conifer wood other than redwood, **6**
6 a. Cap dull orange-brown; taste bitter; PNW only (see 85. *X. campanella*),
       X. brunneola
    b. Cap yellow-brown; taste mild; throughout North America, *85. X. campanella*

83 XEROMPHALINA FULVIPES   Nonpoisonous/Infrequent

(Murr.) A. H. Smith

Cap flat, yellow-brown; gills attached

Cap 1–3 cm broad, convex, nearly flat in age, dry, yellow-brown. Flesh pliant, brown. Gills attached, close, broad, pale yellowish. Stalk 5–8 cm long, 3–8 mm thick, equal, pliant, tough, dry, fine velvety brown hairs over the entire surface.

Spores 4–5 x 1–2 μ allantoid (sausage-shaped), smooth, amyloid (blue) reaction in Melzer's solution, white spore print.

Single to scattered on wood and needles of conifers in Sp., S., and F., in SW, PNW, and A.

This fungus and *X. tenuipes* (Schw.) A. H. Smith have adnate (attached) gills, but *X. tenuipes* is found on hardwood logs and limbs and has larger, elliptical spores (6.5–9 μ long).

84 XEROMPHALINA CAUTICINALIS   Nonpoisonous/Common

(Fr.) Kuehn.

Cap yellow-brown, convex; gills extend down stalk; stalk straight

Cap 5–25 mm broad, convex, nearly plane in age, moist, honey-yellow, margin faintly striate. Flesh pliant, thin, pallid, turns red in 3% KOH (caustic potash solution). Gills extend partially down stalk, fairly well separated, veined, pale yellow. Stalk 3–8 cm long, 1–2.5 mm thick, equal, pliant, tough, straight, yellow-brown to dark brown at base.

Spores 4–7 x 2.5–3.5 μ elliptical, smooth, amyloid (blue) reaction in Melzer's solution, white spore print.

Solitary or several, occasionally in groups on conifer seeds, needles, sticks, or sometimes on aspen leaves, in S. and F.; widely distributed. They are tall, straight, and not on recently fallen logs or stumps. If curving from and directly on logs, see *X. campanella.*

85 XEROMPHALINA CAMPANELLA   Nonpoisonous/Common

(Fr.) Kuehn. and Maire

Cap orange-brown, convex; gills extend down stalk; stalk curved

Cap 3–25 mm broad, convex to broadly convex with a central depression, moist, smooth, without hair, yellow-brown to orange-brown, margin striate, inrolled at first. Flesh thin, pliant, yellowish. Gills extend down stalk, fairly well separated, veins between, yellowish to dull orange. Stalk 1–4 cm long, 0.5–2 mm thick, equal, pliant, tough, yellowish at apex, red-brown at base, which is surrounded by yellow-brown hairs, usually curved from the host.
Spores 5–9 x 3–4 μ elliptical, thin-walled, turns blue in Melzer's solution, buff spore print. Cystidia (sterile cells) present on gill edge, thin-walled, tapering at each end.
Groups to cespitose clusters on decayed conifer wood, in S. and F. or early W.; widely distributed. Large clusters often cover a sizable area along a fallen log with the stalks characteristically curving from the host.
X. kauffmanii A. H. Smith has a yellow cap, smaller spores (4–5 μ long), yellow hairs around stalk base and is found on hardwood logs and stumps. *Xeromphalina brunneola* O. K. Miller has a dull orange cap, disagreeable taste, and grows on western conifers. *Xeromphalina orickiana* (Smith) Singer has a red-brown cap with hairs over the central portion and grows only on redwood logs.

86 SCHIZOPHYLLUM COMMUNE Fr.   Nonpoisonous/Common

Cap small, gray, hairy, fan-shaped; gills with hairy edges; stalk absent

Cap 9–30 mm broad, fan-shaped, dry, densely hairy, whitish gray when dry to brownish gray when moist, margin lobed and also hairy. Flesh tough, leathery, gray. Gills well separated, split and hairy, cuplike in cross section, white to gray. Stalk absent, cap attached directly to the wood.
Spores 3–4 x 1–1.5 μ cylindrical, white spore print.
Several or in dense clusters, sometimes from a common base on a wide variety of hardwood sticks, stumps, and logs, in Sp., S., F., and in warm spells in W.; widely distributed.
This very odd species has been used for many years for experimental genetics research. It fruits readily in culture in the laboratory, but is too small and tough to be edible.

LENTINELLUS

All species grow on wood or on the ground and attached to buried wood. The stipe is either absent or off center. All have coarsely toothed to ragged gill edges (serrate) (Color pl. 87). The spores are subglobose (almost round) (Fig. 62) with fine spines (Fig. 64), and have an amyloid (blue) reaction in Melzer's solution. Most of the seven species in North America (Miller and Stewart, 1971) fruit in the late S. and F., but one western species *(L. montanus)* fruits in the Sp. near snow.

Edibility.   Most are much too bitter; some people have tried *L. montanus* and find it somewhat tough but with fairly good flavor.

Key to Lentinellus

1 a. Cap completely sessile (stalk absent), **2**
   b. Cap with a short or long stalk, **3**
2 a. Widely distributed, common; cap tissue has amyloid (blue) cells, *87. L. ursinus*
   b. Near or under melting snowbanks in spring in Rocky Mts. (see 87. *L. ursinus*),
      L. montanus
3 a. Cap glabrous (smooth, without hairs), pinkish buff; stalk narrow (0.5–4 mm
      thick), *88. L. omphalodes*
   b. Cap hairy, **4**
4 a. Cap not robust, with upraised hairs; stalks slender, fluted and fused; found on
      wood on the ground (see 88. *L. omphalodes*), *L. cochleatus*
   b. Cap robust, with dense white to buff hairs, stalks thick, fused; often on wounds
      of living hardwoods (see *L. ursinus*), *L. vulpinus*

87 LENTINELLUS URSINUS (Fr.) Kuehn.   Nonpoisonous/Common

Cap densely hairy; stalk absent; not found near melting snow; gills toothed

Cap 2–10 cm broad, 2–5 cm wide, convex to nearly flat, dry, sparse hairs to densely hairy with stiff rigid hairs over the center, dark brown to cinnamon, margin incurved and minutely hairy, sometimes without hairs. Flesh firm, white to pinkish buff. Gills close, broad, *coarsely toothed,* often with *ragged edges,* light pink to pinkish buff. Stalk absent. Veil absent. Odor fruity. Taste strongly bitter.
   Spores 3–4.5 x 2–3.5 μ almost round, with minute spines that have an amyloid (blue) reaction, white spore print. Tissue in cap mixed with amyloid and non-amyloid cells.
   Several to overlapping clusters on logs and stumps of hardwoods and conifers, during moist, cool periods in S., F., and W.; widely distributed.
   L. montanus O. K. Miller is similar but has no amyloid reaction in tissue in cap, larger spores (4.5–6.5 x 4–5 μ), and is always found fruiting at high elevations in the Western mountains near melting snow. *L. vulpinus* (Fr.) Kuehn. and Maire has dense white to yellowish hairs on the cap, short fused stalks, and is often found on the wounds of living trees, especially elm. *L. montanus* is edible but I have no reports on other closely related species.

88 LENTINELLUS OMPHALODES (Fr.) Karst.   Edibility unknown/Numerous

Cap smooth, without hairs; stalk thin, slightly off center; gills toothed

Cap 1–5 cm broad, broadly convex, nearly plane with a central depression in age, moist, smooth, without hairs, pinkish buff, cinnamon to wood-brown in age. Flesh soft, white. Gills attached, subdistant, *edges toothed or ragged,* pinkish cinnamon. Stalk 5–50 mm long, 0.5–4 mm thick, equal, sometimes slightly off center, ridged and furrowed, dry, smooth, without hairs, red-brown above, wood-brown toward the base. Veil absent. Taste soon peppery.

Spores 5–6.5 x 3.5–4.5 μ short elliptical, with minute spines that have an amyloid (blue) reaction, buff spore print. Tissue in cap not blue in Melzer's solution.

Solitary or in small groups on wood debris, sticks or logs of conifers and hardwoods, or occasionally on the ground, in S. or F.; widely distributed.

L. cochleatus (Fr.) Karst. is rarely found but has several fused stalks, a deeply depressed cap with conspicuous upraised hairs on it and brown, thick-walled spores on the stalk.

89 RHODOTUS PALMATUS (Bull. ex Fr.) Maire   Edibility unknown/Rare

Cap pink with distinctive ridges and shallow pits; off-center stalk; on wood

Cap 2–5 cm broad, broadly convex, dry, *covered with ridges and pits,* brick-red to flesh color. Flesh firm, flesh color. Gills attached, close, broad, veined, pink. Stalk 15–30 mm long, 4–6 mm thick, off center, enlarging slightly toward base, dry, light pinkish. Veil absent.

Spores 5–7 x 4.5–6.5 μ round, *tuberculate* (strongly warted), cream color spore print.

Single or several on logs and branches of hardwoods during cool, wet weather in Sp., S., and F.

I have collected this rare species in Michigan and Maryland ten years apart. The unique ridges and pits on the cap enable one to instantly identify this rare species. The cap ornamentation, combined with the tuberculate cream-colored spores, makes this a unique species.

PLEUROTUS AND ALLIES

The genera included here are all found on wood or sometimes on buried wood. They have an off-center stipe or no stipe at all (sessile). The gill edges are always smooth unlike *Lentinus* and *Lentinellus.* None of the species has ornamented spores (with spines or tuberculations; Figs. 63–64). They all have white to buff spores except *Pleurotus sapidus,* which has a lilac-tinted spore print. The spores are almost round to elliptical, entire, and smooth. Spores with amyloid (blue) reactions are found in the genus *Panellus,* but all the other genera have non-amyloid spores. One must be careful to obtain a spore print because *Crepidotus,* which has no stalk (sessile), and *Gymnopilus* with an off-center stalk, in the Cortinariaceae, are also found on wood and could be confused with the species described here. Both, however, have brown and yellow-brown spore prints. The majority of the species described here fruit in F. or in S. *Pleurotus ostreatus* and *P. sapidus* may also be found in early Sp.

Edibility.   Many of the species are edible and no toxins are reported in this group. Always be sure that you do not have the orange *Omphalotus olearius.* Some of the species are bitter or the flesh is too tough to make them desirable. However, *P. ostreatus* and *P. sapidus* are both choice edibles. Certain black beetles live in these mushrooms in hot weather. One must always check the flesh to make sure that they have not invaded the fungus. Often great quantities of both species will be encountered in F. or S., sometimes in sufficient quantity to be canned for future use. Both *P. ostreatus* and *P. sapidus* are excellent in the "Mushroom and Cheese Casserole" (see Recipes). *Pleurotus ulmarius* and the tough species of *Panus* should be parboiled to soften them. Young fresh specimens are preferable.

Key to Pleurotus and Allies

1 a. Fruiting body minute (2–6 mm broad), grayish blue; without stalk (sessile), *90. Resupinatus applicatus*

    b. Fruiting body larger; eccentric (off-center) stalk or sometimes sessile, **2**

2 a. Fleshy and easily broken, **3**

    b. Leathery to tough, not easily broken, **10**

3 a. Stalk absent, **4**

    b. Stalk present, **9**

4 a. Cap white, **5**

    b. Cap colored, **7**

5 a. Cap small (3–20 mm broad), *91. Pleurotus candidissimus*

    b. Cap larger (2–30 cm broad), **6**

6 a. Usually fan-shaped with very thin flesh and crowded, thin gills, *92. Pleurocybella porrigens*

    b. Usually convex with thick flesh and fairly well separated gills, *93. Pleurotus ostreatus*

7 a. Cap green to greenish yellow, viscid, *94. Panellus serotinus*

    b. Cap some other color, **8**

8 a. Cap orange to orange-buff, dry, hairy, *95. Phyllotopsis nidulans*

    b. Cap gray to dark grayish brown, *93. Pleurotus ostreatus*

9 a. Cap with mottled water spots over the center, usually in cespitose clusters at base or on logs and stumps; stalk very long or short; spores white, *96. Pleurotus elongatipes* (if orange and cespitose, see 119. *Omphalotus olearius*)

    b. Cap not mottled; stalk very short; pleasant odor; spores lilac (see 93. *P. ostreatus*), *P. sapidus*

10 a. Cap small (12–32 mm broad); in imbricate (overlapping) clusters; stalk short (6–12 mm long), *97. Panellus stipticus*

    b. Cap larger (1.5–40 cm broad); not imbricate, **11**

11 a. Cap densely hairy, **12**

    b. Cap minutely pubescent (fine, short hairs) to glabrous (smooth, without hairs), **13**

12 a. Cap 10–40 cm broad, cream to buff *(orange when dry);* found on wounds in living trees, *98. Panus strigosus*

    b. Cap 1.5–7 cm broad, pinkish to light reddish brown, *99. Panus rudis*

13 a. Gills decurrent (extend down stalk), violet to tan with violet tints, *100. Panus torulosus*

    b. Gills adnexed (notched), white to cream; often on wounds of living trees, *101. Pleurotus ulmarius*

90 RESUPINATUS APPLICATUS   **Edibility unknown**/Common

(Fr.) S. F. Gray

Cap minute, cuplike, grayish blue; jellylike tissue layer in cap

Cap 2–6 cm broad, *minute,* dry, cuplike to convex, grayish blue to grayish black, covered with fine, minute hairs. Flesh firm, gelatinous. Gills fairly well separated, broad, whitish at first to gray sometimes very dark.

Spores 4–5 μ round, smooth, white spore print.

Scattered or often in great numbers on hardwoods including vines and shrubs, in Sp., S. and F.

A cross section reveals a gelatinous flesh, and the gill edges are sterile. This along with the combination of grayish blue to grayish black are the distinguishing characteristics of this minute but common fungus.

91 PLEUROTUS CANDIDISSIMUS   Edibility unknown/Numerous

Berk, and Curt.

Cap small, shell-shaped, white, soft; on wood

Cap 3–20 mm broad, shell-shaped, dry, covered with soft white hairs, inrolled at first. Flesh thin, white. Gills attached to stalk, well separated, broad, white, minutely hairy edges. Stalk absent. Veil absent.

Spores 6–7 x 5.5–7 μ round, smooth, white spore print.

On dead wood of hardwoods and conifers, in F.; widely distributed. There are several small species with minute white caps. *Panellus mitis* (Pers. ex Fr.) Singer is very similar but has amyloid reacting, sausage-shaped spores (3.5–6 x 0.8–1.2 μ).

92 PLEUROCYBELLA PORRIGENS   Nonpoisonous/Numerous

(Pers. ex Fr.) Singer

Cap large, fan-shaped, white, margin inrolled; flesh very thin, white

Cap 5–10 cm broad, convex, petal-like to fan-shaped, dry, minutely hairy, white, margin inrolled at first. Flesh very thin, pliant, white. Gills extend to stalk base, crowded, narrow, white to cream color. Stalk absent, only a pluglike attachment to the host. Veil absent.

Spores 5–7 x 4.5–6.5 μ almost round, smooth, white spore print.

In overlapping clusters from the sides of conifer logs and stumps in late S. and F.; widely distributed. This has often been noted on hemlocks. The shiny white caps attract attention against the somber browns of the forest floor. The very thin, sessile (lacking a stalk), pliant or flexible, white fruiting bodies with very narrow crowded gills are distinctly different from the more fleshy species of *Pleurotus;* the latter also have broader gills and are not pliant.

93 PLEUROTUS OSTREATUS   Edible, choice/Common

Fries ["Oyster Mushroom"]

Cap white to yellow-brown, moist, oyster-shell shape; stalk absent; flesh thick

Cap 2–30 cm broad, 0.8–8 cm wide, oyster-shell to fan-shaped to broadly convex, moist, smooth, without hairs, white, pale pink to light yellow-brown, margin sometimes lobed or wavy. Flesh firm, thick, dull white. Gills extending to base on the blunt point of attachment, fairly well separated, broad, thick, veined, white. Stalk usually absent; if present 5–15 mm long, 5–10 mm thick, off center, equal, dry, white, covered with a downy white pubescence, and often dense with white hairs around base. Odor of anise or fragrant and fruity.

Spores 8–12 x 3.5–4.5 μ elliptical, smooth, white to buff spore print.

Single but usually in large overlapping (imbricate, (Fig. 2) clusters on branches, logs, and stumps of hardwoods and conifers in Sp., early S., F. and in the W.; widely distributed. I have found it commonly on aspens, willow, beech, and pines, but it occurs on many other trees.

The distinctive characteristics include white, oyster-shell-like caps, white to buff spores, and usually a sessile growth habit. A closely related species, *P. sapidus* (Schulzer) Kalch., has a dull white to brown cap, lilac-colored, narrow spores (2.5–3.5 μ wide), and often has an off-center, short stalk. *Pleurotus columbinus* Quel. apud Bres. is a name which was once used for the brown variant of *P. sapidus* and is not a separate species. A shiny black beetle will very frequently be found in numbers between the gills of both species; it lays eggs in the tissue of the cap where the grubs feed until they mature and reproduce the adult. Both *P. ostreatus* and *P. sapidus* are delicious edibles, so you are fortunate if you find this fungus before the beetles do. A check of the flesh during preparation will reveal if insect tunnels are present.

94 PANELLUS SEROTINUS (Fr.) Kuehn.   **Edible**/Numerous

Cap green to yellow-green, convex, viscid; gills pale orange; stalk absent

Cap 2.5–6 cm broad, 2–3.5 cm wide, convex, fan-shaped to shell-shaped, viscid, smooth, lacking hair, green to yellowish green to olive-buff, margin wavy. Flesh thick, firm, white. Gills adnate, fairly well separated, pale orange to light yellowish tan. Stalk absent, attached by a basal plug covered with dense white woolly hairs.

Spores 4–5.5 x 1–1.5 μ sausage-shaped, smooth, amyloid in Melzer's solution, yellow spore print. Cystidia (sterile cells) on gill edge 40–60 x 6–12 μ, cylindric, thin-walled.

Solitary to overlapping on logs and sticks on hardwoods and occasionally on hemlock in the F.; widely distributed.

Cross section of the flesh reveals a gelatinous layer near the cuticle.

95 PHYLLOTOPSIS NIDULANS (Pers. ex Fr.) Singer   **Nonpoisonous**/Common

Cap orange, convex, dry, densely hairy; odor strong and disagreeable

Cap 3–8 cm broad, broadly convex, dry, dense minute hairs over surface, light orange to orange-buff, margin inrolled at first. Flesh duplex (in two layers), upper layer orange-buff, lower layer very pale orange-buff. Gills adnate, close, narrow, orange-buff. Stalk absent. Veil absent. Odor often *strong and disagreeable.*

Spores 5–7 x 2–2.5 μ elliptical to slightly sausage-shaped, smooth, light reddish cinnamon spore print.

In groups on hardwoods and conifer logs and stumps, in early S. through F.; most commonly found in eastern North America but widely distributed.

96 PLEUROTUS ELONGATIPES Pk.   **Edible**/Common

Cap cream color, convex, water spots in center; stalk often long, curved, and usually in clusters attached at base; on wood

Cap 2–14 cm broad, convex, nearly flat at maturity, moist, smooth, without hairs, pinkish cream with distinctive darker, round water spots over the center. Flesh firm but not tough, white to pinkish buff with a soft cottony stalk center. Gills adnexed (notched) with a thin line on the upper stalk, subdistant, broad, veined, buff to pinkish buff. Stalk 4–22 cm long, 4–20 mm thick, nearly equal or gradually tapering toward base, often curved or bent, white down near the apex, white and smooth, without hairs over the rest except the base, which has white stiff hairs. Veil absent.

Spores 4–5.5 x 3.5–4.5 μ almost round, smooth, buff spore print.

Single but usually densely cespitose (several stalks from a common base) on logs and limbs of hardwood trees or shrubs in late S. or F.; widely distributed.

The smooth stalk surface, mottled cap, soft stalk center, and growth habit on fallen dead material separate this from *P. ulmarius,* which it resembles. It is also edible but I have no information on its taste.

97 PANELLUS STIPTICUS (Fr.) Karst   **Nonpoisonous**/Common

Cap orange-buff to brown, dry, densely hairy; gills glow in dark; taste bitter

Cap 12–32 mm long, 12–25 mm wide, convex, dry, covered with woolly hairs, orange-buff, tan to brown. Flesh tough, firm, pinkish tan. Gills close, narrow, salmon to orange-buff. Stalk 6–12 mm long, 3–8 mm thick, off center, with clusters of hairs, dull whitish. Veil absent. Taste very bitter.

Spores 3–4.6 x 1.2–2.2 μ sausage-shaped, smooth, amyloid in Melzer's solution, white spore print.

Crowded, overlapping clusters on logs and stumps in S. and F.; widely distributed.

This common *Panellus* is one of the 6 species in North America (Miller, 1970). It is an important rot of fallen hardwood limbs. The gills of fresh specimens are luminescent. It requires several minutes in the dark to let the eyes adjust to the lower level of illumination.

98 PANUS STRIGOSUS Berk. and Curt.   **Edible**/Rare

Cap large, covered with dense, short hairs, cream but drying orange; stalk off center, hairy like cap; flesh tough

Cap 10–40 cm broad, broadly convex, depressed in center, dry, covered with dense, short, stiff hairs (1–2 mm long), cream to buff turning mustard-yellow when dry, margin inrolled with fine, dense hairs. Flesh tough, thick, yellowish. Gills extend down stalk, well separated, thick, broad, white to lilac-brown with tints of gray or yellow when dried; edges smooth, not toothed. Stalk 2–15 cm long, 2–4 cm thick, enlarging toward base, off center to lateral, hairy like cap, white to buff but drying mustard-yellow like cap. Veil absent.

Spores 10–13 x 3–4 μ nearly oblong, smooth, white spore print. Cystidia (sterile cells) present on gill sides and edge, tapering at end to club-shaped, thin-walled.

Alone or several together from wounds on various living hardwoods, in late S. and F.; found in eastern North America but also on Arizona walnut in Cochise Co., Arizona.

This is the largest *Panus* in the genus. It is reported to be edible when soft and young but I know of no one personally who has tried it. The stiff erect hairs on the cap and stalk and the orange color of both when dry are the chief field characteristics.

99 PANUS RUDIS Fr.   **Edible**/Numerous

Cap medium size, dense curled hairs, tan to red-brown, not drying orange; flesh tough

Cap 1.5–7 cm broad, convex, dry, velvety from dense, long, stiff hairs over the entire surface, pinkish tan to light reddish brown, margin inrolled, often irregularly lobed. Flesh tough, thin, white. Gills extend down stalk, close, narrow, edges smooth (not serrate), white to pallid. Stalk a lateral, short, stout plug of tissue, hairy and colored like the cap. Taste slightly bitter.

Spores 4.5–7 x 2.5–3 μ short elliptical, smooth, white spore print. Cystidia (sterile cells) on side and edge of gill, club-shaped, thick-walled.

Single but usually several together on logs and stumps of hardwoods in Sp., S., and F.; widely distributed.

P. fragilis O. K. Miller is a smaller species with a lighter cap which is known only from the S and SE. Both are tough, which would tend to disqualify them as edibles.

100 PANUS TORULOSUS Fr.   **Edible**/Numerous

Cap tan tinted lilac to reddish brown, smooth, without hairs; gills violet fading to buffy tan; flesh leathery

Cap 1.5–8 cm broad, broadly convex, somewhat depressed in center in age, smooth, without hairs, dry, color variable, tan, often with a light lilac hue to reddish brown, margin inrolled, minutely downy at first. Flesh firm, leathery, white. Gills extend down stalk, fairly well separated, edges even, violet at first fading to tan or buffy tan. Stalk 2–6 cm long, 2–3 cm thick, off center, thick, stout, tough, covered with fine violet-tinted hairs. Veil absent.

Spores 5–7.5 x 2.5–3 μ elliptical, thin-walled, white spore print. Cystidia (sterile cells) on sides and edge of gills, narrowly club-shaped.

Usually several to cespitose clusters on hardwood stumps and logs in Sp., S., and F.; widely distributed.

This species is also known as *P. conchatus* Fries. I would rate it as inedible because it is so tough.

101 PLEUROTUS ULMARIUS (Fr.) Kummer   **Edible**/Numerous

Cap large, cream color with small scales; stalk often off center; flesh tough; from wounds on live trees

Cap 7–12 cm broad, broadly convex to shallow depressed, dry, covered with very fine flattened scales, white, cream or pinkish buff in age. Flesh firm, hard, tough, white throughout. Gills adnexed (notched) to adnate, fairly well separated, broad, white to cream. Stalk 6–9 cm long, 15–25 mm thick, centered or off center, equal or swollen near base, dry, minute hairs at apex, the rest is smooth, without hairs, white, pinkish in age with longitudinal ropelike cords of tissue revealing a white flesh beneath. Veil absent.

Spores 5–8 x 4–5 μ almost round, smooth, buff to white spore print.

Single to several on hardwoods, usually on living trees in wounds or decayed branch stubs, often high up in the tree, in F. and early W.; mostly in eastern North America.

It is tough and would have to be cooked thoroughly before eating. It is much tougher than *P. sapidus* or *P. ostreatus,* is never overlapping, and has almost round spores. It does not have the mottled cap surface and soft cottony stalk center of *P. elongatipes* Peck, which is found in western as well as eastern North America.

THE ARMILLARIAS AND ALLIES

These are usually robust mushrooms with non-waxy, short decurrent (extending partially down stalk) to notched gills. They have, with one exception, a single or double annulus (ring). The cap may be covered with fibrils but it never appears granulose. Most of the species are found on the ground, but *Armillariella mellea* and *A. tabescens* will grow either on the ground from buried wood or directly on wood. All the species have white, entire, smooth spores which are fusiform (tapered at the ends) to elliptical. The genus *Catathelasma* has spores that have an amyloid reaction and a double veil, but the other species have spores that are non-amyloid (yellowish) in Melzer's solution, and a single veil. Most of the species are probably mycorrhiza forming except the Armillariellas which are parasites on the roots of hardwoods and conifers (see Introduction).

One should also be aware of other white spored mushrooms which have annulus. If the gills are free, it must be a mushroom in the Lepiotaceae or Amanitaceae. The presence of a volva (Fig. 9) would place it in the Amanitaceae; the Lepiotaceae lack a volva. If the cap and lower stipe (stalk) appear conspicuously granulose and the gills are not free, it is probably a *Cystoderma*. Waxy gills would indicate a *Hygrophorus*.

Edibility. There are no reports of toxins in this group. Some of the species are tough and tasteless but others can be counted among the really good edible mushrooms. *Armillariella mellea* and *A. tabescens* are in this group. *Armillaria caligata* is sought after in Japan but has received no particular attention in North America. *Armillaria ponderosa* is a large, thick, fleshy species which is collected for the table in western North America. The Armillarias are a good basic group of edibles that can serve in a wide variety of recipes. *Armillariella mellea,* the Honey Mushroom, is particularly good when sautééd in butter and served on steak or mixed with scrambled eggs. It should also be used for the "Hot Mushroom Dip" (see Recipes).

Key to the Armillarias and Allies

1 a. On wood, in cespitose (served from one base) clusters or on soil from buried wood, **2**
   b. On the ground, if cespitose, then only two or three together, **3**
2 a. Veil present and persistent, *102. Armillariella mellea*
   b. Veil absent at all stages, *103. Armillariella tabescens*
3 a. Double veil; spores with amyloid reaction; cap often very large (8–40 cm), **4**
   b. Single veil; non-amyloid spores; cap medium size (usually 5–15 cm), **5**
4 a. Cap 15–40 cm broad; olive-brown; western North America,
     104. Catathelasma imperialis
   b. Cap 8–10 cm broad; light ash-gray; eastern North America,
     105. Catathelasma ventricosa
5 a. Fruiting body basically white with reddish brown stains; sweet odor,
     106. Armillaria ponderosa
   b. Fruiting body colored; no odor or of almonds, **6**
6 a. Cap bright yellow in age, with cinnamon scales over the center; white scales on stalk below ring; western North America, *107. Armillaria albolanaripes*
   b. Cap colored differently; scales colored below ring, **7**
7 a. Cap viscid, orange over center with a white margin; scales on stalk with orange-brown tips; western North America, *108. Armillaria zelleri*
   b. Cap dry, center covered with reddish brown scales; brownish red hairs on stalk; eastern North America, *109. Armillaria caligata*

102 ARMILLARIELLA MELLEA

Edible, choice/Common

(Fr.) Karst. ["Honey Mushroom"]

Cap viscid, erect hairs over center; has ring; cespitose clusters on wood

Cap 3–12 cm broad, convex, convex-knobbed, sticky to viscid or even glutinous in wet weather, *fine, erect hairs over the center,* very finely striate on the margin, variable in color from yellow, honey-yellow, pinkish brown to brown. Flesh thin, white with brown to rusty brown discoloration in age. Gills adnate to extending partially down stalk in age, fairly well separated, brown, white to dingy cream color with rusty brown to pinkish brown stains in age. Stalk 4–15 cm long, 6–20 mm thick, nearly equal, enlarging slowly toward apex, dry, white with downy flattened hairs above ring, white to buff, becoming brown, cinnamon-pink to reddish brown below ring. Veil cottony, with clusters of hairs, white to buff or mustard-yellow, leaving a superior, persistent ring which is dry but in wet weather often has a viscid margin.

Spores 6–9.5 x 5–6 μ elliptical, smooth, white spore print.

Small or large cespitose clusters on living and dead trees or on soil from buried wood, in late S. and F.; widely distributed. Often fruiting year after year from the same log or stump.

This is probably one of the most variable mushrooms in North America and one of the best edibles. Several color variants and numerous growth forms have prompted mycologists to recognize several varieties. The cespitose growth habit on wood, hairs at the cap center, white spore print, and the persistent ring are its most reliable combination of characteristics. A yellow cap variant is most often observed on hardwoods, while a pinkish brown variant is common on conifer wood. Under certain, as yet unknown, conditions, this fungus is capable of becoming a virulent parasite. The name "shoestring root rot" is often used to label the disease which is characterized by proliferating rootlike strands (rhizomorphs)that blacken with age. Economic loss to commercial and shade trees is great but the extent of the damage is not fully known. The fungus will also be found under *Armillaria mellea* Fries.

103 ARMILLARIELLA TABESCENS (Scop. ex Fr.) Singer **Edible,good**/Common

Cap dry, erect hairs over center; no ring; cespitose clusters on wood

Cap 3–10 cm broad, convex, flat or slightly depressed in age, dry, with dense, minute, erect hairs in center, with the rest flattened hairs, nearly yellow, honey-yellow to pinkish brown in age. Flesh thick near center, white to tinted or stained brown in age, brown at stalk base. Gills extend partially down stalk, fairly well separated, white to light pinkish, stained brown in age. Stalk 5–20 cm long, 4–16 mm thick, equal or tapering toward base, dry, silky white near apex gradually darkening to brown over the base. *Veil absent.*

Spores 6–8 x 5–5.5 μ broadly elliptical, smooth, white spore print.

Grows densely in large clusters attached at base on living and dead tree roots, stumps or on the ground from buried wood or roots. Often found as a parasite on oaks and other trees in the SE, but also found in CS, NE, and So., in S. and F.

This fungus is similar to *A. mellea* but even the most minute buttons do not possess a veil, and the cap is dry. Both species produce rootlike structures (rhizomorphs) in nature as well as in laboratory cultures. *A. tabescens* can also act as a parasite but I have always found it on oak and never noted it on any conifer wood. It has also been called *Clitocybe monadelpha* Morg. and *Clitocybe tabescens* Bres. in earlier American publications. Like *A. mellea* it is a choice edible.

104 CATATHELASMA IMPERIALIS (Fr.) Singer   **Nonpoisonous**/Infrequent

Cap large, viscid, olive-brown; veil double; gills extend down stalk

Cap 15–40 cm broad, convex, broadly convex to plane in age, viscid when wet, soon dry, somewhat cracked at the center, cinnamon-buff, sometimes flushed olive to dark brown in age, margin incurved at first and outer veil continuous with the margin in buttons. Flesh very thick, firm, white. Gills extend down stalk, close, broad, forked, buff to olive-gray. Stalk 12–18 cm long, *6–8 cm thick,* tapering to a dull point at base, dry, surface covered at first to the base by a membranous, pinkish buff to brownish veil. Veil double; outer veil membranous and attached to upper cap margin; inner veil softer, with clusters of hairs attached on lower margin of cap, leaving a persistent, superior, double ring which flares upward at first.

Spores 10–15 x 4–5.5 μ long, tapering at ends, smooth, *amyloid reaction in Melzer's solution,* white spore print.

Single or several together under conifers in the mountains of the PNW in S. and F.

This is an extremely large fungus with a double veil and spores that have an amyloid reaction. The buttons are often six inches (15 cm) thick and very robust. In North America, it seems to be confined to the dense forests of the western mountains.

105 CATATHELASMA VENTRICOSA   Edibility unknown/Infrequent

(Pk.) Singer

Cap dry, ash-gray; veil double; gills extend down stalk

Cap 8–10 cm broad, convex, broadly convex in age, dry, smooth, without hairs, pallid to *ash-gray.* Flesh firm, white. Gills extend partially down stalk, close to fairly well separated, often forked, pinkish buff. Stalk 6–8 cm long, 3–4 cm thick, tapering to a narrow base, dry, white above the ring, dull yellow-brown below. Veil pallid with a white, hairy inner veil, leaving a superior, persistent, double ring which flares upward at first.

Spores 8.5–11 x 4–6 μ elliptical, smooth, *amyloid reaction in Melzer's solution,* white spore print.

Alone or several, under mixed woods in late S. and F., in eastern North America.

The even, ash-gray color and smooth appearance, combined with the double veil and amyloid (blue) reaction of the spores in Melzer's solution are the distinguishing characteristics.

106 ARMILLARIA PONDEROSA (Pk.) Sacc.   Edible,good/Common

Cap white with brown scales, tacky or dry; ring white, single; odor spicy

Cap 8–20 cm broad, broadly convex, flat in age, tacky to moist, white with flattened brownish scales developing over the center and streaked with brown hairs elsewhere, margin a narrow, white, inrolled flap. Flesh firm, white. Gills adnexed (notched), crowded, broad, white, bruising pinkish to reddish brown. Stalk 6–18 cm long, 2–4 cm thick, tapering somewhat toward base, sticky when wet to dry, smooth and white above the ring, below covered with reddish brown scales and hairs in age. Veil membranous, white, leaving a superior, persistent, skirtlike ring soon stained reddish brown on the edge and outside. Odor spicy, with a sweet smell even in button stage.

Spores 5.5–7 x 4.5–5.5 μ broadly elliptical, smooth, white spore print.

Scattered to abundant under mixed conifers in F.; widely distributed but abundant and common only along the Pacific Coast.

This robust, handsome fungus is distinguished by the thick stalk and cottony ring, non-waxy gills, reddish brown scales, and sweet odor. It is frequently collected for the table in the Far West. *Hygrophorus subalpinus* is similar but fruits in the Sp. near snowbanks. However, they are both edible.

107 ARMILLARIA ALBOLANARIPES Atk.   Edibility unknown/Numerous

Cap viscid, bright yellow with brown scales; ring and stalk shiny white

Cap 5–12 cm broad, convex, broadly convex to plane, sometimes with an obscure low knob, margin viscid to moist over center, bright yellow to mustard-yellow with rich reddish to cinnamon-brown flattened scales and hairs over the center which darken in age, margin inrolled somewhat at first and covered with ragged, pure white veil remnants. Flesh firm, white to yellow beneath cap cuticle. Gills adnexed (notched), fairly well separated, broad, white to cream or yellowish in age, edges often toothlike. Stalk 3–8 cm long, 10–25 mm thick, equal, smooth and shiny white above cottony ring, covered with shiny, ragged, white scales below, staining yellowish or even dingy yellow-brown in age. Veil fragile, cottony, white leaving a ragged, short, superior ring which hangs skirtlike.

Spores 5–7 x 3–4.5 μ elliptical, smooth, white spore print.

Solitary to several under western conifers in S. and F., in PNW and RM.

This handsome fungus with its bright yellow cap, cinnamon-brown scales and shiny, white stalk and veil is not likely to be confused with any other fungus. I have often found it on hard ground beside roads or on roadbanks. I have not tried to eat it and I have seen no reports by anyone who has.

108 ARMILLARIA ZELLERI Stuntz and Smith   Edibility unknown/Numerous

Cap viscid, orange-brown with white margin; ring white; stalk orange-brown

Cap 5–15 cm broad, nearly conic, convex with a broad knob to nearly flat in age, viscid when fresh with moist, minute flattened hairs, solid, bright orange-brown over the center becoming scaly only in age, abruptly white over the inrolled margin which has clinging pieces of veil at first. Flesh thick, white, and bruises orange-brown. Gills adnate (attached), crowded, white to buff and staining rusty brown in age. Stalk 4–13 cm long, 1–3 cm thick, tapering to a blunt pointed base, dry, nearly smooth, white, with flattened, downy hairs above the ring, covered below with loose scales which have orange-brown edges or stained orange-brown generally near base. Veil white, membranous, leaving a persistent, superior ring that flares upward at first.

Spores 4–5.5 x 3–4 μ broadly elliptical, smooth, white spore print.

Scattered to numerous or even in partial fairy rings under western conifers in F. in the PNW. I have collected this species in quantity under pure western white pine plantations near the west side of Priest Lake, Idaho. *Tricholoma aurantia* (Schaeff. ex Fr.) Ricken has somewhat similar coloration but the nearly hyaline (transparent) veil separates early and usually leaves no trace of a ring.

109 ARMILLARIA CALIGATA Viv. and Bres.   Edible, choice/Common

Cap dry, flattened hairs, reddish brown; ring dull white, flares upward

Cap 6–12 cm broad, convex, broadly convex to flat, sometimes slightly depressed, dry, center covered with *flattened, hairy, reddish brown to brownish drab scales* over a pallid to pinkish cinnamon surface, margin silky with few scales. Flesh firm, white. Gills wavy to a fine line, close, broad, white, sometimes conspicuously veined. Stalk 4–9 cm long, 1.5–4 cm thick, equal or somewhat larger in center, dry, smooth and very light tan above ring, covered with brownish red hairs below. Veil soft, membranous, dull white, leaving persistent, superior ring which flares upward at first, soon dingy brownish like stalk. Odor none or occasionally reported as almond-like.

Spores 6–8 x 4.5–5.5 μ elliptical, smooth, white spore print.

Single to scattered or numerous under mixed hardwood and conifers in F. in eastern North America, on rare occasions along the Pacific Coast.

This species often fruits in large numbers under mixed conifers and hardwoods in the Northeast and under mixed oaks and pine in the Southeast. The typical red-brown cap scales, upward-flaring ring, white spore print, and hairy lower stalk set this apart from other Armillarias. Described in Europe as *Tricholoma caligatum* (Viv.) Ricken and known in Japan as *Armillaria matsutake* Ito and Imai, it is sought after as an excellent edible in Europe and Asia but not often in North America.

CYSTODERMA

Cystoderma is made up of a group of relatively small species (1.5–8 cm broad), growing on ground or well-decayed wood, with *granulose appearing dry caps.* The stipe is *densely granulose below the annulus.* A partial veil is present in all species. The gills are attached, non-waxy and the spores are entire, smooth and have amyloid to non-amyloid reactions. There are fourteen species in North America (Smith and Singer, 1945); the widespread species are described here.

Edibility.   I have no knowledge of any toxic species in this group. Several are reported to be edible but I have not eaten them. They are firm and fleshy and often found in great numbers so they could serve as good edibles.

Key to Cystoderma

1 a. Cap yellow-brown to yellow-orange; in moss under conifers; veil leaves poorly formed ring, *110. C. amianthinum*
  b. Cap another color; not as above, **2**
2 a. Cap rusty brown; ring persistent (see 110. *C. amianthinum*) *C. fallax*
  b. Not as above, **3**
3 a. Cap bright orange; on rotten hardwoods in western North America (see 110. *C. amianthinum*), *C. granosum*
  b. Cap orange-brown, covered with erect, pointed scales; in conifer duff and on well-decayed wood, *111. C. cinnabarinum*

---

110 CYSTODERMA AMIANTHINUM   Nonpoisonous/Common

(Scop. ex Fr.) Fayod

Cap dry, granulose, yellow-brown; ring poorly formed; in moss

Cap 1.5–4.5 cm broad, convex to broadly convex, dry, radially rugulose (finely wrinkled), granulose often forming granular scales, yellow-brown to yellow-orange, margin with conspicuous flaps of veil tissue. Flesh firm, white, thin. Gills adnate to notched, close to crowded in age, edges even, orange-yellow to buff. Stalk 25–60 mm long, 2–7 mm thick, equal or slightly enlarged downward, has a powdery cover, and pallid to buff above ring, densely granulose below and same color as cap. Veil silky above, granulose below, fragile, soon breaking to form a superior, poorly formed ring and leaves pieces hanging from cap margin.
Spores 4.5–7.5 x 2.5–3.5 μ elliptical, thin-walled, amyloid (blue) reaction, white spore print.
Single or in dense troops, usually in moss, under conifers, in S. and F.; widely distributed.
Cystoderma fallax Smith and Singer is a PNW species with a persistent ring and a more rusty cap color. *C. granosum* (Morg.) Smith and Singer has a bright orange cap and is found on the rotten wood of deciduous trees in eastern North America.

111 CYSTODERMA CINNABARINUM (Secr.) Fayod   **Nonpoisonous**/Common

Cap dry, granulose, orange-brown; ring soon gone; on conifer wood

Cap 3–8 cm broad, egg-shaped, convex, often with a broad knob, dry, densely granular and covered with small sharp-pointed scales, dense at center, margin incurved with hanging veil fragments at first, orange-brown to cinnamon-brown. Flesh thin, pallid near cap cuticle, light orange to light rusty brown. Gills adnate to nearly free in age, close to crowded, thin, white, edges minutely roughened. Stalk 3–6 cm long, 6–15 mm thick, nearly equal, with a powdery cover above ring, covered below with granules similar to and with the same color as cap. Veil fragile, granulose leaving a thin ring which soon disappears and hanging pieces on cap margin.

Spores 3.5–5 x 2.5–3.5 μ elliptical, smooth, non-amyloid, white spore print.

In humus, needle duff, or in well-decayed conifer wood in conifer or hardwood stands, in late S. and F.; widely distributed.

The large size, cinnamon color and non-amyloid reaction in the spores are the important characteristics.

MARASMIUS

The species of *Marasmius* are small, rather tough, and have pliant, thin stipes (stalks). The entire fruiting body and stalk will revive and regain a fresh appearance when placed in water. This ability to revive when moistened is also possessed by the species of *Xeromphalina* and *Crinipellis*. *Marasmius* is also typified by white, smooth, non-amyloid reacting spores and a pileus cuticle composed of club-shaped cells similar to that shown in Figure 80 but not as rounded. The gills are adnate, adnexed (notched), or nearly free but do not extend down stalk. Collybias are similar but have brittle stalks and are often larger with broader caps. Mycenas have conic caps and also have brittle stalks. There are a number of species of *Marasmius* not treated here, and a definitive work on this group in North America has yet to appear. The species are generally found on leaves, needles, and grass.

Edibility.   I have no reports of poisons or toxic compounds in these mushrooms. Only *M. oreades,* however, is commonly eaten.

Key to Marasmius

1 a. Stalk clothed in fine hairs, **2**
   b. Stalk glabrous (smooth, without hairs), often shiny blackish brown, **4**
2 a. Cap campanulate (bell-shaped); scattered in grass, *112. M. oreades*
   b. Cap plane or slightly depressed; densely cespitose (several from one base) in leaves and humus, **3**
3 a. Odor and taste mild, 113. *M. confluens*
   b. Odor and taste of garlic (see 113. *M. confluens*), *M. polyphyllus*
4 a. Odor, when crushed, of garlic, *114. M. scorodonius*
   b. No garlic odor, **5**
5 a. Cap white; gills attached to a collar around stalk; stalk shiny black, 115. *M. rotula*
   b. Cap rose to brown; gills adnate, adnexed (notched), or free, **6**

80

6 a. Cap campanulate, delicate rose-colored, parasol-like; gills adnexed to nearly free, *116. M. siccus*

   b. Cap convex, depressed in center, brown with a pinkish tint; gills attached to stalk, *117. M. androsaceus*

112 MARASMIUS OREADES Fr.   Edible, choice/Common

Cap dry, bell-shaped, light tan; stalk hairy; in grass

Cap 2–6 cm broad, convex, bell-shaped in age, dry, smooth, without hairs, light tan, light brown to reddish brown usually faded in age, margin somewhat striate in age. Flesh thin, watery, white. Gills adnexed (notched), nearly free, fairly well separated, veined, broad, light buff. Stalk 3–7 cm long, 3–5 mm thick, equal, tough, dry, buff at apex to reddish brown at base, minute hairs especially dense over the base. Veil absent. Odor fragrant.

Spores 7–11 x 4–7 μ pip-shaped (like an apple seed), smooth, white to buff spore print.

Grows in groups or in fairy rings on lawns, pastures, and grasslands of various kinds, in Sp., S., and F.; widely distributed.

The tough stalks and characteristic bell-shaped caps, along with its occurrence in grass, distinguish *M. oreades* from other common species of *Marasmius.* The fairy rings increase in size each year as the mycelium (or spawn) grows out into new grass. The fungus plant dies within the circle, leaving an advancing ring of mycelium from which the fruiting bodies arise. It is a very good edible with a flavor that enhances vegetables or gravies. "Marasmius Cookies" (see Recipes) are also enjoyed by mushroom fanciers.

113 MARASMIUS CONFLUENS (Fr.) Pers.   Nonpoisonous/Common

Cap dry, reddish brown to pinkish gray, plane in age; stalk hairy; cespitose clusters on wood debris

Cap 18–50 mm broad, convex, flat sometimes with a small central depression, dry, smooth, without hairs, reddish brown fading to grayish pink when moist, margin faintly striate. Flesh thin, tough, pliant, white. Gills free, crowded, narrow, white. Stalk 5–10 cm long, 2–5 mm thick, tough, pliant, equal or slightly larger toward base, sometimes grooved or ridged, dry, pinkish to reddish under a dense covering of very short hairs.

Spores 4–6 x 3–4 μ pip-shaped, thin-walled, white spore print.

Found in dense cespitose clusters among fallen leaves and debris on the ground under hardwoods and conifers, in S. and F.; widely distributed.

Known as *Collybia confluens* (Pers. ex Fr.) Kummer by many authors. *Marasmius polyphyllus* Peck is somewhat similar, not as common, and has the odor and taste of garlic.

114 MARASMIUS SCORODONIUS Fr.   Nonpoisonous/Numerous

Cap dry, flat in age, brown; narrow gills; tapering stalk; garlic smell

Cap 5–12 mm broad, dry, convex then flat, pliant, brown but tinted reddish, margin whitish, wrinkled, wavy. Flesh very thin, whitish. Gills attached, close, narrow, whitish. Stalk 20–30 mm long, 1–2 mm thick, tapering toward base, tough, smooth, hairless, white at apex, blackish brown below. Veil absent. Odor of strong garlic when crushed (crush cap material).

Spores 6–8 x 3–4 μ elliptical, smooth, white spore print.

Grows in groups on twigs and debris in grassy places and under forest trees in Sp., S., and F.; widely distributed.

The garlic smell, tapering stalk, and narrow gills are the distinguishing characteristics.

115 MARASMIUS ROTULA (Fr.) Kummer   Nonpoisonous/Common

["Horsehair Fungus"]

Cap dry, white, striate margin; gills join collar at apex of stalk

Cap 3–12 mm broad, convex, with a central, naval-like depression, radially striate, plicate, white, margin wavy. Flesh very thin, membranous, white. Gills joined to a collar which encircles the stalk, well separated, broad, white. Stalk 12–50 mm long, 0.2–1 mm thick, tough, equal, smooth, hairless, shiny black, hollow. Veil absent.

Spores 7–10 x 3–5 μ pip-shaped, smooth, white sport print.

Several or in troops on twigs, leaves, and debris on the ground, in Sp., S., and F. Very common in eastern North America especially in New England on west to the Lake States.

116 MARASMIUS SICCUS (Schw.) Fries   Edibility unknown/Common

Cap dry, delicate rose color, striate, like a parasol; veil absent; on leaves

Cap 10–25 mm broad, convex, bell-shaped, often depressed in center, striate almost to the top, dry, *rose colored,* darker at center. Flesh very thin, white. Gills adnexed (notched) to free, well separated, white. Stalk, 3–8 cm long, 0.2–1 mm thick, equal, tough, smooth, hairless, dry, white at apex, brown to dark brown at the base, which is covered with white mycelium and is hollow. Veil absent.

Spores 13–18 x 3–4.5 μ pip-shaped, smooth, white spore print.

Several or numerous on leaves, twigs, and debris in S. and early F.; in eastern North America.

This beautiful *Marasmius* resembles a rose-colored parasol (Fig. 7).

117 MARASMIUS ANDROSACEUS (L.) Fries   Nonpoisonous/Common

Cap dry, pinkish brown, minute; stalk shiny black; veil absent

Cap 4–12 mm broad, conic, broadly convex, depressed in center, smooth, without hairs, dry, brown with a pinkish tint, margin striate to radially wrinkled. Flesh very thin, dull white. Gills adnate (attached), fairly well separated, whitish to flesh color. Stalk 2–5 cm long, 1–3 mm thick, equal, tough, dry, hairless, pale at apex, rest shiny black, hollow. Veil absent.

Spores 6–9 x 2.5–3.5 μ broadly elliptical, smooth, white spore print.

Grows in groups, often in great numbers on conifer needles during wet periods in Sp., S., and F.; widely distributed. I have seen great numbers of this species in the PNW and A during and following heavy rain.

118 ASTEROPHORA LYCOPERDOIDES   Nonpoisonous/Numerous

(Bull.) Ditmar ex S. F. Gray

Cap dry, powdery white; on dead mushrooms

Cap 1–2 cm broad, nearly round, dry, powdery, white. Flesh firm, pallid. Gills adnate, well separated, narrow, white. Stalk 2–3 cm long, 3–8 mm thick, equal, minutely downy, white stained brown in age. Odor and taste of fresh meal.

Spores 12–18 μ round, spiny, brownish spore print.

Several together on other mushrooms, especially *Russula* and *Lactarius,* in S. and F.; widely distributed.

This curious fungus is also found on other agarics but they are often well rotted and beyond recognition. The cap is covered with thick-walled warted, spores (which are oval to elliptical) giving it a powdery appearance.

Asterophora parasitica (Fr.) Sing. is also parasitic on mushrooms but the cap is covered with flattened, gray hairs.

CLITOCYBE, OMPHALINA, AND OMPHALOTUS

All three genera have decurrent gills and are treated together because they are difficult to tell apart. The margin of the cap is inrolled at first and the cap center is often depressed at maturity. The spores are white to buff, smooth or minutely spiny, non-amyloid reaction, and entire. The stipe is brittle or fibrous but not pliant and there is no annulus.

Omphalotus consists of one species, *O. olearius,* and is a large, distinctive, orange species with luminescent orange gills. In contrast, the Omphalinas are a group of small, delicate, waxy-appearing species in moss or surrounded by lichens. All species of *Clitocybe* are medium to large, often fleshy; on wood, needles, or directly on the ground.

Bigelow's study (1969, 1970) of the latter genera has revealed that a good many species in North America are restricted to specific habitats or narrow ranges. The total number of species is not yet known, but there are many.

Edibility. Several species, including *Omphalotus olearius, Clitocybe dealbata,* and *C. aurantiaca,* are poisonous. The toxic effects include gastrointestinal symptoms that can be quite serious and are discussed under Group VI Toxins.

There are a number of good edibles in *Clitocybe* which are found in quantity and possess good flavor. *Clitocybe irina* and *C. nuda* both have strong, distinctive flavors which make them well suited for casseroles and excellent in turkey stuffing. They are excellent for canning and should be on the list of those that can be preserved for off-season use. A little goes a long way in most dishes because of the strong, distinctive flavor.

In any event, in a group that includes both edible and poisonous species it is important to read the identifications and study the illustrations with care.

Key to Clitocybe, Omphalina, and Omphalotus

1 a. Cap 7–11 cm broad, orange to orange-yellow; in cespitose clusters on the ground, usually around hardwood stumps, *119. Omphalotus olearius*
  b. Not as above, **2**
2 a. Cap small, waxy appearing, 3–40 mm broad; in moss or surrounded by lichens, usually on ground or on moss-covered, well-decayed wood or conifer logs and stumps, **3**
  b. Cap large, dry to felted appearing, 2–20 cm broad; in humus, on lawns or on wood, **8**
3 a. Cap minute (3–8 mm broad), orange; in tufts of moss, *120. Omphalina fibula*
  b. Cap larger, yellow, straw-yellow to yellow-brown or even tinted gray, **4**

4 a. Gills orange to orange-yellow; on conifer wood (logs, stumps, etc.), **5**

   b. Gills pale to bright yellow; in moss and algae (sometimes old moss covered logs), **6**

5 a. Cap yellow-brown, tinted gray, 121. *Omphalina chrysophylla*

   b. Cap orange, (see 122. *O. ericetorum*), *Omphalina luteicolor*

6 a. In tundra, usually in the arctic or in subarctic tundra areas, **7**

   b. Widely distributed in North America, 122. *Omphalina ericetorum*

7 a. Cap and stalk bright yellow (see 122. *Omphalina ericetorum*)—O. luteovitellina

   b. Cap orange-yellow; stalk pale yellow (see 122. *O. erictorum*), *Omphalina hudsoniana*

8 a. Cap large (9–20 cm broad); stalk robust (2.5–4 cm thick), **9**

   b. Cap smaller (1.5–10 cm broad); stalk slender (4–25 mm thick), **10**

9 a. Cap gray to drab, *123. Clitocybe nebularis*

   b. Cap dull white, (see *C. 123. nebularis*), *C. robusta*

10 a. Cap, gills, and stalk white or nearly white overall, **11**

   b. Cap, gills, and stalk not white throughout, **12**

11 a. Cap 15–33 mm broad; spores 4–5 x 2.5–3 μ, *124. C. dealbata*

   b. Cap 4–13 cm broad; spores 7–10 x 4–5 μ, *125. C. irina*

12 a. Cap light gray to chalky white in age; in cespitose clusters, *126. C. dilatata*

   b. Cap another color; not in cespitose clusters, **13**

13 a. Odor strong and fragrant; cap pale bluish green, *127. C. odora*

   b. Odor mild; cap another color, **14**

14 a. Cap violet to violet-gray; gills and stalk pale violet, *128. C. nuda*

   b. Cap, gills, and stalk not colored as above, **15**

15 a. Gills close, forked, bright orange to orange-yellow, *129. C. aurantiaca*

   b. Gills white to buff, **16**

16 a. Stalk base forming a basal bulb; cap convex-depressed, *130. C. clavipes*

   b. Stalk not forming a basal bulb; cap deeply depressed, *131. C. gibba*

119 OMPHALOTUS OLEARIUS   **Poisonous**/Common

(D. C. ex Fr.) Singer ["Jack-O'Lantern"]

Cap dry, orange; gills orange, extend down stalk, luminescent; in clusters on wood

Cap 7–11 cm broad, convex to flat, depressed in center, often with a shallow knob, dry, smooth, streaked with flat-lying hairs, *bright orange* to *orange-yellow,* margin inrolled, nearly even in age. Flesh thin, firm, white tinted orange. Gills extend down stalk, close, *yellow-orange,* luminescent in the dark. Stalk 5–18 cm long, 5–22 mm thick, tapering to a narrow base, dry, minutely downy or somewhat scaly in age, even, *light orange.* Odor unpleasantly sweet.

Spores 3–5 μ round, smooth, creamy white spore print.

Large dense cespitose clusters at the base of hardwood stumps or on the ground from buried roots (very often oak) in late S. and F.; eastern North America and across the southern United States to the Pacific Coast.

The old name in most books is *Clitocybe illudens* Schw. The luminescence of the gills will often last 40–50 hours after collecting. Luminescence may be seen after sitting in the dark facing the gills for about 2–4 minutes and is usually bright enough to enable one to see each gill clearly. If it does not luminesce, wrap in wax paper, place indoors, and repeat the test in two to three hours. This feature combined with the orange color and growth in cespitose clusters on wood distinguishes this fungus from all others. It is a poisonous species but the toxins are unknown. It is placed in Group VI toxins and should be avoided.

120 OMPHALINA FIBULA Fr.   **Nonpoisonous**/Numerous

Cap minute, orange; gills extend down stalk; stalk white; in moss

Cap 3–8 mm broad, conic to convex depressed with an inrolled margin, dry to moist, orange to reddish orange, with a dark center, and finely striate margin. Flesh thin, whitish. Gills extend down stalk, close, narrow, white to buff. Stalk 2–5 cm long, 0.5 mm thick, equal, dry, white tinted yellow, minutely downy, tough. Veil absent.
Spores 5–6 x 2–3 μ elliptical, smooth, white spore print.
This minute species usually grows in groups in tufts of moss in moist areas in hardwood and hardwood evergreen (conifer) forests, in Sp., S., and F.; widely distributed. I have collected this most often in the Lake States and New England.

121 OMPHALINA CHRYSOPHYLLA (Fr.) Murr.   **Edibility unknown**/Numerous

Cap buff-brown; gills yellow-orange; on conifer wood

Cap 5–40 mm broad, convex depressed to flat depressed, dry to moist, minutely hairy, scaly, buff-brown with tints of gray or drab, margin incurved at first. Flesh thin, pliant, reddish orange. Gills extend down stalk fairly well separated, orange-yellow to apricot. Stalk 2–4 cm long, 1.5–3 mm thick, equal, often curved, smooth, without hairs, moist, or with a few scattered hairs, yellow to smoky orange-red. Veil absent.
Spores 8.5–15.5 x 4.5–6 μ elliptical, smooth, yellow spore print.
Numerous on conifer logs in Sp., S., and F.; widely distributed.
The buff-brown scaly cap, yellow-orange gills, and conifer wood habitat are distinctive characteristics of this species.

122 OMPHALINA ERICETORUM (Fr.) M. Lange   **Nonpoisonous**/Common

Cap straw color, dry, striate margin; gills extend down stalk; surrounded by minute round lichens

Cap 5–35 mm broad, flat with slightly incurved margin, deeply depressed in age, moist to dry, deeply striate to striate, margin wavy, light brown at first, soon straw to yellowish or fading to whitish in age. Flesh very thin, pliant, color of cap. Gills extend down stalk, well separated, at times veined, pale yellowish. Stalk 10–30 mm long, 1–2 mm thick, equal, often curved, dry, smooth, without hairs with fine white mycelium over the base, light reddish brown at apex to light brown below, fading to pale yellowish in age. Veil absent.
Spores 7–9 x 4–6 μ elliptical, smooth, white spore print.
Scattered in low moss and algae, often on moss-covered logs, in Sp., S. but F. and W. along the Pacific Coast; widely distributed in temperate and Arctic regions. The small round lichen, *Botrydina vulgaris* Breb. is usually clustered around the fruiting bodies and can be easily seen with a hand lens. *Omphalina luteovitellina* (Pilat and Nannf.) Lange has a bright yellow cap and stalk, is found in Arctic regions of A and C and is also associated with *Botrydina vulgaris. Omphalina hudsoniana* (Jenn.) Bigelow is bright orange-yellow and associated with a small leaflike lichen, *Coriscium viride* (Ach.) Vain in tundra habitats. A small completely orange fungus, *O. luteicolor* Murr., is found on conifer logs and stumps in the PNW. Its waxy-appearing gills suggest *Hygrophorus,* but it is always found on wood unlike the members of *Hygrophorus* which grow on the ground.

123 CLITOCYBE NEBULARIS (Fr.) Kummer   **Edible**/Common

Cap large, gray, dry; gills extending partially down stalk; stalk large, silky, white

Cap 9–15 cm broad, convex, flat with incurved margin and a shallow depression, dry, radiating flat-lying hairs, gray to drab. Flesh tough, thick, white. Gills attached to short decurrent, close, whitish. Stalk 8–10 cm long, 2.5–4 cm thick, equal with enlarged base, curved, sometimes off center, silky, white, covered with scattered buff-brown hairs. Veil absent. Odor and taste disagreeable.
    Spores 5.5–8 x 3.5–4.5 μ elliptical, smooth, pale yellow spore print.
    Solitary, sometimes several in humus and soil under conifers, S. and F., in PSW, PNW, and adjacent C.
    This large robust *Clitocybe* is recognizeable by the gray to drab cap color. *Clitocybe robusta* Peck (also called *C. alba* [Bat.] Singer) is also a robust species but the cap is white and the stalk white to buff without buff-brown hairs. *C. robusta* is distributed throughout North America. *C. nubularis* is sometimes found with the mushroom parasite, *Volvariella surrecta* (Knapp) Singer, growing from its cap.

124 CLITOCYBE DEALBATA   **Deadly poisonous**/Numerous

(Sow. ex Fr.) Kummer

Cap dry, dull white; gills and stalk white; in grass

Cap 15–33 mm broad, convex to flat, depressed in center in age, dry, pallid to dull whitish, smooth, without hairs, margin incurved, nearly even in age. Flesh thin, whitish, gills extending down stalk, close, narrow, dull white to nearly buff. Stalk 1–5 cm long, 2–6 mm thick, equal or slightly larger at apex, smooth, without hairs to minutely downy, sometimes curved, dull white color of cap.
    Spores 4–5 x 2.5–3 μ short elliptical, thin-walled, white spore print.
    Single, scattered, or even numerous in lawns and grass, on leaves in open woods, in late S. and F.; widely distributed.
    This small poisonous *Clitocybe* may be found growing among edible species in lawns or occasionally in mushroom beds. It is, therefore, important *to be aware of* the *decurrent, close gills and dull white overall appearance.* The poisonous component is not known. It is found in Group VI toxins.

125 CLITOCYBE IRINA (Fr.) Bigelow and Smith   **Edible, choice**/Common

Cap dry, white, plane; gills pale pink in age; often in fairy rings

Cap 4–13 cm broad, broadly convex, flat with a broad low knob, smooth, sticky but soon dry, whitish but soon dingy buff with a white, cottony margin which is inrolled at first. Flesh thick, soft, white or tinted pinkish. Gills adnate to extending partially down stalk, crowded, white at first to pale pinkish in age. Stalk 4–8 cm long, 10–25 mm thick, equal or even thicker in the middle, dry, thin, hairy, dull whitish sometimes dingy in age. Odor very faint, fragrant.
    Spores 7–10 x 4–5 μ elliptical, roughened (some smooth), pale pinkish buff spore print.
    Grows in groups, often in fairy rings under conifers or mixed hardwoods, in late S. or F.; widely distributed. Bigelow and Smith (1969) report this fungus from Idaho in April on infrequent occasions. My wife and I have collected large quantities of it under Norway spruce, canned it, and enjoy eating it in various casseroles, on meat, or in turkey stuffing throughout the winter months. In 1958 we encountered the largest fairy

ring we had seen, 82 feet in diameter, under a pure Norway spruce stand in Ann Arbor, Michigan. Known also as *Lepista irina* (Fr.) Bigelow, *Tricholoma irinus* (Fr.) Kummer, and *Rhodopaxillus irinus* (Fr.) Metrod.

126 CLITOCYBE DILATATA Pers. ex Karst.   Edibility unknown/Common

Cap gray to chalky white; gills attached to stalk, close, buff; stalk dry, dull white, in cespitose clusters

Cap 1.5–15 cm broad, convex, flat in age, dry to moist, fine matted hairs, gray at first, soon mostly chalky white, margin inrolled at first, smooth. Flesh thick, white to gray. Gills adnate to extending partially down stalk, close, narrow, forked sometimes, buff to pinkish buff. Stalk 5–12 cm long, 7–28 mm thick, equal or slightly larger at base, dry, somewhat fibrous, dull white. Taste sour.

Spores 4.5–6.5 x 3–3.5 μ elliptical, smooth, white spore print.

In dense, cespitose clusters on hard soil, along log roads usually after cool, wet weather in early Sp., S., and F., PNW and adjacent C.

This fungus is often seen in wet September weather at every turn along back roads. The caps are often misshapen from being crowded together in large clusters. *Lyophyllum decastes* (Fr. ex Fr.) Sing. looks similar but has a gray-brown to brown cap.

127 CLITOCYBE ODORA (Fr.) Kummer   Edible/Common

Cap convex, moist, bluish green; odor fragrant, anise-like

Cap 3–10 cm broad, convex, broadly convex in age, moist, radially streaked with hairs, bluish green to pale bluish green covered with a whitish cast young. Flesh thin, soft, white to pale buff. Gills adnate to extending partially down stalk, close, whitish to pale buff. Stalk 2–9 cm long, 4–16 mm thick, equal, enlarged just at base, moist, flat-lying hairs, white to buff. Odor strong, fragrant, of anise.

Spores 5–9 x 3.5–5 μ elliptical, smooth, pinkish to pinkish cream spore print.

Scattered or several under hardwoods or mixed conifer-hardwood stands, in S. and F.; widely distributed.

The cap is quite variable in color but the odor of anise is consistently present in young fresh specimens. *C. olida.* Quel. is a small, dull orange-colored fungus with a strong, fragrant odor but is seldom encountered.

128 CLITOCYBE NUDA (Fr.) Bigelow and Smith   Edible, choice/Common

Cap convex, dry, violet-gray; gills pale lilac; stalk dull lavender; odor fragrant

Cap 4–15 cm broad, broadly convex, flat with uplifted margin in age, sometimes with a low knob, smooth, without hairs, dry, various shades of violet to violet-gray to cinnamon with buff color in age, margin inrolled at first. Flesh usually thin, soft, light lilac-buff. Gills adnexed (notched), pale violet to pale lilac to sometimes brownish in age. Stalk 3–6 (up to 10) cm long, 10–25 mm thick, equal, often with an oval basal bulb, dry, pale violet, dull lavender covered below with scattered white, flat-lying hairs. Odor pleasant, faintly fragrant.

Spores 5.5–8 x 3.5–5 μ elliptical, roughened, pinkish buff spore print.

Single but more often numerous, occasionally cespitose in needle duff or deep leaf litter under hardwoods and conifers or even in piles of leaves, lawn grass, or around compost piles, in S. and F.; widely distributed.

Known also as *Tricholoma nudum* (Fr.) Kummer, *Lepista nuda* (Fr.) Cooke, and the name *T. personatum* (Fr. ex Fr.) Kummer has also been used for some color varients of this species. This is a delicious edible fungus and is often found in great quantity.

129 CLITOCYBE AURANTIACA (Fr.) Studer   Poisonous/Common

Cap convex, brownish orange; gills bright orange; stalk orange; on decayed wood

Cap 2.5–6 cm broad, convex, flat to depressed in age, dry, minutely hairy with a velvety appearance, orange-yellow to brownish orange, margin often lobed, inrolled at first, but elevated in age. Flesh thin, soft, whitish to buff or tinted orange. Gills extend down stalk, crowded, forked repeatedly, narrow, bright orange to orange-yellow. Stalk 2.5–9 cm long, 5–12 mm thick, enlarging toward the base, sometimes off center, sometimes curved or twisted, dry, minutely woolly, orange to orange-brown, often color of cap margin.

Spores 5–7 x 3–4 μ elliptical, smooth, turns yellow to rusty brown in Melzer's solution, white spore print.

Single or several on the ground or on partially decayed wood, even on standing dead snags, in S. and F., under hardwoods and conifers; widely distributed.

Known also as *Cantharellus aurantiacus* Fries and *Hygrophoropsis aurantiacus* Schroet., this colorful orange fungus does have narrow, forked gills which remind one of a *Cantharellus.* Reports of the toxic effects of this fungus lead me to consider it in Group VI toxins, but many have tried it and consider it good.

130 CLITOCYBE CLAVIPES (Fr.) Kummer   Edible/Common

Cap depressed in center, gray-brown; stalk with a large basal bulb; odor fragrant

Cap 2–9 cm broad, convex, soon flat with a slightly depressed center, often with a low, pointed knob, moist, gray-brown to olive-brown, margin pale. Flesh thick, watery, whitish. Gills extend down stalk, close, often forked, veins between, pale cream. Stalk 3.5–6 cm long, 4–10 mm thick, equal then abruptly *widening to a basal bulb,* moist, olive-buff to pale ash-colored hairs over a white stalk. Odor often fragrant.

Spores 6–10 x 3.5–5 μ egg-shaped, smooth, white spore print.

Single, scattered to abundant, usually under conifers or in mixed woods, in S. and F.; widely distributed. A most common and often very abundant *Clitocybe* which is usually long lived and does not decay readily.

It is edible and, according to reports, has a good flavor.

131 CLITOCYBE GIBBA (Fr.) Kummer   Edible/Common

Cap deeply depressed, pinkish tan; gills extend down stalk; stalk dry, sometimes enlarged

Cap 3–9 cm broad, plane to shallow depressed, soon deeply depressed, (Fig. 11) moist, smooth, without hairs, pinkish tan to flesh color, margin inrolled at first, sometimes striate. Flesh thin, white, fragile. Gills extend down stalk, crowded, forked, narrow, white to pale buff. Stalk 3–7 cm long, 4–12 mm thick, equal, sometimes with an enlarged base, dry, smooth, without hairs, white, dense woolliness over the base.

Spores 5–10 x 3.5–5.5 μ elliptical, smooth, white spore print.

Solitary to scattered under woods of all kinds in S. and F.; widely distributed.

Also called *C. infundibuliformis* (Fr.) Quel. in many mushroom books. The pinkish tan, deeply depressed cap is very distinctive.

COLLYBIA

The Collybias have attached or adnexed (notched) but never decurrent gills and flat or convex caps which are at first inrolled. The flesh is quite thin and the stalk flesh often has a very soft center. The stipe is brittle or fibrous but not pliant, and the fruiting body will not revive when moistened. There is no ring or partial veil. Most of the species are found in humus, on needles and leaves, or on rotten wood. Fairy rings and exceedingly large fruitings of some species are not uncommon. Both *Flammulina* and *Baeospora* cannot be distinguished from *Collybia* by field characteristics and so are included here.

Edibility. Most species have very thin flesh or are bland and tasteless. *Collybia dryophila* has caused severe gastrointestinal disturbance, but most of the species are not toxic. *Flammulina velutipes,* Winter Mushroom, is one of the edibles in this group. It fruits in the late F., W., and Sp. when few edible mushrooms are out. Its flavor is brought out when it is sauteed in butter on or with meat.

Key to Collybia

1 a. Small, delicate, in troops on decaying mushrooms, *132. C. tuberosa*
   b. Not on decaying mushrooms, **2**
2 a. Growing from deep in the soil by means of a long rootlike base, **3**
   b. No rootlike base present, **4**
3 a. Cap small 12–25 mm broad; stalk 2–7 cm long, often on cones, *133. C. albipilata*
   b. Cap large 30–120 mm broad; stalk 5–25 cm long, not on cones, *134. C. radicata*
4 a. On the cones of conifers or magnolia, **5**
   b. Not on cones, **6**
5 a. Spores turn blue in Melzer's solution (amyloid), *135. Baeospora myosura*
   b. Spores yellow in Melzer's solution (non-amyloid), *136. C. conigenoides*
6 a. Stalk clothed in dense velvety brown hairs; usually found in late fall or winter on living trees, *137. Flammulina velutipes*
   b. Not as above, **7**
7 a. In cespitose clusters (often large and dense) on decaying wood, 138. *C. acervata* and related species
   b. Gregarious (groups) at times but not really cespitose, **8**
8 a. Cap white to cream with mustard-yellow stains in age, *139. C. maculata*
   b. Cap pinkish buff, yellow-brown to gray-brown, **9**
9 a. Cap and stalk gray-brown; edges of gills often brown, *140. C. exculpta*
   b. Cap and stalk without gray coloration; edges of gills not brown, **10**
10 a. Stalk with fine longitudinal striations; gills with ragged edges; usually in needle duff under conifers, *141. C. butyracea*
   b. Stalk smooth; gills even; under hardwoods and sometimes conifers, *142. C. dryophila*

132 COLLYBIA TUBEROSA (Bull. ex Fr.) Kummer   Nonpoisonous/Common

Cap small, white, hairless, found in dense troops on decaying mushrooms

Cap 2–10 (up to 15) mm broad, convex, soon flat, sometimes with a small knob, dry, smooth, without hairs, white but often tinted yellowish to reddish. Flesh very thin, white. Gills adnate, close, white. Stalk 4–20 mm long, 0.5–1.5 mm thick, nearly equal, dry, minutely downy, white tinted reddish brown, attached to a red-brown to blackish brown sclerotium (rounded tough body), 2–4 mm diameter.

Spores 4–5.5 x 2–3 μ elliptical, smooth, white spore print.

Small, dense clusters on decaying mushrooms, in S. and F.; widely distributed.

Collybia cookei (Bres.) Arnold is very similar but the sclerotium is yellow if it develops at all. Both of the above species are too small to be edible, but they are very conspicuous during periods of prolific fruiting of the mushrooms on which they develop, (see 118. *Asterophora*).

133 COLLYBIA ALBIPILATA Pk.   Edibility unknown/Infrequent

Cap small, dark brown, striate margin; on buried conifer debris and cones near snow

Cap 12–25 mm broad, convex, dry, rich brown, striate margin. Flesh firm, white. Gills adnate, close, white. Stalk 2–7 cm long, 2–4 mm thick, equal, white above, tan below covered with fine white hairs, a thin rootlike extension below ground. Veil absent.

Spores 5–6 x 3.5–4 μ short elliptical, smooth, white spore print. Cystidia (sterile cells) present on gill edge.

Gregarious, from buried conifer debris and cones, usually near melting snowbanks in the PNW in Sp.

The small, brown caps and rooting bases separate this from the other Collybias.

134 COLLYBIA RADICATA (Fr.) Quel.   Edible/Common

Cap large, brown, viscid; gills and stalk white; with underground rootlike base

Cap 3–12 cm broad, convex, nearly flat, often with a low knob, viscid or moist in dry weather, smooth, lacking hairs, sometimes wrinkled over the center, dark brown to gray-brown. Flesh thin, white. Gills notched, fairly well separated, broad, shiny, white. Stalk 5–25 cm long above ground, 5–15 mm thick, enlarging slightly toward base, dry, white above, brownish below, twisted-striate, with a long tapering underground rootlike base (Fig. 12). Veil absent.

Spores 14–17 x 9–11 μ broadly elliptical, smooth, white spore print.

Alone or scattered in hardwood forests in late Sp., S., and early F.; widely distributed but most common in eastern North America.

The long rootlike underground extension of the stalk is the distinctive characteristic of this tall, slender, graceful agaric.

135 BAEOSPORA MYOSURA (Fr.) Singer   Nonpoisonous/Infrequent

Cap small, dry, cinnamon color; on conifer cones

Cap 5–20 mm broad, convex, nearly flat in age, dry, pinkish brown to cinnamon-pink, margin slightly inrolled at first, striate. Flesh soft, thin, white. Gills adnexed (notched), close, light pinkish buff. Stalk 10–20 mm long, 1–2 mm thick, equal, dry, cinnamon-pink, clothed with minute, white hairs, attached to conifer cones by short, white filaments. Veil absent.

Spores 3.5–4.5 x 1.5–2.5 μ egg-shaped, smooth, is amyloid (dark blue) in Melzer's solution, white spore print. Sterile cells on gill edge club to bowling-pin shaped.

Numerous on *conifer cones,* especially Norway spruce and eastern white pine in F.; in eastern North America, but exact distribution is unknown.

Collybia albipilata should be compared with this fungus because it is also found on cones.

136 COLLYBIA CONIGENOIDES Ellis   Edibility unknown/Numerous

Cap minute, dry, tan; on conifer cones

Cap 1–5 mm broad, convex, dry, minutely downy, tan, margin cream-colored and striate. Flesh thin, white. Gills adnexed (notched) to nearly free, fairly well separated, whitish. Stalk 20–30 mm long, 0.5–1 mm thick, nearly equal, finely downy, white to tan with white mycelium over the base. Veil absent.

Spores 4–5 x 2–3 μ short elliptical, smooth, white spore print. Cystidia (sterile cells) present on gill edge.

Numerous on cones of white pine, Norway spruce, and *Magnolia fraseria.* I have found it on Norway spruce in early F. in Maryland, but its distribution is imperfectly known.

137 FLAMMULINA VELUTIPES (Fr.) Singer ["Winter Mushroom"] Edible/Common

Cap viscid, reddish orange; stalk short, velvety hairs; it fruits in winter

Cap 1.5–6 cm broad, convex, nearly flat in age, viscid, reddish yellow, reddish orange to reddish brown, margin often irregular. Flesh thick, white to yellowish. Gills adnexed (notched), fairly well separated, broad, cream color to yellow, edges minutely hairy. Stalk 2–6 cm long, 3–8 mm thick, slightly narrowed at base, yellowish at apex, rest *dense, velvety, short, brown to blackish brown hairs,* tough, hollow in age. Veil absent.

Spores 7–9 x 3–4 μ narrowly elliptical, smooth, white spore print.

Single but most often in cespitose clusters (attached at base) on living trees in late F., W., and Sp., but in the high mountains of the PNW and RM in Sp. and S. as well, widely distributed. I have found this cold-loving species in every month of the year from Alaska to New Hampshire, but it always favors cool periods of weather. A thaw of any length during the W. is a good time to look for this species, especially in southern exposures.

The viscid cap, velvety brown stalk, cespitose habit, and occurrence in cold weather readily distinguish this from all other white-spored species. This species tastes quite good, but the stalk is extremely tough and should be discarded. (Also called *Collybia velutipes* Fries.)

138 COLLYBIA ACERVATA (Fr.) Kummer   Edibility unknown/Numerous

Cap reddish brown, dry, striate; stalk red-brown; in cespitose clusters on wood

Cap 2–5 cm broad, convex to broadly convex in age, smooth, without hairs, dry, light reddish brown, slightly striate over the margin when wet. Flesh thin, whitish to dull pinkish. Gills adnexed (notched) to nearly free, close, narrow, white sometimes tinged pinkish. Stalk 4–10 cm long, 3–6 mm thick, equal, dry, reddish brown to wine-brown, usually darker than the cap. Veil absent.

Spores 5–7 x 2–3 μ elliptical, smooth, white spore print.

In large, dense, cespitose clusters on well-decayed logs and sticks or from buried wood, in late S. and F.; widely distributed.

Collybia familia (Pk.) Sacc. has a whitish to buff-colored stalk and amyloid reaction in spores but is also densely cespitose.

139 COLLYBIA MACULATA (Fr.) Quel.   Nonpoisonous/Common

Cap white with yellow stains in age; gills attached, white; on humus or wood debris

Cap 3–8 (up to 15) cm broad, convex, broadly convex to nearly flat in age, dry, smooth, without hairs, feltlike to touch, white to cream color with mustard-yellow to yellow stains which usually develop in age, margin inrolled at first to wavy in age. Flesh firm, white. Gills attached at first to notched and appearing almost free, crowded, white to buff, sometimes tinted light pinkish in age. Stalk 4–15 cm long, 6–12 mm thick, equal, dry, white, sometimes with mustard-yellow stains over the base which is clothed in a dense white mycelium. Veil absent.

Spores 5–7 x 4–5 μ almost round, smooth, buff spore print.

Single or in small clusters on rich humus, wood debris, or even well-decayed logs in Sp., S., and F.; widely distributed.

This is basically a white species and can be recognized by the yellowish stains which develop as it ages. The stalk is not striate or twisted, tends to be more fleshy than most Collybias, and the crowded gills are almost free. Though edible, this species does not taste good.

140 COLLYBIA EXCULPTA Fr. Edibility unknown/Common

Cap small gray-brown; stalk color of cap; gills yellowish

Cap 9–40 mm broad, broadly convex, sometimes depressed somewhat in age, moist to dry, minutely hairy but soon smooth and hairless, gray with buff, gray-brown, to brown. Flesh thin, pallid. Gills adnate to adnexed (notched), fairly well, separated, dull white sometimes with brown to blackish brown edges or tinted yellow-brown overall. Stalk 12–36 mm long, 1–4 mm thick, equal, dry, color of the cap, scattered fine white hairs, base surrounded by white mycelium. Veil absent.

Spores 4.5–5.5 x 3.5–4 μ elliptical, smooth, white spore print.

Scattered to numerous on needles, leaves, and well-decayed conifer logs, in Sp., S., and F.; widely distributed.

This is a nondescript fungus, but the cap and stalk are of the same color and the gills frequently have brownish margins. I suspect that several species are included under this name in North America.

141 COLLYBIA BUTYRACEA Fr. Nonpoisonous/Common

Cap pinkish buff; gills with ragged edges; stalk with lines or fine ridges (Fig. 7)

Cap 2–8 cm broad, broadly convex with a low knob, dry, feltlike to touch, pinkish buff at first to brown or reddish brown. Flesh soft, white. Gills adnate, adnexed (notched), to almost free, crowded, thin, *toothed or ragged edges,* white but soon stained lilac to buff. Stalk 3–12 cm long, 4–12 mm thick, enlarging slightly at base, dry, *longitudinally lined* (striate), sometimes twisted, dull white to light tan, covered with downy white mycelium over the base, often brown stained in age. Veil absent.

Spores 5–7 x 3–3.4 μ short elliptical, smooth, white to buff spore print.

Scattered to numerous, *usually in needle duff under conifers* in late S. and F.; widely distributed.

The striate stalk, ragged gill edges, and growth in conifer needle duff are the important field characteristics. Atypical forms are encountered, creating a problem which will be solved only by a critical study of the genus *Collybia.*

142 COLLYBIA DRYOPHILA (Bull. ex Fr.) Kummer Poisonous/Common

Cap flat, yellow-brown; gills crowded, narrow, white; stalk smooth

Cap 2.5–6 cm broad, convex, flat in age, pliant, smooth, without hairs, dry to moist when fresh, color variable from yellow or yellow-brown to reddish brown, usually one color overall, margin often upturned in age, sometimes wavy. Flesh thin but pliant, watery, white. Gills adnate (attached to stalk) to adnexed (notched), narrow, crowded, white to dull white. Stalk 3–8 cm long, 2–6 mm thick, equal or nearly so, smooth, white with a tint of yellow below to yellow-brown or even reddish brown toward the base which is covered with white mycelium often in strands. Veil absent.

Spores 5–7 x 3–3.5 μ elliptical, smooth, white spore print. Cystidia (sterile cells) absent.

Several to numerous in deep humus or leaf litter usually under hardwoods, mixed woods, or alder and willow, but under conifers in the PNW; fruiting in Sp., S. and F. during or after wet periods; widely distributed.

This is an extremely variable species, but the stalk is not twisted striate as in *C. butyracea* Fries, nor is it as fleshy as in *C. maculata* (A and S ex Fr.) Quel. It also does not form the dense, large, cespitose clusters typical of *C. acervata* (Fr.) Kummer or *C. familia* Peck. I know of persons who suffered *severe* gastrointestinal disturbance shortly after eating this species (Group VI toxins). Although it is eaten without ill effect in many areas I would consider it poisonous pending a thorough study of this genus in North America.

MYCENA

The Mycenas are a dainty, fragile group with thin, brittle to pliant stipes (stalks) and small to minute conic to campanulate (bell-shaped) caps. They have white spore prints. The gills are most often adnate or adnexed (notched) and the cap margin is striate at first and *appressed against the stalk. Collybia.* Omphalina and other genera similar in other respects have inrolled cap margins at first or sometimes gills which extend down the stalk (decurrent). The Mycenas are single to cespitose on wood or in troops on needles, leaves, and humus. They are fleshy and do not revive when moistened like *Marasmius* and *Xeromphalina.* Alexander Smith's book (1947) on the North American species of *Mycena* includes more than 230 species, and no doubt more will be discovered on this continent.

Edibility. I have not discovered any references to poisonous species, but most of them are untested because they are much too small to be regarded as edible. Only a few of the distinctive or abundant species are included here.

Key to Mycena

1 a. Cap minute (3–10 mm broad); on bark of living or dead trees *143. M. corticola*
   b. Cap larger; on ground or on decayed wood, **2**
2 a. Stalk covered with a thick, transparent, gelatinous sheath, *144. M. rorida*
   b. Stalk viscid or dry but not as above, **3**
3 a. Broken or cut stalk exuding a white or red latex, **4**
   b. Broken or cut stalk may be watery but not exuding a latex, **5**
4 a. Latex blood-red; cap red-brown, *145. M. haematopus*
   b. Latex white; cap black with white margin (see *145. M. haemotopus*), *M. galopus*
5 a. Cap bright orange, viscid; cespitose on hardwoods, *146. M. leaiana*
   b. Cap another color, **6**
6 a. Cap 2–5 cm broad, rosy red, purplish to lilac-gray; on ground, *147. M. pura*
   b. Cap usually smaller, another color, **7**
7 a. Cap yellow, moist to viscid; stalk dry to viscid, **8**
   b. Cap gray-brown to brown; stalk moist to dry, **9**

8 a. Cap viscid, mustard-yellow; stalk viscid, lemon-yellow, *148. M. epipterygia*

   b. Cap moist, orange to yellow center; stalk dry, white above, yellow below, *149. M. aurantiidisca*

9 a. Fruiting near or in snow on conifer wood in mountains of western North America in Sp., *150. M. overholtzii*

   b. Not fruiting near snow, widely distributed, **10**

10 a. Odor strongly alkaline; cap 5–30 mm broad; on decaying conifer wood, *151. M. alcalina*

   b. Odor not distinctive; cap 2–4 cm broad; on decaying hardwoods, *152. M. galericulata*

143 MYCENA CORTICOLA (Fr.) S. F. Gray　　　Nonpoisonous/Common

Cap minute, brown, dry, hairy; stalk curved from wood, hairy

Cap 3–10 mm broad, convex, broadly convex in age, dry, densely hairy, dark purplish fading to brown in age, margin straight, striate. Flesh thin, color of cap. Gills attached to stalk, well separated, color of cap. Stalk 3–10 mm long, 0.5 mm thick, often curved from host, equal, densely hairy. Veil absent.

Spores 9–11 μ round, smooth, amyloid reaction in Melzer's solution, white spore print. Cystidia (sterile cells) on gill edge only, club-shaped, top covered with finger-like projections.

Grows in groups in large numbers on living or dead hardwoods and some conifers in early Sp. and F.; eastern North America.

144 MYCENA RORIDA (Fr.) Quel.　　　Edibility unknown/Numerous

Cap small, yellowish brown, dry; stalk covered with a thick glutin

Cap 5–15 mm broad, broadly convex, often depressed in the center, dry, brownish to yellowish white, margin striate. Flesh thin, pallid. Gills attached to stalk, extending down stalk in age, subdistant, white. Stalk 2–5 cm long, 1 mm thick, equal, *covered with a thick, glutinous transparent sheath,* beneath glutin bluish black near apex, turning whitish overall in age, glutin sliding to base in age. Veil absent.

Spores 8–10 x 4–5 μ elliptical, smooth, darkly amyloid reaction in Melzer's solution. Cystidia (sterile cells) on gill edge only, club-shaped with a round head.

Several to numerous on needles or conifer wood in Sp., S., and F.; widely distributed. I have found it in Glacier Park, Montana. The extremely thick gelatinous sheath makes this one of the most distinctive of all Mycenas.

145 MYCENA HAEMATOPUS (Fr.) Quel.　　　Edibility unknown/Common

Cap red-brown; exuding a blood-red latex from broken stalk or flesh

Cap 1–4 cm broad, conic to bell-shaped in age, moist, red-brown center, reddish gray margin, covered at first by a frosted white downiness, margin striate, straight, often torn and ragged. Flesh thin, grayish red, exuding a *blood-red latex* when cut. Gills attached to stalk, close, whitish to grayish red, staining reddish brown. Stalk 3–8 cm long, 1–2 mm thick, equal, dry, covered with dense minute hairs, pale cinnamon, exuding a *blood-red latex* when broken, brittle, hollow. Veil absent.

Spores 8–11 x 5–7 μ elliptical, smooth, amyloid reaction in Melzer's solution, white spore print. Cystidia (sterile cells) abundant on gill edge, broad in center but tapered at both ends.

Single but most often in cespitose clusters on well-decayed wood, in Sp., S., and F.; widely distributed.

There are a number of species of *Mycena* which have a white or colored latex. They are too numerous to be included here, but they are well described and illustrated by Smith (1947) in section *Lactipedes*. *M. galopus* (Fr.) Quel. has a black cap with a white margin and exudes a white latex when broken. The latex should be checked in *Mycena* by breaking the stalk near the base in one of the specimens in each collection when they are first collected.

146 MYCENA LEAIANA (Berk.) Sacc.   Nonpoisonous/Common

Cap and stalk orange, viscid; gills salmon color; in cespitose clusters on wood

Cap 1–4 cm broad, broadly convex, bell-shaped in age, viscid, smooth, without hairs, shining, bright orange to orange-red, fading in age, eventually whitish, margin slightly striate when moist. Flesh soft, watery, white. Gills adnate, close, salmon, edges reddish orange, bruise orange-yellow. Stalk 3–7 cm long, 2–4 mm thick, equal or enlarging below, viscid, orange, covered with fine hairs, with densely hairy base. Veil absent. Odor somewhat mealy.

Spores 7–10 x 5–6 μ elliptical, smooth, blue (amyloid) reaction in Melzer's solution, white spore print. Cystidia (sterile cells) on edge and sides of gills, variously shaped.

In cespitose clusters on hardwood logs and limbs in SE, CS, NE, and adjacent C., in S. and F.

The distinctive field characters are the conspicuous bright orange color, viscid cap and stalk, and cespitose habit on the wood of hardwoods.

147 MYCENA PURA (Fr.) Quel.   Nonpoisonous/Common

Cap purplish, dry; gills lilac; stalk white, tinted color of cap

Cap 2–5 cm broad, conic, broadly convex with a broad low knob in age, moist, smooth, without hairs, color quite variable, rosy-red, purplish to lilac-gray, margin straight, sometimes yellowish to white, striate. Flesh firm, purplish to lilac fading to whitish in age. Gills adnate (attached) to adnexed (notched), close, broad, veins between, bluish to lilac. Stalk 3–10 cm long, 2–6 mm thick, equal or enlarged below, sometimes twisted striate, dry, smooth, without hairs, color of cap to whitish, hollow. Veil absent.

Spores 6–10 x 3–3.5 μ narrowly elliptical, smooth, amyloid reaction in Melzer's solution, white spore print. Cystidia (sterile cells) abundant on gill edge, 40–100 x 10-25 μ, flask-shaped with long necks.

Single to several on humus under hardwoods and conifers, in Sp. and early F.; widely distributed.

The attractive, variable colors of this small fungus are quite distinctive. It is found in great abundance throughout Canada and the United States, but it is too small to be of food value.

148 MYCENA EPIPTERYGIA (Fr.) S. F. Gray   Edibility unknown/Common

Cap small, mustard-yellow with olive tones, viscid; stalk viscid, white to yellowish

Cap 8–20 mm broad, conic, broadly conic with knob in age, smooth, without hairs, viscid, mustard-yellow in center often with faint greenish tone, margin straight, striate, whitish. Flesh thin, yellowish. Gills attached to stalk, fairly well separated, white to buff. Stalk 6–8 cm long, 1–2 mm thick, equal, pliant (bends easily), viscid, smooth, white to lemon-yellow, base with a few coarse hairs. Veil absent.

Spores 8–10 x 5–6 μ elliptical, smooth, amyloid reaction in Melzer's solution, white spore print. Cystidia (sterile cells) on gill edge club-shaped, covered with short finger-like projections.

Scattered to many in humus under conifers in northern conifer forests in the F.

The viscid cap and long, thin, viscid stalk combined with the mustard-yellow cap with faint greenish tones are the distinguishing characteristics. A variety which grows on wood has also been described by A. H. Smith (1947).

149 MYCENA AURANTIIDISCA Murr. Edibility unknown/Numerous

Cap minute, moist, orange center, white margin; under conifers

Cap 7–20 mm broad, conic, broadly conic, moist, smooth, without hairs, center orange to yellow, margin straight, white. Flesh fragile, orange. Gills attached to stalk (adnate), close, white, yellowish in age. Stalk 20–30 mm long, 1 mm thick, equal, dry, white above, yellow below, hollow. Veil absent.

Spores 7–8 x 3.5–4 μ elliptical, smooth, white spore print.

Grows in groups, often in great numbers under conifers in the PNW in Sp. and F.

Mycena adonis (Fr.) S. F. Gray is similar but has a scarlet cap center and is also found in the PNW under conifers. *Mycena amabilissima* (Pk.) Sacc. has a coral-red cap and is more widely distributed under conifers throughout the northern forests.

150 MYCENA OVERHOLTZII Smith and Solheim Edibility unknown/Numerous

Cap large, brown, moist; stalk base covered with dense white hairs; near melting snow.

Cap 1.5–5 cm broad, convex with a knob in age, moist, smooth, hairless, brown to tan in age, margin with minute radial ridges. Flesh soft, white. Gills almost distant, adnate or extending a short distance down the stalk, light gray. Stalk 4–15 cm long, 3–10 mm thick, enlarging toward base, dry, light pinkish brown and smooth near the gills, lower part covered with dense, cottony white hairs. Veil absent.

Spores 5–7.5 x 3.5–4 μ, elliptical, thin-walled, amyloid (blue) in Melzer's solution, white spore print.

In large clusters on conifer logs and stumps near melting snow in Sp. in the PNW, PSW, RM, and adjacent C. This large *Mycena* is a conspicuous part of the spring fungi throughout much of the western mountains.

151 MYCENA ALCALINA (Fr.) Quel. Nonpoisonous/Common

Cap minute, gray-brown, dry; odor strongly alkaline

Cap 5–30 mm broad, conic, bell-shaped in age, dry, covered at first with a white fuzz (bloom), brown to gray-brown, margin white to pallid, straight, lightly striate at first. Flesh thin, white to grayish. Gills adnate, close, narrow, white to grayish, staining red-brown in age. Stalk 1.5–11 cm long, 1–3 mm thick, enlarging toward base, brittle, dry, also covered with a white bloom at first, brown to gray-brown. Veil absent. Odor strongly alkaline. Taste mildly acrid. The odor should be noted immediately after collection or when collection is opened again.

Spores 7.5–11 x 4.5–7 μ elliptical, smooth, amyloid reaction in Melzer's solution, white spore print. Cystidia (sterile cells) abundant on sides and edges of gills, flask-shaped with long necks.

Grows in groups, often in cespitose clusters, on decaying conifer logs and stumps, in Sp. and F.; one of the common Mycenas in North America.

152 MYCENA GALERICULATA (Fr.) S. F. Gray　　　　Edible/Common

Cap moist, buffy brown, striate; stalk with rootlike base; no odor; on wood

Cap 2–4 cm broad, conic, bell-shaped in age, moist to greasy feeling but not viscid, buff-brown, dingy cinnamon-brown, margin gray, striate, often spliting. Flesh thick, gray. Gills adnexed (notched), close, veined, grayish white to pale pink. Stalk 5–10 cm long, 2–4 mm thick, equal, dry, smooth, without hairs, twisted-striate, pale grayish white, often with a rootlike base penetrating the host. Veil absent.

Spores 8–10 x 5.5–7 μ elliptical, smooth, amyloid reaction in Melzer's solution, white spore print.

Single or in cespitose clusters on decayed logs and stumps of hardwoods in Sp. and F.; widely distributed. Cystidia (sterile cells) only on gill edge, club-shaped with finger-like projections over the top.

Mycena elegantula Peck is somewhat similar but the cap is blackish-brown to red-brown and it is widely distributed on decayed conifer wood. There are a series of closely related species that are too numerous to include here.

LACCARIA

The Laccarias have thick, fleshy, pink to violet, adnate to short decurrent gills (extending down stalk). The pileus is convex to nearly flat, and there is no partial veil or annulus. The spores are elliptical to subglobose (almost round) with minute spines to smooth in one species. The spore print is white to pink in one species.

Edibility. They are fleshy but tasteless and would not make desirable edibles. I have no reports of toxins in this group.

Key to Laccaria

1 a. Cap and stalk covered with sand, usually buried partially or up to the cap, 153. *L. trullisata*
　b. Cap and stalk not covered with sand or buried, **2**
2 a. Cap brown or pinkish brown; gills thick, well separated, pink, 154. *L. laccata*
　b. Cap dull white to purple; gills purple, **3**
3 a. Fruiting body deep violet overall (see 154. *L. laccata*), *L. amethystina*
　b. Cap and stalk dull white to gray; gills dark purple, 155. *L. ochropurpurea*

153 LACCARIA TRULLISATA (Ellis) Pk.　　　　Nonpoisonous/Infrequent

Cap dingy brown; gills purple; in sand or sand dunes

Cap 2.5–6 cm broad, convex to broadly convex, moist to tacky, flattened small scales, dingy light brown, with dark brown spots and stains, always covered with sand, margin inrolled at first but straight and split in age, sometimes deeply. Flesh firm, light brown. Gills adnate, well separated, thick, pink to *purple*. Stalk 3–8 cm long, 8–20 mm thick, enlarging toward base, curved, twisted, pinkish at first to dingy dull brown, covered with embedded sand particles. Mycelium (hyphae) around base is violet and also covered with sand. Veil absent.

Spores 16–21 x 6–7.5 μ slightly tapered at ends, long elliptical, *smooth,* white spore print.

Alone or several in one place, sometimes cespitose in clusters of two or three in sand dunes or very sandy soil, in S. and F. after wet weather. I have collected this species several years in a row in nearly pure sand near a log road. The caps are sometimes the only part of the fungus which shows and the rest is buried in the soft sand which clings to all but the gills; widely distributed.

154 LACCARIA LACCATA (Fr.) Berk. and Br.   Nonpoisonous/Common

Cap pinkish brown, dry; gills thick, flesh color; very common; on soil

Cap 1–4 (up to 8) cm broad, convex, flat or with upraised margin in age, dry to moist, smooth, without hairs, flesh color, pinkish brown to pinkish orange, margin sometimes wavy and uneven. Flesh thin, color of cap. Gills extend partially down stalk, well separated, thick, flesh color to pink. Stalk 2–10 cm long, 4–8 mm thick, nearly equal, dry, smooth, without hairs but with scattered loose scales in age, colored like cap. Veil absent.
Spores 7.5–10 x 7–8.5 μ broadly elliptical to almost round, minutely spiny, white spore print.
Single, scattered to abundant on damp soil, moss and in extremely variable habitats, in Sp., S., and F.; very widely distributed.
One of the most variable and common mushrooms in North America, *L. laccata* will be most often recognized by the thick, well-separated, flesh colored gills and minutely spiny spores. *L. amethystina* (Bolt. ex Hooker) Murr. is similar but colored deep violet overall. It is so variable that it can easily be confused with other nonedible species. It is not rated as tasting very good.

155 LACCARIA OCHROPURPUREA   Edible/Common

(Berk.) Peck

Cap dull white, dry; gills purple; stalk white, dry

Cap 4–12 cm broad, convex, nearly flat in age, dry, nearly hairless, dull whitish to light gray. Flesh tough, dull whitish. Gills adnate, broad, well separated, thick, *light to dark purple*. Stalk 5–12 cm long, 10–25 mm thick, equal or variable in shape, often curved, dry, smooth, without hairs (glabrous) color of cap. Veil absent. Taste slightly acrid, disagreeable.
Spores 8–10 μ round, spiny, pale lilac spore print.
Alone or several under hardwoods or in openings in forests in S. and early F.; widespread in eastern North America.
This large, coarse fungus has distinctive purple, well-separated gills and an overall dull white to gray cap color which sets it apart from other Laccarias. I have often collected it in September in moist lowland hardwood stands along the Patuxent River in Maryland. This fleshy mushroom can provide a good meal, but I have no direct information on its taste.

LYOPHYLLUM

The convex to flat caps are always dry to moist and usually brown to gray-brown with an incurved margin at first. The gills are adnate to short decurrent (extending down stalk) and stain dingy gray, bluish to black in some species. The stipe is usually brittle to fleshy, dry, and no veil is present. The fruiting bodies remind one of a *Collybia* with the more robust species resembling a *Tricholoma.* They are found on the ground and may be saprophytes or decaying material or form mycorrhiza. The spores are non-amyloid, broadly elliptical, and white.

Edibility. The group has not been well studied in North America. I have seen no reports of toxins in European species.

Key to Lyophyllum

1 a. Caps large (6–14 cm broad); in cespitose clusters, 156. *L. decastes*
   b. Caps smaller; not cespitose, **2**
2 a. Gills staining black, 157. *L. infumatum*
   b. Gills not as above, **3**
3 a. Foul smelling; in summer along the Pacific Coast (see 157. *L. infumatum*), *L. rancidum*
   b. Mild smelling; near melting snowbanks in the RM in Sp. (see 157. *L. infumatum*, *L. montanum*

156 LYOPHYLLUM DECASTES (Fr.) Singer   Edibility unknown/Numerous

Cap gray-brown, moist, lobed; gills and stalk white; in cespitose clusters on ground

Cap 6–14 cm broad, broadly convex, flat, smooth, without hairs, moist not viscid, metallic gray to gray-brown, margin lobed to crenate, incurved at first. Flesh firm, not thick, white, not staining. Gills adnate, fairly well separated, broad, uneven, white to pallid in age. Stalk 3–6 cm long, 15–35 mm thick, equal or smaller toward base, moist to dry, smooth, without hairs or with minute, scattered fibrils, white sometimes with brownish stains near base in age. Veil absent.
Spores 5–7 x 5–6 μ broadly elliptical to almost round, smooth, white spore print.
In cespitose clusters on the ground in early S. through F.; widely distributed.
I have observed it in Alaska and the Yukon in late July and August. In Virginia it fruits in late November but always during cool moist weather. It is noted as edible and good in Europe, and it is found in sufficient quantity to make a good edible. I have no reports on its edibility in North America.

157 LYOPHYLLUM INFUMATUM (Bres.) Kuehn.   Edibility unknown/Infrequent

Cap gray; gills dull white, blackening when bruised

Cap 1–6 cm broad, convex to nearly flat, dry, smooth, without hairs, gray to pinkish gray. Flesh firm, white. Gills extend partially down stalk, subdistant, dull white, blackening irregularly. Stalk 2–5 cm long, 2–9 mm thick, equal, dry, smooth, without hairs, pallid to very light gray. Veil absent.
Spores 9–11 x 5–6 μ elliptical, smooth, white spore print.
Single to several on ground under conifers in early Sp. and S.; widely distributed.
The white gills readily bruise black and this is the diagnostic feature of the species. *Lyophyllum montanum* A. H. Smith is a metallic gray species found only near melting snow in the RM in late Sp. It does not bruise as above and is never cespitose. A rare species, *L. rancidum* (Fr.) Kuehn, and Romaq., is found along the Pacific Coast north to Alaska, where I have collected it. It has a very foul smell and a gray to brown cap.

LEUCOPAXILLUS

These are robust, large, fleshy mushrooms with attached gills and no partial veil. The cap is dry and convex to nearly flat in age. The outstanding field characteristic is the chalky white appearance of gills and stalk. This characteristic, combined with the tuber-culate (Fig. 63), amyloid-reacting spores, and the presence of clamp connections (Fig. 73), distinguishes this genus from other white-spored genera. All species are found on the ground and are probably all mycorrhizal with hardwoods and conifers.

Edibility. Some species are reported to be edible. There are no reports of toxins in the genus. The bitter to slightly disagreeable taste fresh is not appealing.

Key to Leucopaxillus

1 a. Cap reddish brown with a pinkish inrolled margin, *158. L. amarus*
   b. Cap pure white to cream or buff, **2**
2 a. Cap large (10–45 cm broad); taste mild; odor fishlike to skunklike, *159. L. giganteus*
   b. Cap smaller (3–12 cm broad); taste bitter; odor of meal or aromatic, **3**
3 a. Cap pure white; odor sweet or aromatic; widely distributed under conifers and hardwoods, *160. L. albissimus* and varieties
   b. Cap dull white, tinted pink to buff; odor of meal, unpleasant; in eastern North America under beech and oaks, *161. L. laterarius*

158 LEUCOPAXILLUS AMARUS   **Edibility unknown**/Common

(Alb. and Schw.) Kuehn.

Cap reddish brown, dry; stalk bulbous; taste bitter

Cap 4–12 cm broad, broadly convex, dry, smooth, without hairs, reddish brown with a pinkish inrolled margin. Flesh firm, white. Gills adnate, close to crowded, chalky white. Stalk 4–6 cm long, 8–45 mm thick, equal to bulbous, minutely powdered, white to dingy brownish over the base. Veil absent. Odor unpleasant. Taste bitter.

Spores 4–6 x 3.5–5 μ almost round, amyloid warts in Melzer's solution, white spore print. Cystidia (sterile cells) abundant on gill edge. Clamp connections present.

Single to numerous under conifers in S. and F., PNW, but under oak and conifers in the PSW; infrequent elsewhere.

The reddish brown cap, bitter taste combined with the numerous cystidia on the gill edge and round spores separate this from other species of *Leucopaxillus.* A number of forms have been described by Singer and Smith (1943).

159 LEUCOPAXILLUS GIGANTEUS (Fr.) Singer   **Edible**/Infrequent

Cap very large, cream color; odor fishy; taste mild

Cap 10–45 cm broad (usually 15–30 cm broad), broadly convex, flat or depressed to deeply depressed in age, moist to dry, smooth, without hairs, cream to buff, margin inrolled, sometimes striate, and finely downy at first. Flesh firm, white. Gills extend down stalk, crowded, white to light buff, sometimes forked. Stalk 4.5–7.5 cm long, 2–5 cm thick, equal, smooth, dry, without hairs, same color as cap. Odor weak, skunk-like or like fish meal. Taste mild. Veil absent.

Spores 5–8 x 3–5.5 μ egg-shaped to elliptical, smooth, amyloid walls in Melzer's solution, white spore print. Cystidia (sterile cells) absent. Clamp connections always present.

Single, several together or in a fairy ring, in grassy areas or open mixed woods, under spruce and fir in late S. and F.; widely distributed but seldom seen.

Clitocybe gigantea Fr. is an older name. *Leucopaxillus candidus* (Bres.) Singer ranges from 6–20 cm broad at maturity and is found in eastern North America but is similar in all other aspects to *L. giganteus.*

160 LEUCOPAXILLUS ALBISSIMUS (Pk.) Singer　　　　Edible/Common

Cap chalky white; odor sweet; taste bitter

Cap 3–9 cm broad, convex to flat in age, dry, smooth, without hairs, pure white, margin inrolled at first and often striate. Flesh soft, white. Gills adnate or with a short line on stalk, crowded, white (even when dried). Stalk 4–7 cm long, 7–15 mm thick, equal or with a basal bulb entangled with leaves or needles, chalky white, smooth, without hairs or with scattered hairs near the apex at times. Veil absent. Odor aromatic and sweet. *Taste bitter.*

Spores 5.5–8.5 x 4–5.5 μ elliptical, amyloid warts in Melzer's solution, white spore print. Cystidia (sterile cells) absent. Clamp connections present.

Single, several, or in fairy rings under conifers and hardwoods in S. and F.; widely distributed. Eight varieties and four forms of this species are described by Singer and Smith (1943). *Leucopaxillus albissimus* var. *piceinus* (Pk.) Singer and Smith has yellow color on old and dried specimens and fused ridges at the apex of the stalk. *Leucopaxillus albissimus* var. *paradoxus* (Cost. and Duf.) Singer and Smith is not bitter but has a disagreeable taste. It is found in PSW and PNW.

161 LEUCOPAXILLUS LATERARIUS　　　　Nonpoisonous/Common
(Pk.) Singer and Smith

Cap white, tinted pink; odor of meal; taste very bitter

Cap 4–12 cm broad, broadly convex often with a low knob, dry, dull, white with a tint of pink to buff, margin inrolled and sometimes striate. Flesh firm, thick, white. Gills adnate to very short decurrent, crowded, narrow, white to pale buff. Stalk 4–11 cm long, 6–20 mm thick, equal or enlarging toward base, smooth, without hairs or minutely downy, dull chalky white, attached to the leaves and ground by a mass of white mycelium. Veil absent. Odor of meal, often unpleasant. Taste very bitter.

Spores 3.5–5.5 x 3.5–4.5 μ round to almost round, amyloid warts in Melzer's solution, white spore print. Cystidia (sterile cells) absent. Clamp connections present.

Single, scattered to numerous in leaves and debris under beech and oaks in S. and F.; in eastern North America.

The bitter taste, enlarged lower stalk, and round spores form a distinctive combination of characteristics.

MELANOLEUCA

The North American species, for the most part, are not thoroughly studied. The tall, thin stalk without any annulus, combined with the convex, umbonate (knobbed), wide cap gives the species a characteristic appearance. Like *Leucopaxillus,* they also have warted (Fig. 63) spores with an amyloid reaction but the stalk and gills differ in that they are never chalky white. In fact, many species of *Melanoleuca* have brown or gray-brown fibrils on the stipe. They occur on the ground in fields and woods and may well form mycorrhiza.

Edibility. I have no information on the edibility of this genus.

162 MELANOLEUCA MELALEUCA (Pers. ex Fr.) Murr.   **Unknown**/Infrequent

Cap convex-knobbed, dark brown; stalk long with long hairs

Cap 2–8 cm broad, broadly convex, flat with a low knob and upturned margin in age, moist to dry, smooth, without hairs, smoky brown to dark brown, margin wavy or straight. Flesh thin, whitish to tinted brown in the lower stalk. Gills adnexed (notched), close, broad in center, white. Stalk 2.5–12 cm long, 4–9 mm thick, equal or slightly swollen at the base, dry, white with brownish longitudinal hairs. Veil absent. Taste mild.

Spores 6–8 x 4–5 μ elliptical, warts have amyloid reaction in Melzer's solution, white spore print. Cystidia (sterile cells) lancelike, incrusted at apex, forming a harpoon-like tip. Clamp connections absent.

Solitary or scattered in pastures and open woods in the F. in eastern North America; in S. and F. in western North America.

It is eaten in Europe and is probably edible here, but I have no specific information on it. *Melanoleuca alboflavida* (Pk.) Murr. is larger, has a buff to white cap and longer spores (7–10 μ long). There are several undescribed species in the RM which occur in the Sp. in great numbers. If they are edible, an abundant food source would be available in Sp. They are smaller than the above species and have dark brown caps.

TRICHOLOMA

There are many species of *Tricholoma,* eight of the most common are included here. The fleshy, brittle stalk with no annulus, convex cap, and non-waxy, attached gills that are usually adnate (attached) or adnexed (notched) are the chief field characteristics. They must also be growing on the ground and have smooth, white spores with a non-amyloid reaction. Such a species found on wood would be a *Tricholomopsis.* If the gills extend down the stalk, *Clitocybe* should be checked, but if they are waxy when rubbed, it may be Hygrophorus. *Tricholoma* species are on the ground and form mycorrhiza.

Edibility. Several species are edible but most are not edible or their edibility is unknown. Since several are reported to be poisonous, including *T. pardinum* (Group VI toxins) and *T. sulphureum* (Group V toxins), I would not recommend trying unknown species in this group.

Key to Tricholoma

1 a. Veil present, hairy and usually gone at maturity; cap orange-red and with flattened reddish brown scales, *163. T. aurantium*
  b. Veil absent, **2**
2 a. Cap, gills, and stalk white, **3**
  b. Cap another color, **5**
3 a. Cap viscid; taste mild, *164. T. resplendens*
  b. Cap dry; taste mild or bitter, **4**

4 a. Taste mild (see 164. *T. resplendens*), *T. columbetta*
   b. Taste bitter (see 164. *T. resplendens*), *T. album*
5 a. Cap sticky to viscid, yellow to olive or olive-brown, **6**
   b. Cap dry, reddish brown to purplish gray or gray-brown, **7**
6 a. Cap viscid, yellow with brownish center; gills yellow, *165. T. flavovirens*
   b. Cap sticky to viscid, grayish olive to olive-brown; gills white, 166. *T. saponaceum*
7 a. Cap gray-brown with prominent, hairy small scales, *167. T. pardinum*
   b. Cap different color with appressed hairs, **8**
8 a. Cap slate-gray to purplish gray with appressed dark hairs, *168. T. virgatum*
   b. Cap reddish brown with appressed fine dark hairs, *169. T. vaccinum*

163 TRICHOLOMA AURANTIUM

Edibility unknown/Common

(Schaeff. ex Fr.) Rick.

Cap with orange-red scales, viscid; fragile veil; disagreeable odor

Cap 4–7 cm broad, convex, flat in age, dry to sticky or viscid in wet weather, flattened-scaly or scattered fibrils, orange-red to orange-brown with reddish brown stains. Flesh firm, pallid to white. Gills adnexed (notched), adnate to short decurrent (extending down stalk), close, white, spotted or stained orange-brown, veined near stipe. Stalk 3–6 cm long, 8–20 mm thick, equal, enlarged near the base, dry, bright white above the ring, orange-brown flattened hairs below. Veil thin, sticky, hairy, colorless to white, leaving a very thin fibrillose ring or not present at maturity. Odor disagreeable, strong, as cornsilk.

Spores 4.5–6 x 3–4 μ almost round to short elliptical, smooth, white spore print.

Single to scattered or in small groups under various conifers in S. and F.; widely distributed. I have found this fungus under pure Engelmann spruce in Alberta. *Armillaria zelleri* Stuntz and Smith is similar but has a persistent, superior ring.

164 TRICHOLOMA RESPLENDENS Fr.

Edibility unknown/Infrequent

Cap white, viscid; taste mild

Cap 3.5–10 cm broad, convex, broadly convex in age, viscid to slightly viscid in dry weather, dull white with light brown spots in age, margin straight. Flesh firm, white. Gills adnate to adnexed (notched), fairly well separated, white, tinted flesh color. Stalk 3–6 cm long, 10–18 mm thick, equal or slightly narrowed just at the base, dry, lacking hairs or sometimes downy near the apex, white overall. Veil absent.

Spores 6–7.5 x 3.5–4.5 μ elliptical, smooth, white spore print.

Single, scattered, or occasionally numerous, in conifers and hardwoods in late S. and F.; eastern North America.

The white, viscid cap sets this apart from *T. columbetta* Fries which has a dry cap. *Tricholoma venenata* Atk. has a dry cap, mild taste, and the gills and stalk bruise brown. It is a poisonous species. *Tricholoma album* Fries has a dry cap but also has a bitter taste. The gills in this group are not waxy which eliminates *Hygrophorus;* there is no latex which eliminates *Lactarius.* The spores are not blue in Melzer's solution which means that it could not be a *Russula.*

165 TRICHOLOMA FLAVOVIRENS (Fr.) Lund.

Edible/Common

Cap center brown with yellow margin, viscid; gills yellow

Cap 5–10 cm broad, convex, nearly flat in age, viscid, smooth, lacking hairs, brownish to reddish margin, pale yellow, even. Flesh firm, white, yellowish beneath cap cuticle. Gills adnexed (notched), close, yellow. Cap 3–8 cm long, 1–2 cm thick, equal to enlarged over the base, dry, white to buff, usually hairless, solid. Veil absent. Smell and taste of new meal.

Spores 6–7 x 4–5 μ elliptical, smooth, white spore print. Cystidia absent.

Single to scattered, sometimes in clumps on the ground under mixed conifers and hardwoods in late S. and F.; in eastern North America.

Known also as *T. equestre* Fries, our variant of this species has more yellow in the cap than is found in the Europe variety but in other respects seems to be the same. I have often found it in sandy loam in late F. in Maryland in Virginia pine-oak woods. The cap often has a faint greenish tint over the yellow areas and is sometimes partially covered by sand and soil. *Tricholoma sulphureum* Fries has a sulfur-yellow cap and stalk, a strong, disagreeable odor, and is poisonous.

166 TRICHOLOMA SAPONACEUM (Fr.) Kummer   Nonpoisonous/Common

Cap olive-gray, viscid; smell of soap

Cap 3–8 cm broad, convex, hairless, sticky to viscid, grayish olive, yellowish olive shaded to olive-brown especially over the center, margin inrolled at first. Flesh thick, white. Gills adnate, fairly well separated, white with a greenish tint, stained reddish near margin. Stalk 3–8 cm long, 7–25 (35) mm thick, equal to enlarged over the base, dry, smooth, with small cottony scales, white with brown stains in age. Veil absent. Smell of soap. Taste mild or slightly soaplike.

Spores 5.5–7 x 3.5–5 μ short elliptical, smooth, white spore print.

Single to groups under hardwoods and conifers in S. and F.; widely distributed.

167 TRICHOLOMA PARDINUM Quel.   Poisonous/Numerous

Cap dry with gray-brown, hairy small scales; white stalk

Cap 6–10 cm broad, convex with a low, broad knob, dry, covered with gray-brown, hairy small scales, white beneath. Flesh firm, white. Gills adnexed (notched), fairly well separated, white to pallid in age. Stalk 5.5–10 cm long, 15–20 mm thick, nearly equal or enlarged toward base, dry, chalky white, smooth, without hairs. Veil absent. Smell of fresh meal.

Spores 7.5–10 x 5.5–6.5 μ short elliptical, smooth, white spore print.

Single to several under conifers in the F., in NE, CS, PNW, RM and adjacent C.

The toxins are unknown and are included under Group VI. The white, gray to gray-brown species of *Tricholoma* should be avoided. I have collected this species in Idaho and eastern Washington, where it is common every F.

168 TRICHOLOMA VIRGATUM (Fr.) Kummer   Edibility unknown/Numerous

Cap gray to purplish gray, dry, streaked with flattened hairs.

Cap 2.5–7 cm broad, conic, bell-shaped in age, dry to sticky, with radiating dark flattened hairs, slate gray, pale gray to purplish gray, with incurved margin. Flesh thin, white. Gills adnexed (notched), close, broad, white. Stalk 3.5–11 cm long, 6–18 mm thick, equal, dry, smooth, without hairs, or with light fibrils, white. Veil absent.

Spores 6–7.5 x 4.5–5 μ egg-shaped to short elliptical, smooth, white spore deposit.

Single but usually in scattered groups in conifer woods in late S. and F.; widely distributed. In the PNW it often seems to have more purple coloration and less gray than eastern populations.

T. subacutum Peck is another name for this fungus. I have little or no information on its edibility in North America and therefore do not recommend it.

(Pers. ex Fr.) Kummer

Cap reddish brown, dry, streaked with flattened hairs

Cap 4–7 cm broad, convex, nearly flat or with a low knob in age, dry, flattened, radially arranged hairs, reddish brown to light reddish brown, margin wooly. Flesh thin, white with a reddish tinge. Gills adnexed (notched), close, white, stained reddish brown. Stalk 5–8 cm long, 6–15 mm thick, dry, hairy, whitish at the apex, reddish brown. Veil absent. Taste unpleasant.

Spores 4–5 μ round, smooth, white spore print.

Scattered to numerous in needle duff under conifers in S. and F.; widely distributed. The distinctive features of this fungus are the radially hairy red-brown cap and stalk.

VOLVARIACEAE

There are only two families of salmon to pink-spored gilled fungi, the Volvariaceae and the Rhodophyllaceae. The most conspicuous characteristic of the Volvariaceae is the free gills which are best seen in mature specimens. In addition, the most common genus, *Pluteus,* is always found on wood and the infrequently encountered Volvariellas all have a volva (Fig. 84). The spores are pink, smooth, and elliptical (Fig. 64), and not angular as in the Rhodophyllaceae. In contrast the other pink-spored family, the Rhodophyllaceae, has attached gills, no volva, and is found on the ground. In addition, the spores are mostly angular (Fig. 69) or occasionally long striate (Fig. 65). If a lilac spore color is encountered, check the genus *Laccaria.* The spores will be round or almost round and spiny. *Pleurotus sapidus* has lilac, elliptical, smooth spores. It also has a short eccentric stalk and is found on wood.

Edibility. I know of no poisonous species in this family. However, many species are either too small or too rarely encountered to be treated as edibles. *Pluteus cervinus* and the closely related species which resemble it are rated as good and are eaten both here and in Europe. The soft flesh of *Pluteus* is subject to rapid deterioration.

Key to the Genera of the Volvariaceae

1 a. Volva (cup) present at stalk base, *Volvariella*
   b. Volva (cup) absent, *Pluteus*

Key to Volvariella

1 a. Growing on other gilled mushrooms, especially *Clitocybe* (see 170.
     V. bombycina), *V. surrecta*
   b. Not growing on other gilled mushrooms, **2**
2 a. On hardwood logs, sticks, and stumps, 170. *V. bombycina*
   b. On rich soil, compost, or humus, **3**
3 a. Cap gray-brown to blackish brown (see 170. *V. bombycina*), *V. volvaceae*
   b. Cap grayish to white (see 170. *V. bombycina*), *V. speciosa*

Key to Pluteus

1 a. Cap and stalk white (see 171. *P. cervinus*), *P. pellitus*
   b. Cap and stalk some other color, **2**
2 a. Cap dark brown to drab brown, *171. P. cervinus*
   b. Cap dark yellow to yellowish olive; stalk thin (2–4 mm), *172. P. admirabilis*

170 VOLVARIELLA BOMBYCINA (Schaeff. ex Fr.) Singer   **Edible**/Rare

Cap dingy yellowish; volva deep and dull white; on wood

Cap 5–20 cm broad, egg-shaped, bell-shaped to convex in age, dry, silky, fibrillose in age, yellowish to dingy yellow, margin white and fringed with fine hairs. Flesh thin, soft, white. Gills free, crowded, broad, white at first to pink. Stalk 6–20 cm long, 1–2 cm thick, dry, without hairs, white enlarging or even bulbous at the base. Volva present as a thick, deep cup around the stalk base, dull whitish to dingy yellowish, persistent.
Spores 6.5–10.5 x 4.5–6.5 μ elliptical, smooth, *salmon to pink* spore print.
Solitary or several together *on hardwood logs, sticks, and stumps* in late S. and F.; widely distributed but most frequently found in eastern North America.
The universal veil surrounding the button has brownish-tipped small scales over the top and the buttons resemble a small covered thumb, but the veil splits and does not fragment, so the cap does not have veil remnants on it. No Amanitas grow on wood or are parasitic on other agarics. The salmon to pink spore print and free gills combined with the volva and the lack of a partial veil clearly indicate a species of *Volvariella*. There are 18 species in North America but most are very rarely encountered. *Volvariella surrecta* (Knapp) Singer is parasitic on agarics and is seldom encountered; it is often found on *Clitocybe nebularis* (Fr.) Kummer. *Volvariella volvacea* (Bull. ex Fr.) Singer is found on soil or compost and has a gray-brown to blackish brown cap. *Volvariella speciosa* (Fr.) Singer is also on rich soil but has a viscid, white to grayish cap. Many of the other species reported by Shaffer (1957) are found on soil or humus under forest stands.

171 PLUTEUS CERVINUS (Schaeff. ex Fr.) Kummer   **Edible**/Common

Cap dark brown; gills free and pink; on wood

Cap 5–14 cm broad, conic, broadly convex to bell-shaped in age, flattened hairs over the center, rest smooth, without hairs, moist but not viscid, often slightly wrinkled, dark brown to drab brown, usually lighter brown in age. Flesh thick, soft, white. Gills free, crowded young to close, broad, white at first, soon pink from the maturing spores. Stalk 5–12 cm long, 6–15 mm thick, enlarging evenly at base, dry, white with tinted brown or dull whitish hairs, sometimes nearly erect.
Spores 5.5–7 x 4–6 μ elliptical, smooth, *salmon to pink* spore print. Cystidia (sterile cells) present, walls thickened, apex with hornlike projections.
Single or several closely positioned on decaying wood of hardwoods and conifers, in Sp. and F.; widely distributed. The soft flesh soon spoils in warm weather. Preparation and refrigeration as soon as possible are a must.
A striking white species, *P. pellitus* (Pers. ex Fr.) Kummer, is occasionally encountered in the PNW. It is eaten in Europe but I have no information on it in North America.

172 PLUTEUS ADMIRABILIS (Pk.) Peck   **Nonpoisonous**/Common

Cap dark yellow; stalk very thin, clear yellow

Cap 13–40 mm broad, bell-shaped to broadly convex with a low knob, moist, without hairs, wrinkled over the center, *dark yellow,* sometimes olive-yellow with a striate margin when fresh. Flesh thin, dull whitish. Gills free, close, broad, buff becoming pinkish in age. Stalk 2–6 cm long, 2–4 mm thick, equal, very thin, without hairs, moist, *clear yellow.*

Spores 5.5–6.5 x 5–6 μ nearly round, smooth, *salmon to pink* spore print.

Single or in small groups on very decayed logs or woody debris of hardwoods in Sp., S., and F.; widely distributed.

The fragile yellow cap and stalk are reliable field characteristics along with the tendency of this species to have a split stalk.

RHODOPHYLLACEAE

Like the Volvariaceae treated previously, this family has salmon to pink spores. The Rhodophyllaceae, however, have attached gills and angular (Fig. 69) to long-ridged (Fig. 65) spores unlike the Volvariaceae. In addition, the species in the Rhodophyllaceae with central stalks are always on the ground and not on wood. There are two genera: *Clitopilus* and *Entoloma.*

Clitopilus has fewer species and is less frequently encountered. It is a distinctive genus with decurrent (extending down the stalk) gills and long-ridged spores. *Entoloma,* on the other hand, is frequently encountered, has many species, adnexed (notched), adnate to short decurrent gills, and has angular spores. The caps are often conic and nearly pointed to convex or flat. The stalks are often brittle, striate, and appear twisted to fleshy. Many Entolomas fruit during rainy, cool weather when few other mushrooms are out. Hesler (1967) reports 200 species in the southeastern United States, but no complete studies have been done elsewhere.

Edibility. The species of *Clitopilus* are edible. In *Entoloma,* only *E. abortivus* is generally reported to be edible. By contrast a number of Entolomas are poisonous. The toxins are not known and are discussed under Group VI toxins. Most species have very little flesh and in view of the history of poisonings from the species I would not recommend eating any other *Entoloma.*

Key to Genera of the Rhodophyllaceae

1 a. Gills long decurrent (Fig. 5); spores long striate (Fig. 65), *Clitopilus*
   b. Gills adnexed (notched) (Fig. 5) to short decurrent (Fig. 5); spores angular (Fig. 69), *Entoloma*

Key to Clitopilus

1 a. Cap gray, dry, margin straight; stalk glabrous (lacking hairs), *173. C. prunulus*
   b. Cap whitish, tacky, margin inrolled; stalk minutely hairy, *174. C. orcellus*

173 CLITOPILUS PRUNULUS Fr.   **Edible**/Common

Cap ashy gray; gills extend down stalk, pale pink

Cap 5–10 cm broad, broadly convex, flat or edges upturned in age, dry felty, dull white to ash-gray, margin even and wavy in age. Flesh firm, white. Gills extend down stalk (long decurrent), fairly well separated, white to pale pink in age. Stalk 4–8 cm long, 4–15 mm thick, nearly equal, dry, without hairs, dull white. Veil absent. Odor and taste mealy.

Spores 10–12 x 5–7 μ elliptical, longitudinal ridges, best seen in end view, salmon-pink spore print.

Scattered in open woods in S. and F.; widely distributed.

174 CLITOPILUS ORCELLUS Fr.   **Edible**/Numerous

Cap yellowish white, flat; gills extend down stalk, light pinkish

Cap 3–8 cm broad, convex, flat in age, tacky, with hairs, whitish to yellowish white, margin inrolled and often lobed. Flesh soft, white. Gills long decurrent (extend down stalk), close, narrow, white to pinkish in age from the spores. Stalk 3–5 cm long, 5–12 mm thick, equal, sometimes slightly eccentric, white, covered with minute hairs. Veil absent.

Spores 7–10 x 4–6 μ tapering at ends, longitudinally striate, pale salmon spore print.

Single to scattered in grassy areas or open woods in S. and F.; widely distributed.

C. prunulus has a dry, gray cap, a stalk lacking hairs and longer spores.

Key to Entoloma

1 a. Gills short decurrent; oval, white, aborted fruiting bodies usually present, *175. E. abortivus*

   b. Gills adnexed (notched); all normal gilled mushrooms, **2**

2 a. Cap conic (Fig. 13) (10–20 mm broad), salmon-pink, stalk 2–3 mm thick, *176. E. salmoneum*

   b. Cap bell-shaped (Fig. 8) to convex (Fig. 5), tan to brown; stalk 3–40 mm thick, **3**

3 a. Stalk thin (3–5 mm thick), appears twisted; cap gray-brown to brown, *177. E. sericeum*

   b. Stalk robust (10–40 mm thick), not obviously twisted; tan to dull white, **4**

4 a. Cap dull white, tacky to viscid (see 178. *E. lividum*), *E. prunuloides*

   b. Cap dingy tan to livid tan, tacky to dry, *178. E. lividum*

175 ENTOLOMA ABORTIVUM (Berk. and Curt.) Donk   **Edible**/Common

Cap large, dry, gray-brown; gills close, pink; stalk white tinted gray

Cap 4–10 cm broad, convex, flat in age, dry, lacking hairs, gray-brown, margin even at first, lobed in age. Flesh soft, fragile, white. Gills extend down stalk, close, pale gray to pink or dull salmon in age. Stalk 3–10 cm long, 5–12 mm thick, equal, dry, minutely fibrillose, grayish white. Veil absent. Odor and taste mealy.

Spores 8–10 x 5–7 μ elliptical-angular, salmon-pink spore print.

Several or in large groups around stumps or on rotten wood in F.; in eastern North America.

Often found with oval, white aborted fruiting bodies. These are often found where no normal fruiting body can be located. They have been tried and are edible. *Clitopilus abortivus* Berk. and Curt. is another name for this curious fungus. I have found it in abundance in the mountains of Virginia. The novice should compare this with *Entoloma lividum,* which is poisonous.

176 ENTOLOMA SALMONEUM (Pk.) Sacc.   **Edibility unknown**/Numerous

Cap conical, salmon-pink; stalk very long and thin, pink; in wet moss

Cap 10–20 mm broad, narrowly conical, bell-shaped in age, moist, without hairs, salmon-pink to orange-salmon, margin even. Flesh fragile, thin, pallid. Gills adnexed (notched), fairly well separated, broad, pink to pinkish buff. Stalk 5–12 cm long, 2–3 mm thick, equal, without hairs, minutely hairy at apex, pink, hollow. Veil absent.

Spores 10–13 μ nearly round but angular (four-sided), salmon-pink spore print.

Grows in troops in moist, moss-covered areas under evergreens (conifers) in late S. and F., in the CS, NE, and adjacent C. I have found it in great numbers in central New Hampshire in August.

This fungus looks like a *Conocybe* but is distinguished at once by the pink spore print and four-sided, angular spores.

177 ENTOLOMA SERICEUM (Bull. ex Fr.) Quel.   Edibility unknown/Common

Cap convex with a knob, brown, dry; stalk gray-brown, appears twisted

Cap 2–6 cm broad, convex with an obscure knob, dry to moist, smooth, without hairs, dark brown to gray-brown, dry, silky appearing, margin with very short striations, wavy in age. Flesh thin, water-soaked, brown. Gills adnexed (notched), close, grayish white to pinkish gray in age. Stalk 2–8 cm long, 3–5 mm thick, enlarging toward base, dry, with fine silky fibrils, *appears twisted,* grayish brown. Veil absent. Odor and taste mealy.

Spores 8–10 x 6–7 μ egg-shaped, strongly warted and angular, deep salmon-pink spore print.

Solitary or several in grassy areas or open woods in Sp. and S.; widely distributed. Often fruiting *during* rainy periods unlike many ground fungi which fruit on the drying cycle.

Through the season there are a large number of Entolomas that resemble each other, all having the general appearance of the one described here. They should be avoided because they are so poorly known.

178 ENTOLOMA LIVIDUM Fr.   Poisonous/Common

Cap large, bell-shaped, dry, tan; gills adnexed (notched); stalk robust, white

Cap 8–15 cm broad, broadly bell-shaped, nearly flat with an obscure knob in age, tacky to dry, without hairs, light dingy tan to livid tan, margin inrolled at first, straight and often split in age. Flesh thick, firm, white. Gills adnexed, fairly well separated, broad, gray to pink at maturity. Stalk 4–12 cm long, 1–4 cm thick, equal, robust, dry, white tinted grayish, with scattered infrequent scales. Veil absent. Odor and taste mealy.

Spores 7–10 μ nearly round but angular (5–6 sided), salmon-pink spore print. Cystidia (sterile cells) on gill edge narrowly club-shaped.

Scattered, sometimes several, under hardwoods and conifers in S. and F.; widely distributed. I have found it in many diverse habitats.

Also known as *Rhodophyllus lividus* (Fr.) Quel. and *R. sinuatus* (Bull. ex Fr.) Sing. This is a poisonous species (Group VI toxins) which has caused fatalities and should be avoided along with all robust, fleshy Entolomas. *Entoloma prunuloides* Fries is a large, white, tacky- to viscid-capped, closely related species. I have encountered it in Glacier National Park in Montana.

GOMPHIDIACEAE

This is a distinctive family with smoky-gray to black, entire spores and thick gills extending down stalk (decurrent). The cap cuticle is dry, viscid to glutinous, and composed of filamentous hyphae (Figs. 81, 82) but never cellular as in Fig. 80. All of the species are found under or near pines, spruces, firs, and other conifers with which they form mutually beneficial (mycorrhizal) relationship with the roots. The genus *Gomphidius* has a gelatinous cap cuticle (Fig. 81) and white flesh in the cap which is not blue in Melzer's solution. *Chroogomphus* has a dry to viscid or tacky cap cuticle (Fig. 82), and colored flesh in the cap which turns blue in Melzer's solution. *Gomphidius* has 9 species while *Chroogomphus* has 7 species (Miller, 1964, 1971).

Edibility.   All of the species in the family are edible. I would not rate them as choice but in certain areas, such as the PNW, large fruitings of many of the species could yield quantities of table food. They would serve well in general-purpose dishes such as casseroles and in stuffing. When cooked, the flesh turns maroon in color and the texture is somewhat slimy.

Key to Gomphidius and Chroogomphus

1 a. Cap dry to viscid; flesh light orange to salmon color, **2**
   b. Cap glutinous; flesh white, **5**
2 a. Cap dry; known only from western North America, **3**
   b. Cap viscid; widely distributed, **4**
3 a. Cap orange to orange-buff, *179. Chroogomphus tomentosus*
   b. Cap reddish brown (see *179. C. tomentosus*), *C. leptocystis*
4 a. Cap reddish brown, *180. C. rutilus*
   b. Cap orange; usually under pines (see *180. C. rutilus*), *C. ochraceus*
5 a. Fruiting body cespitose, usually partially buried; spores small (10–14 μ long); known only from western North America, *181. Gomphidius oregonensis*
   b. Fruiting body not cespitose; more widely distributed; spores large (16–26 μ long), **6**
6 a. Cap pink to red, *182. G. subroseus*
   b. Cap reddish brown, orange to purple, **7**
7 a. Veil absent; stalk covered with purplish black fibrils over the lower half, *183. G. maculatus*
   b. Veil present and glutinous; stalk yellow over the lower half, *184. G. glutinosus*

179 CHROOGOMPHUS TOMENTOSUS   **Edible**/Common
(Murr.) O. K. Miller

Cap dry, orange; gills extend down stalk, yellow-orange; stalk dry, orange

Cap 2–6 cm broad, broadly convex, flat in age, dry, downy fibrillose, light orange to orange-buff. Flesh firm, light orange. Gills extend down stalk, well separated, yellow-orange to smoky at maturity. Stalk 4–17 cm long, 9–14 mm thick, tapering to a narrow base, dry, smooth with scattered hair, color of cap. Veil thin, hairy, leaving a few fibrils but no ring.

Spores 15–25 x 6–9 μ long elliptical, smooth, smoke gray to black spore print. Cystidia (sterile cells) long cylindric, thick-walled, conspicuous above basidia. Cap tissue has amyloid reaction in Melzer's solution.

Single to numerous or scattered under western conifers (especially western hemlock) in F. in PNW, PSW, and RM.

This attractive orange mushroom stands out in the dark green hemlock forests of the western mountains. *Chroogomphus leptocystis* (Sing.) O. K. Miller is a reddish brown, dry, hairy-capped species also found in the same habitats as *C. tomentosus.*

110

180 CHROOGOMPHUS RUTILUS (Fr.) O. K. Miller   Edible/Common

Cap viscid, reddish brown; gills extend down stalk, cinnamon; stalk dry, orange-buff

Cap 2.5–12 cm broad, convex, convex with a small pointed knob, viscid, reddish brown, margin incurved at first. Flesh firm, *light salmon* to pinkish near the cuticle to light yellowish near stalk base. Gills extend down stalk, fairly well separated, broad, buff to dull cinnamon. Stalk 4.5–18 cm long, 5–25 mm thick, tapering toward base, dry to moist, covered in center with thin fibrils, orange-buff flushed reddish. Veil hairy, dry, leaving a thin, superior, annular zone of hairs which soon disappears.

Spores 14–22 x 6–7.5 μ long elliptical, smooth, *smoke gray to black* spore print. Cystidia (sterile cells) long cylindrical, thin-walled, conspicuous above basidia. Cap tissue has amyloid reaction in Melzer's solution.

Single to abundant only under conifers, especially pines S. and F.; widely distributed.

This species turns a deep maroon when cooked and I would not rate it as an outstanding edible. *Chroogomphus ochraceus* (Kauff.) O. K. Miller is closely related but has an orange pileus and stalk. *Chroogomphus vinicolor* (Pk.) O. K. Miller is a very dark red-brown nearly conic species with a pointed knob, viscid cap, and thick-walled cystidia (sterile cells).

181 GOMPHIDIUS OREGONENSIS Pk.   Edible/Numerous

Cap glutinous, dingy salmon; gills white at first; cespitose, stalks deep in ground

Cap 2–15 cm broad, broadly convex, glutinous, without hairs, salmon-pink to reddish brown usually covered with soil and appearing dingy. Flesh thick, soft, white near cap cuticle tinted color of cap, yellow in stalk base. Gills extend down stalk, fairly well separated, thick, white to smoke gray in age. Stalk 6–12 cm long, 1–5 cm thick, equal or either enlarged or tapered at base, without hairs, white above the ring, yellowish to deep yellow below, attached deep in soil to an enlarged fleshy base. Veil glutinous, with a white hairy layer beneath leaving a glutinous, superior ring, which may be a blackish ring from the spores.

Spores 10–13 x 4.5–8 μ nearly tapering at ends, smooth, entire, black spore print.

Single but usually cespitose (Fig. 3) under conifers in S. and F. in PNW, PSW, and RM. This western species grows under redwood and Douglas fir as well as many other conifer species.

This robust fungus typically grows in cespitose fashion from a large fleshy mass buried deeply in the soil. This combined with the small spores and the glutinous veil sets *G. oregonensis* apart from all other species. A small-spored species without a glutinous veil, *G. pseudomaculatus* O. K. Miller, is known only from Idaho.

182 GOMPHIDIUS SUBROSEUS Kauf.   Edible, good/Common

Cap glutinous, red; gills white at first; stalk with yellow base

Cap 4–6 cm broad, convex, flat in age with upturned margin, glutinous, without hairs, pink to red. Flesh thick, firm, white. Gills extend down stalk, subdistant (fairly well separated), broad, white to smoky gray in age. Stalk 3.5–7 cm long, 6–18 mm thick, tapering somewhat toward base, white and silky above ring, cream to yellow below, deep yellow at base. Veil glutinous, thin, leaving a superior, colorless, glutinous ring which is soon blackened from the spores.

Spores 15–21 x 4.5–7 μ nearly tapering at ends, smooth, black spore print.

Solitary but more often abundant under conifers in late Sp., S., and F. (most often in F.); widely distributed.

This is a rather small species compared to *G. glutinosus*. In addition, *G. subroseus* has only pink to red cap coloration. It is a good edible but like other species of *Gomphidius* the glutinous cuticle must be removed for the best taste.

183 GOMPHIDIUS MACULATUS (Scop. ex Fr.) Fries   **Edible**/Infrequent

Cap glutinous, reddish brown; gills white at first; stalk with purple black fibrils

Cap 1–11 cm broad, convex, broadly convex to flat in age, glutinous, smooth, without hairs, or sometimes with fine flattened hairs, light cinnamon to reddish brown appearing streaked and dingy. Flesh firm, white tinted wine color near cap cuticle or dingy reddish to yellow just at stalk base. Gills extend down stalk, well separated, sometimes forked, veins between, white to smoky gray in age. Stalk 1–8 cm long, 3–35 mm thick, tapering somewhat toward base, white above, covered below with *dark ochre to purplish black hairs,* yellowish brown to bright yellow over the base. Veil absent.

Spores 14–22 x 6–8 μ nearly tapering at ends, smooth, entire, black spore print.

Single to numerous under mixed conifer stands, especially noted near or under larch in late S. and F., widely distributed in northern North America.

The lack of a veil and the purplish black fibrils on the stalk, combined with a dingy cinnamon cap color, are the distinctive characteristics.

184 GOMPHIDIUS GLUTINOSUS (Schaeff. ex Fr.) Fries   **Edible**/Common

Cap glutinous, purple-brown; gills pale drab; stalk yellow over base

Cap 2–10 cm broad, broadly convex to flat with an upturned margin in age, glutinous, smooth, without hairs, gray-brown to purple-gray or reddish brown often spotted or stained blackish. Flesh thick, soft, white with a pinkish tint near the cap cuticle, yellow in the stalk base. Gills extend down stalk, close, pale drab to smoky gray in age. Stalk 4–10 cm long, 7–20 mm thick, tapering toward base, shiny white above glutinous ring, yellow over the lower part. Veil a thin, glutinous outer layer with a white hairy inner layer leaving a superior ring in the form of a glutinous band which soon darkens from the spores.

Spores 15–21 x 4–7.5 μ nearly tapered at ends, smooth, smoke-gray to blackish spore print.

Seldom solitary, usually abundant under many different conifers in Sp., S., and F.; throughout North America but most abundant in the west.

There are two closely related varieties: *G. glutinosus* var. *salmoneus* O. K. Miller has an orange-colored cap and *G. glutinosus* var. *purpureus* O. K. Miller with a dark purple cap. Both are found only in PNW and at first glance seem to have no relationship to *G. glutinosus*. *Gomphidius smithii* Singer has a grayish red to cinnamon-gray cap and little or no yellow flesh in the stalk. *Gomphidius largus* O. K. Miller is a robust western species with a cap up to 21 cm broad which is dull pinkish brown to dingy reddish drab. All these species and varieties are edible.

COPRINACEAE

This family has black to dark purple-brown spores which have a pore at the apex (Fig. 64). The gills are usually adnate (attached) to adnexed (notched) and are deliquescent (inky) in one genus. The cap is most often conic and the stalks thin and fragile.

The cap cuticle is dry to viscid and *always cellular* (Fig. 80). The species are common saprophytes on dung, humus, and in grassy areas. Three genera are included here although sometimes more than three names will be encountered.

Edibility. The genus *Panaeolus* has a sinister reputation, possessing poisons that often produce hallucinogenic reactions and visual distortion. The toxins are found in Groups III and IV and sufficient quantities could certainly prove fatal. The presence of low levels of psilocybin and psilocin (Group IV poisons) have been reported by Ol'ah (1969) and Robbers *et al* (1969). In *Coprinus* some species produce gastrointestinal disturbance (Group VI toxins) and the discussion under *C. atramentarius* should be consulted. I have no information on poisons in *Psathyrella,* but there are a large number of North American species.

Some of the truly good edible species are found in this family. The distinctive appearance of several species make them easily recognizable and the large numbers which fruit at one time make them very desirable. Most of the good edibles are fragile species which deteriorate quickly.

Key to the Genera of the Coprinaceae

1. a. Gills black and inky in age and/or cap with deep striations, often folded in umbrella-like fashion, *Coprinus*
   b. Gills not black or if black not inky in age; cap not folded as above, **2**
2. a. Narrowly conic with a thin rigid stipe (stalk); on dung or grass; single to numerous but not cespitose, *Panaeolus*
   b. Campanulate to convex cap; usually on humus or wood; often numerous or cespitose, *Psathyrella*

COPRINUS
"Inky Caps"

The conic cap is striate and often folded in umbrella-like fashion. This characteristic is combined with thin flesh and a very thin stipe (Fig. 10). The gills are often white or light gray at first but soon turn black as the spores mature. At maturity, in many species, the gills start to dissolve, producing an inky fluid. This process is called *deliquescence* or autodigestion of the gill and is caused by an enzyme. If the tissue is cooked, the enzyme is inactivated and the process of deliquescence halted. Several species of *Coprinus* do not undergo deliquescence and are therefore placed in the genus *Pseudocoprinus* by some authors. The spore print is black and the spores are smooth with a pore at apex (Fig. 64). These mushrooms are found on wood, wood debris, humus, dung, or on the ground and are nearly always saprophytes living on decaying matter.

Edibility. Most of the species are edible. A toxic reaction may result from the ingestion of *C. atramentarius* after an alcoholic beverage (see the discussion following the description of this species). *C. comatus* is excellent with eggs and as a garnish for meats.

Key to Coprinus

1. a. Cap pure white, **2**
   b. Cap colored or with colored scales, **3**

2 a. Veil absent; white fibrils on cap rub off revealing grayish black surface (see 185. *C. niveus*), *C. narcoticus*
   b. Veil present with faint, hairy ring; cap cuticle white, *185. C. niveus*
3 a. Cap small (4–20 mm broad, 5–20 mm tall), conic, striate to deeply striate, **4**
   b. Cap large (2–8 cm broad, 3–12 cm tall), conic, if striate or wrinkled just over the margin, **7**
4 a. Grows in groups in great numbers on well-decayed wood, *186. C. disseminatus*
   b. Single to several on grass, humus, or dung, **5**
5 a. In grass; gills free but attached to a collar, *187. C. plicatilis*
   b. On dung or well-decayed humus; gills not attached to a collar, **6**
6 a. Cap yellow-brown with fine white hairs over the center; spores 11–14 μ long (see *187. C. plicatilis*), *C. radiatus*
   b. Cap yellow-brown without white hairs; spores 15–18 μ long (see *187. C. plicatilus*), *C. ephemerus*
7 a. Cap narrowly conical, white with light reddish brown recurved scales, *188. C. comatus*
   b. Cap conic to campanulate (bell-shaped), gray-brown to reddish brown, scaly just at the center or covered with glistening particles, **8**
8 a. Cap gray-brown, scaly just over the center; usually in grass, *189. C. atramentarius*
   b. Cap reddish brown, covered over center with glistening particles, *190. C. micaceus*

185 COPRINUS NIVEUS Fr.   Nonpoisonous/Numerous

Cap conic, pure white; stalk white, veil present

Cap 10–30 mm tall, 10–20 mm broad, narrowly conic, bell-shaped in age, dry, covered with pure white, soft hairs. Flesh soft, thin. Gills adnexed (notched), nearly free, crowded, narrow, white then black and inky in age. Stalk 3–8 cm long, 4–9 mm thick, equal or enlarged toward the base, dry, white, covered with fine fibrils. Veil hairy and mealy sometimes leaving a faint hairy, superior ring.
   Spores 15–17 x 8–11 μ elliptical, smooth, pore at apex, black spore print.
   Single or several on manure and compost, usually cow and horse dung, in Sp., S., and F.; widely distributed.
   C. narcoticus (Pers. ex Fr.) Fries is white but the powdery tufts of hairs rub off revealing a grayish black surface, and has no veil.

186 COPRINUS DISSEMINATUS (Pers.) Fr.   Nonpoisonous/Numerous

Cap minute, conic, gray-brown; gills never inky; in large troops on wood debris

Cap 6–15 mm broad, conic, bell-shaped, dry, yellow-brown to gray-brown, striate with folds over the margin. Flesh soft, thin, pallid. Gills attached, fairly well separated, narrow, white, gray to black in age but not becoming inky. Stalk 25–40 mm long, 1–2 mm thick, equal, without hairs, white, curved, hollow. Veil absent.
   Spores 7–10 x 4–5 μ elliptical, smooth, pore at apex, black spore print.
   Gregarious *in great numbers* on well-decayed wood debris or buried wood, in Sp., S., and F.; widely distributed in eastern North America.
   Sometimes known as *Pseudocoprinus dissmeninatus* (Pers. ex Fr.) Kuehn.

187 COPRINUS PLICATILIS Fr.   **Nonpoisonous**/Numerous

Cap conic, deeply striate; gills free, attached to a collar; in grass

Cap 5–20 mm tall, 5–10 mm broad, narrowly conic, upturned in age, dry, center brown, margin brown but soon gray, deeply striate to nearly fragile. Flesh fragile, very thin. Gills free but attached to a collar, well separated, gray to black, gills slowly turn inky (autodigest). Stalk 5–7 cm long, 1–2 mm thick, equal, dry, without hairs, brittle, white, hollow with a small basal bulb. Veil absent.

Spores 10–12 x 7.5–9.5 μ broadly elliptical, apical pore, black spore print.

Numerous in grass in Sp., S., and F. during wet cool weather; widely distributed.

C. ephemerus Fries is similar but the cap is yellow-brown, the gills are not attached to a collar, the spores are larger (15–18 μ long), and it fruits on dung. *C. radiatus* Fries is also found on dung but it is very small (2–10 mm broad) with a narrowly conic cap which is covered with fragile white hairs especially dense over the center. Its spores measure 11–14 x 6–7 μ. There are a number of other small species found on grass, humus, and dung.

188 COPRINUS COMATUS   **Edible, choice**/Common

(Mull. ex Fr.) S. F. Gray ["Shaggy Mane"]

Cap tall and cylindrical with many scales, large; inferior ring; on hard ground

Cap 5–10 cm tall, narrowly cylindric, expanding to bell-shaped, dry, covered with flattened scales which recurve in age, white with light reddish brown scales. Flesh soft, white. Gills notched, crowded, white to black becoming inky in age. Stalk 8–20 cm long, 10–15 mm thick, equal, dry, white, bulbous. Veil fibrous. soon free, leaving an inferior ring.

Spores 13–18 x 7–8 μ elliptical, smooth, apical pore, black spore print.

Scattered or more often in groups on hard ground or grassy areas, often along road-sides, in late S. and F.; widely distributed.

One of the most delightful of all edible mushrooms. It should be prepared for the table as quickly as possible after being collected. It is excellent with eggs. The size of the elongated, conical cap and the curved scales near and over the top make it a very distinctive species.

189 COPRINUS ATRAMENTARIUS (Bull. ex Fr.) Fries   **Edible**/Common

Cap conic, gray-brown, striate; inferior ring; clusters in grass

Cap 2–8 cm broad, conic, bell-shaped in age, dry, hairless to scaly over the central portion, light gray-brown, margin striate to wrinkled. Flesh thin, pallid. Gills nearly free, crowded, broad, white, gray to black and inky in age. Stalk 4–10 cm long, 6–10 mm thick, equal or enlarged toward base, dry, white, covered with minute flattened white hairs. Veil membrane-like, leaving an inferior, fibrous ring (Fig. 10).

Spores 8–12 x 4.5–6.5 μ elliptical, smooth, pore at apex, black spore print.

In tight clusters of three or more in grass, wood debris, or from buried wood, in S. and F.; widely distributed.

The autodigestion (inky gills) occurs after the spores mature. Should not be eaten along with or followed by alcoholic drinks. The toxic reaction is a flushed hot feeling, nausea, and gastrointestinal disturbance as described in toxic Group VI.

190 COPRINUS MICACEUS (Bull. ex Fr.) Fries   **Edible**/Common

Cap conic, red-brown, glistening particles over the top; cespitose on wood

Cap 3–6 cm broad, conic, bell-shaped in age, dry, reddish brown, covered with glistening white particles which disappear in age, margin striate. Flesh soft, watery, pallid. Gills notched, crowded, white to black becoming inky in age. Stalk 15–70 mm long, 3–6 mm thick, equal, silky white, hollow.

Spores 7–10 x 3.5–5 μ elliptical, smooth with apical pore, blackish brown spore print.

In dense cespitose clusters often in great numbers, on wood debris or around old stumps on buried wood, during cool, moist weather in Sp., S., and F.; widely distributed.

This is a good edible. It is small but often found in great quantity, which makes collecting it worthwhile. The youngest specimens, with white to very light gray gills, are much the best.

PANAEOLUS

The conic caps are grayish white to gray-brown or red-brown and not folded striate as *Coprinus*. The gills are mottled, deep purple-brown to purple-black but do not become inky as *Coprinus*. The stipe (stalk) is thin and rigid and only one species, *P. separatus*, has a ring. All the species except *P. foenisecii* on grass, are found directly on dung, unlike species of *Psathyrella* which are found on wood and humus. In addition, *Panaeolus* spores retain their color in concentrated H_2SO_4 (sulphuric acid) while those of Psathyrella discolor in H_2SO_4. *Panaeolus* spores are smooth or warted but have pore at apex and a black to purple-black spore print.

Edibility. The species in this genus are poisonous (Groups III and IV) and should be completely avoided (see discussion under the Coprinaceae).

Key to Panaeolus

1 a. Veil present leaving a persistent annulus (ring) in center of stipe (stalk); cap conic, viscid, dull grayish white; on horse dung, *191. P. separatus*
　 b. Veil absent or not leaving an annulus, **2**
2 a. Scattered to abundant in lawns and grassy areas; cap brown to reddish brown, *192. P. foenisecii*
　 b. Cap smooth or wrinkled over the center, cap margin with toothlike, white hanging veil remnants (Fig. 13), **3**
3 a. Cap smooth over the center, *193. P. campanulatus*
　 b. Cap wrinkled over the center (see *193. P. campanulatus*), *P. retirugis*

191 PANAEOLUS SEPARATUS   **Poisonous** (hallucinogenic)/Numerous
(L. ex Fr.) Quel.

Cap tall, conic, viscid, grayish white; ring; on horse dung

Cap 8–45 mm broad, tall, conic, bell-shaped, light gray to dull grayish white, viscid, smooth, often splitting at margin in age. Flesh thin, soft, white. Gills adnate, close,

116

broad, gray, mottled blackish brown, nearly black in age. Stalk 7–18 cm long, 2–6 mm thick, nearly equal, enlarged just at base, smooth or somewhat striate above ring, smooth whitish below. Veil membranous leaving a *persistent ring* near the center of the stalk, and soon blackened by falling spores.

Spores 15–20 x 8–11 μ elliptical, smooth with pore at apex, black spore print.

Single or several on horse dung during cool, moist weather in Sp., S., and F.; widely distributed.

Also known as *P. semiovatus* (Fr.) Lundell, or *Anellaria separata* (L. ex Fr.) Karst., this fungus is the only large *Panaeolus* with a persistent ring. *Panaeolus solidipes* Peck is larger, also on manure, but has a scaly to wrinkled cap, (4–10 cm broad), no ring, and it has a frequently twisted, striate stipe. *P. solidipes* fruits well in the cool moist Yukon and Alaskan summers.

192 PANAEOLUS FOENISECII   Poisonous(hallucinogenic)/Common

(Fr.) Kuehn.

Cap small, dry, brown drying tan; stalk very thin, whitish; abundant in grass

Cap 10–25 mm broad, narrowly conic, bell-shaped in age, dry, without hairs, dark smoky brown to reddish brown when moist and fresh, soon drying in sun to tan or light gray-brown, margin smooth or faintly striate in age. Flesh thin, watery, tan to light brown. Gills attached, close, broad, deep purple-brown to chocolate, often mottled with whitish edges. Stalk 4–10 cm long, 1.5–3 mm thick, nearly equal or slightly enlarged toward base, brittle, minute hairs at apex, rest without hairs, whitish to pinkish brown. Veil absent.

Spores 11–18 x 6–9 μ broadly elliptical, ornamented with warts, pore at apex, dark purple-brown spore print.

Scattered to abundant on lawns and grassy areas, in the dew of cool early mornings, wilted and gone by midday; most frequently found in Sp. and early S. but also S. and F.; widely distributed.

The presence of low levels of psilocybin and psilocin (Group IV poisons) has been reported by Ola'h (1969) and Robbers, et al. (1969). In 1966 I was consulted on a case of poisoning in which this fungus was eaten by a four-year-old boy. He was rendered comatose for a short time. I now suspect that the toxins reported by Ola'h were responsible for what I considered an unusual reaction to the fungus at the time. The warted spores set this nondescript fungus apart from all others with purple-brown spores.

193 PANAEOLUS CAMPANULATUS   Poisonous (hallucinogenic)/Common

(L.) Quel.

Cap conic, dry, gray brown tinted olive with white toothlike veil pieces on margin; on dung

Cap 8–40 mm broad, conic, narrowly bell-shaped in age, without hairs, dry, gray-brown to yellowish gray-brown, often with a tint of olive, margin white at edge with *small toothlike, white, hanging veil remnants when young* (Fig. 13). Flesh soft, white. Gills attached, close, broad, gray, speckled dark purple-brown with white edges at maturity. Stalk 4–11 cm long, 1–3 mm thick, equal, brittle, rigid and straight, enlarged somewhat at base, light gray and striate at apex, rest reddish brown to gray-brown. Veil soft, white, not leaving a ring.

Spores 14–18 x 8–13 μ broadly elliptical, smooth with pore at apex, black spore print.

Alone or several on horse or cow dung, following moist weather in Sp. and S., occasionally in F.; widely distributed.

Another name for it is *P. sphinctrinus* (Fr.) Quel. *Panaeolus retirugis* (Fr.) Gillet is similar but has a wrinkled cap especially near center and is also found on dung. Both of the above species contain pantherin and other related toxins discussed in Group III toxins. A combination of sickness and hallucinations is experienced by those who have mistakenly eaten these fungi.

PSATHYRELLA

This is a large genus with approximately 450 species. It is currently being studied and revised by Alexander H. Smith. The convex to nearly flat cap ranges from hairy to smooth and hairless, but is usually not folded striate. The stalk sometimes has a ring. The species are most often gregarious to cespitose on humus, grass, or wood which may be buried. The spores are warted or smooth and have pore at apex (Fig. 64). The spore print is purple-brown to deep purple-brown, but the spores discolor to red, dingy blue, or some other color in the presence of concentrated H_2SO_4 (sulphuric acid). The species in this genus could be mistaken for Agaricus, but the gills in *Psathyrella* are not free. They might also be confused with the Strophariaceae, and this family should be checked if one is in doubt. *Psathyrella* has a cellular cuticle (Fig. 80) and the cuticle in the Strophariaceae is composed of threadlike, long narrow (filamentous) hyphae (Fig. 82).

Edibility. I have no reports of toxins in the genus, but I recommend caution because of reported toxins in related genera and families. Several are reported as edible, but I have no information concerning any special preparation for them. They are not fleshy mushrooms and large fruitings would have to be found to provide enough for a meal.

Key to Psathyrella

1 a. Cap covered with dark brown dense fibrils (hairs), *194. P. velutina*
  b. Cap light brown, glabrous (lacking hairs) or with scattered white hairs, **2**
2 a. Cap pale cinnamon to honey-yellow, glabrous (without hairs) with scattered white fibrils at first; in lawns and grassy areas, *195. P. candolleana*
  b. Cap light brown to dark brown, without hairs; in cespitose clusters on well-decayed hardwood stumps and logs, *196. P. hydrophila*

194 PSATHYRELLA VELUTINA (Fr.) Sing.   **Edible**/Common

Cap convex, dry, with dense dark brown hairs; on grass or humus

Cap 1–5 cm broad, convex, convex with knob in age, dry, covered with dense hairs, dark brown, sometimes with an olive hue, margin lighter to pinkish buff. Flesh soft, brownish. Gills adnexed (notched) to nearly free, light brown to dark brown, mottled, edges white, minutely hairy. Stalk 3–7 cm long, 4–10 mm thick, equal, dry, with hairs and whitish above ring, light brown flattened hairs below. Veil fibrous, white leaving an obscure, superior, hairy ring.

Spores 8.5–12 x 5.5–7 μ elliptical, warted, dark purple-brown to blackish brown spore print. Cystidia (sterile cells) on gill edge narrowly club-shaped, thin-walled.

Single to abundant on lawns or grassy areas or humus, in Sp., S., and F.; widely distributed.

Reported as edible, but I have no personal information on it.

195 PSATHYRELLA CANDOLLEANA (Fr.) A. H. Smith   **Edible**/Common

Cap broadly convex, dry, light brown, white scales at first; stalk shiny-white; on grass

Cap 2–8 cm broad, convex, broadly convex in age, dry, light brown, pale cinnamon to honey-yellow, with scattered small white scales at first, margin at first with hanging veil remnants. Flesh thin, fragile, white. Gills attached, close to crowded, white, grayish to purplish brown. Stalk 2–5 cm long, 2–4 mm thick, smooth, shiny-white, minute fibrils at the apex, hollow. Veil membranous, white, rarely leaving ring.

Spores 6.5–8 x 4–5 μ elliptical, smooth with pore at apex, purple-brown spore print.

Single to scattered or numerous in grassy areas, lawns, or forest openings, in Sp., S., and F.; widely distributed. Often in grassy areas in open forest in the PNW.

Also known as *Hypholoma appendiculatum* Fries and *Hypholoma incertum* Pk. This fungus is one of the few members of the genus on whose edibility we have information. However, there is not really much flesh and a large number would be required for a meal although the flavor is reported to be excellent.

196 PSATHYRELLA HYDROPHILA (Fr.) A. H. Smith   **Unknown**/Common

Cap convex, brown, lacking hairs, viscid; stalk shiny-white; cespitose on wood

Cap 15–40 mm broad, convex to broadly convex with low knob, moist but not viscid, lacking hairs, light brown to dark brown, margin faintly striate when moist. Flesh thin, fragile, light brownish. Gills adnate (attached) to adnexed (notched), close, narrow, brown to purple-brown. Stalk 2–8 cm long, 2–5 mm thick, equal or enlarging toward base, dry, shiny-white, streaked with inconspicuous hairs, hollow. Veil usually not seen, delicate, thin, soon disappears.

Spores 5–7 x 3–3.5 μ kidney-shaped, smooth with pore at apex, purple-brown spore print.

In cespitose clusters, often in great numbers, on well-decayed hardwood stumps and logs in S. and F.; widely distributed.

Not enough is known about closely related species of *Psathyrella* to recommend these as edibles. In addition, they have relatively little flesh. *P. spadicea* (Schaeff. ex Fr.) Singer looks similar but lacks a veil. The cap is deep red-brown and it has larger spores (8–9 x 3.5–4.5 μ) and a reddish to red-brown spore print. I have encountered this species in the Yukon Territory of Canada in abundance but it is infrequent further south.

STROPHARIACEAE

This family includes those mushrooms with purple-brown spore prints, the spores have a pore at the apex (Fig. 64), and the cap cuticle is filamentous (Fig. 82). The species have a central stipe (stalk) with the gills attached to it. Members of the Agaricaceae have a chocolate-brown spore print but the gills are free from the stipe. The Cortinariaceae have yellow-brown to rusty-brown spore prints and the spores are entire but lack a pore at apex (Fig. 63).

The Stropharias have convex, viscid caps and a persistent annulus (ring). They are not cespitose and are usually found in grass, on wood chips or mulch, and on dung. The Naematolomas are also usually convex but often cespitose on wood and often have olive-colored gills. Species of *Psilocybe* are less frequently encountered and are quite small; they usually have conic caps and some have blue-staining stalks.

Edibility. Much caution must be exercised in collecting and eating members of this family. Only two species of *Naematoloma* are good edibles and even here several closely related species should be avoided. *Stropharia* and *Psilocybe* should be avoided. The toxins in *Stropharia* are not known (see Group VI toxins). Psilocybes have not been totally investigated but those in section *Caerulescentes* have blue-staining stalks and several also have hallucinogenic toxins.

Key to the Genera of the Strophariaceae

1. a. A persistent, superior annulus (ring) *Stropharia*
   b. Ring absent or a thin zone of fibrils which soon disappear, **2**
2. a. On dung, grass, or humus; thin stipe (stalk); conic cap; never cespitose, *Psilocybe*
   b. On wood of conifers and hardwoods; thickened stalk; often cespitose, *Naematoloma*

Key to Stropharia

1. a. Cap bright green, viscid, *197. S. aeruginosa*
   b. Cap another color, **2**
2. a. Cap straw-yellow, viscid; on dung, usually horse dung, *198. S. semiglobata*
   b. Cap some other color or not on dung, **3**
3. a. Cap straw-yellow to yellow-brown; in grass (esp. lawns), *199. S. coronilla*
   b. Cap not colored as above, **4**
4. a. Cap deep maroon-red to red with or without scales, **5**
   b. Cap cinnamon-gray, brown to tan, **6**
5. a. Cap deep maroon-red; yellow patches on surface of unopened veil (see 200. *S. squamosa*), *S. rugosoannulata*
   b. Cap cinnamon-red with scattered buff hairs; no yellow patches on surface of unopened veil (see 200. *S. squamosa*), *S. squamosa* var. *thrausta*
6. a. Stalk tall and thin (6–8 mm thick), *200. S. squamosa*
   b. Stalk thicker (8–25 mm thick) with or without cottony scales, **7**
7. a. Dense cottony scales on stipe; persistent skirtlike annulus, *201. S. hornemanni*
   b. Few to no scales on stipe; annulus is fragile and soon disappears (see 201. *S. hornemanni*), *S. ambigua*

197 STROPHARIA AERUGINOSA (Curt. ex Fr.) Quel.   **Poisonous**/Infrequent

Cap viscid, bright green; in grassy areas

Cap 1–5 cm broad, convex, convex bell-shaped in age, viscid, *bright green,* yellowish green in age. Flesh soft, white often tinted bluish. Gills attached, close, broad, gray, purplish brown in age, edges minutely hairy and white. Stalk 3–8 cm long, 3–10 mm thick, equal, viscid, smooth, white above ring, greenish blue and viscid below. Veil soft, membranous, leaving a superior ring which may disappear in age.

Spores 7–10 x 4–5 μ elliptical, smooth with a pore at apex, purple-brown spore print. Cystidia (sterile cells) on gill edge club-shaped with an elongated neck.

Single or several in grassy areas, gardens, or in damp areas in forests, late S. and F.; widely distributed but infrequently encountered.

The striking green, viscid cap and stalk, along with the purple-brown spore print set this fungus apart from all others. It is reported in many works on fungi to be poisonous, but I have not been able to find any specific cases of poison attributed to it.

198 STROPHARIA SEMIGLOBATA (Fr.) Quel.   Nonpoisonous/Common

Cap yellow, viscid; ring present; on dung, usually horse or cow

Cap 12–35 mm broad, conic to convex, without hairs, viscid, light yellow to straw-yellow. Flesh firm, watery, buff. Gills attached, close to fairly well separated in age, broad, gray, soon dark purple-brown. Stalk 3–12 cm long, 2–4 mm thick, equal or enlarged slightly toward base, minutely hairy above ring, viscid below, white to buff. Veil delicate, membranous, leaving a superior, fragile ring, sometimes absent, soon purple-brown from accumulated spores. Taste sometimes bitter.

Spores 15–22 x 8.5–11 μ elliptical, smooth with a pore at apex, purple-brown spore print.

Single or several on dung, often horse dung, or in manured fields in Sp., S., and F.; widely distributed. Predictable on dung and in corrals following or during long periods of moist weather.

Psilocybe coprophila looks similar, is also on dung, but has a gray-brown cap, no ring, and smaller purple-brown spores (12–14 x 6.5–8 μ).

199 STROPHARIA CORONILLA (Bull. ex Fr.) Quel.   Poisonous (?)/Numerous

Cap yellow-brown, viscid, small; short, thin stalk; in lawns or grassy areas

Cap 1–4 cm broad, convex, broadly convex in age, without hairs, slimy, viscid to nearly viscid in moist weather, buffy tan, yellowish brown to straw yellow. Flesh firm, thick, white. Gills adnate, close, broad, brownish violet to purple-black in age. Stalk 3–4 cm long, 3–7 mm thick, equal, dry, white, minute fibrils above ring, shiny with longer fibrils below. Veil membranous, white, leaving a persistent ring grooved striate above, about in center of stalk, soon colored purple-brown above from falling spores. Odor slightly disagreeable.

Spores 7–9.5 x 4–5 μ ovoid, smooth with pore at apex, purple-brown spore print. Cap cuticle of filamentous hyphae.

Several to abundant on lawns, meadows, under open stands of trees, in late S. and F.; widely distributed. I have found this in great numbers on lawns in central Idaho. *Stropharia hardii* Atk. is larger and usually has dark spots on the cap and is common in the SE.

200 STROPHARIA SQUAMOSA (Pers. ex Fr.) Quel.   Poisonous/Infrequent

Cap brown with buff hairs; long thin stalk; on wood, wood chips, etc.

Cap 3–6.5 cm broad, convex, broadly convex in age, viscid with scattered buff fibrils, especially near margin, over a yellow-brown to orange-brown surface. Flesh firm, white. Gills attached, fairly well separated, buff to violet-brown with white, minutely hairy margins. Stalk 7–12 cm long, 6–8 mm thick, equal or enlarging somewhat toward base, dry, above ring white with minute white scales, below with dense buff to orange-buff small scales, base surrounded with long, often stiff, orange-buff hairs. Veil fragile, membranous, buff, leaving a superior ring which hangs, often torn, around the stalk.

Spores 11–15 x 6–8 μ broadly elliptical, smooth, with a pore at apex, purple-brown spore print. Cystidia (sterile cells) present on gill edge, broadly club-shaped.

Single or several on buried wood, wood chips of hardwoods, in late S. and F.; in eastern North America, other distribution unknown.

Stropharia squamosa var. *thrausta* Kalch. has red cap but in all other ways resembles *S. squamosa.* I have found this red-capped variety more frequently, especially on compost piles made up of wood chips and from buried wood in Maryland. *Stropharia rugosoannulata* Farlow is a very beautiful, robust species with a deep maroon-red cap and radially arranged yellow, cottony patches on the lower surface of the unopened veil. The stalk is thick and equal (13–20 mm thick). Mr. Andrew Norman of New York has eaten *S. rugosoanulata* and reports that it has an excellent flavor.

201 STROPHARIA HORNEMANNI (Fr.) Lund. and Nannf.   **Unknown**/Numerous

Robust with cottony scales on cap and stalk; skirtlike ring; under conifers

Cap 2–14 cm broad, convex, broadly convex often with a broad low knob, slimy, viscid, without hairs with white cottony scales over the margin at first, cinnamon-gray, reddish brown to light purple-brown. Flesh firm, white, light yellowish near stalk base. Gills attached, close, broad, whitish, light gray to smoky violet in age. Stalk 5–15 cm long, 7–25 mm thick, equal, silky, smooth, white above ring, cottony, abundant, soft, pure white scales below. Veil soft, white, leaving a persistent, superior ring which flares upward at first, later hanging skirtlike around stalk. Taste disagreeable.

Spores 10.5–13 x 5.5–7 μ elliptical, smooth, with a pore at apex, purple-brown spore print.

Single, several, sometimes numerous under conifers or on well-decayed conifer wood in F.; widely distributed.

The robust fruiting body with cottony scales on the lower stalk, the persistent, skirtlike ring, and the habitat under conifers separate this beautiful *Stropharia* from all others. *Stropharia ambigua* (Pk.) Zeller is similar in size but has few scales on cap, its ring soon disappears, and it is most abundant in the PNW. We have conflicting reports of poison cases in *Stropharia*. I would not recommend this group as edible.

PSILOCYBE

This is mainly a group of small species with conic to campanulate caps and rather long, thin, central stipes (stalks), sometimes with an annulus (ring). The gills are attached and they have purple-brown to chocolate-brown spore prints. The spores are smooth and have a pore at the apex. The stipe is white to reddish brown. However, in section *Caerulescentes,* the stipe stains blue when bruised. This blue-staining reaction is correlated with the presence of hallucinogenic compounds described in Group IV toxins. There are 18 species in this section alone, but only 7 are known in the United States and Canada. Most of them are encountered infrequently.

Edibility. There are no edible species (see Group IV toxins), but only a few of the species have been assayed for toxins.

Key to Psilocybe

1 a. Stalk bruises blue, **3**
   b. Stalk does not bruise blue, **2**
2 a. On dung, especially horse dung, *202. P. coprophila*
   b. On moss, usually at high elevations, *203. P. montana*
3 a. Ring membranous and persistent, *204. P. cubensis*
   b. Ring absent or leaving a thin fibrillose zone, *205. P. caerulipes*

202 PSILOCYBE COPROPHILA (Bull. ex Fr.) Kummer   **Nonpoisonous**/Common

Cap gray-brown, viscid; no ring; on dung

Cap 1–2 cm broad, convex, tacky to viscid, smooth with fine hairs over the margin, gray-brown to dark brown. Flesh thin, brown, not staining. Gills attached, fairly separated, gray-brown to deep purple-brown in age. Stalk 2–6 cm long, 1–3 mm thick, dry, nearly equal, yellow to yellowish brown with scattered fibrils. Veil absent. Taste of fresh meal.

Spores 11–15 x 6.5–9 μ broadly elliptical, smooth with a pore at apex, purple-brown spore print.

Several or in groups on horse dung or cow dung; widely distributed in Sp., S., and F.

Stropharia semiglobata (Fr.) Quel. is similar and also occurs on dung. It has a yellow, viscid cap; persistent ring; viscid, yellowish stalk; and its spores are also much larger.

203 PSILOCYBE MONTANA (Pers. ex Fr.) Kummer   Edibility unknown/Common

Cap red-brown, moist; no ring; in moss

Cap 5–15 mm broad, conic to convex with a low knob, moist, dark red-brown with lighter striate lines over the margin. Flesh thin, brown. Gills attached, fairly well separated, light brown to red-brown in age. Stalk 20–40 mm long, 1–2 mm thick, slightly larger toward base, dry, smooth or with a few scattered hairs, reddish brown. Veil very thin, leaving no ring.

Spores 5.5–8 x 4–5 μ short elliptical, smooth with a pore at apex, purple-brown spore print. Cystidia (sterile cells) club-shaped with a rounded head, abundant on the gill edge.

Several to numerous, usually in deep moss, sometimes buried almost to the cap; widely distributed, common in Sp. at high elevations in the RM. This minute agaric is also known as *P. atrorufa* (Fr.) Quel.

204 PSILOCYBE CUBENSIS (Earle) Singer   Poisonous, hallucinogenic/ Rare

Cap pale yellowish, viscid; persistent ring; blue-staining stalk

Cap 1.5–8 cm broad, conic, bell-shaped, convex in age, viscid, without hairs, whitish to pale yellow, light brownish in age, stains bluish in age. Flesh firm, white bruises blue. Gills adnate (attached) to adnexed (notched), close, gray to violet-gray in age with white edges. Stalk 4–15 cm long, 4–14 mm thick, enlarging somewhat toward the base, dry, without hairs, white staining blue when bruised. Veil white, leaving a superior membranous ring.

Spores 10–17 x 7–10 μ elliptical to oval in side view, thick-walled, with a large pore at apex, purple-brown spore print. Cystidia (sterile cells) on gill edge club-shaped with rounded heads.

Several or in groups on cow dung, horse dung, or on manured soil. Fruiting from late W. to late F. Known only from Florida in the United States, but widely distributed southward. Also called *Stropharia cubensis* Earle and *Stropharia cyanescens* Murr. The presence of psilocybin and psilocin (Group IV toxins) has been reported for this species. Several varieties with cap colors ranging from nearly white to clear yellow have been described.

205 PSILOCYBE CAERULIPES (Pk.) Sacc.   Edibility Unknown/Infrequent

Cap cinnamon-brown, viscid; fibrillose annular zone; blue-staining stalk

Cap 10–35 mm broad, conic to convex with a low knob, viscid, without hairs, cinnamon-brown, margin often tinted green, incurved at first. Flesh thin, dull white, bluish when bruised. Gills attached, close, cinnamon to cinnamon-brown, edges whitish. Stalk 3–6 cm long, 2–3 mm thick, nearly equal, dry, minutely hairy, white to buff, blue when bruised. Veil very thin, white, ring a thin hairy zone or absent.

123

Spores 7–10 x 4–5.5 μ elliptical to broadly elliptical in face view, smooth with a pore at apex, dark purple-brown spore print. Cystidia (sterile cells) on gill edge numerous, club-shaped with a narrow neck (see right cystidium, Fig. 75).

Solitary to cespitose clusters on debris under hardwood trees and well-decayed hardwood logs. It is found in SE, CS, and adjacent C., in S. and F.

Psilocybe caerulescens Murr. is somewhat similar but has a large white veil at first, is larger (cap 2–8 cm broad), and is reported only from Alabama in the United States. It is frequently encountered south of the Mexican border. Though they have not specifically investigated both of these species there is little doubt that they contain psilocybin or psilocin (Group IV toxins).

NAEMATOLOMA

There are 21 species of *Naematoloma* in North America (Smith, 1951). The most common species occur in cespitose clusters on wood. They resemble Pholiotas to some extent but their purple-brown spore print, filamentous cap cuticle (Fig. 82), and spores with a pore at apex (Fig. 64) place them in the Strophariaceae. The dense clusters usually have a cap or two with a good spore print on the top from a taller neighboring cap. This provides a quick way in which the typical spore color of the genus can be spotted in the field. Remember that Naematolomas have no rings and that the edible ones are found on wood.

Edibility. I have no reports of severe poisonings in this genus, some species have a very bitter taste while others are quite good. When conditions are right one can usually find many cespitose clusters of one species fruiting on wood. *Naematolomas capnoides* and *N. sublateritium* are of good flavor and can be used as all-purpose edibles.

Key to Naematoloma

1 a. Cap conic; stalk thin (2–5 mm thick); single to numerous but not in cespitose clusters, *206. No. dispersum* and *N. udum*
  b. Cap convex; stalk thicker (5 mm +); cespitose, **2**
2 a. Gills white, then gray finally purple-brown but not greenish; cap orange to cinnamon; usually on conifer wood, *107 N. capnoides*
  b. Gills sulphur-yellow to yellow with greenish hue; cap brick-red to citron-yellow; on hardwoods and conifers, **3**
3 a. Cap brick-red; gills sometimes tinted olive; on hardwood stumps and logs, *208. N. sublateritium*
  b. Cap citron-yellow to orange-yellow; gills conspicuously green; on hardwood and conifer logs and stumps, *209. N. fasciculare*

206 NAEMATOLOMA DISPERSUM (Fr.) Karst.   Edibility unknown/Common

Cap conic, orange-brown; stalk thin, dry; never in cespitose clusters; on wood debris

Cap 1–4 cm broad, conic, convex in age, waxy, moist, knobbed in age, glabrous over the central portion, thinly silky over the margin, tawny to orange-tawny, yellowish to dingy olive over the margin. Flesh thin, dingy white. Gills attached, close, broad, white to dingy olive, purplish brown in age with whitish edges. Stalk 6–10 cm long, *2–5 mm thick,* equal, dry, pruinose (with a powdery cover) and buff to whitish above the obscure, hairy annular zone, below with scattered hairy patches, reddish brown over the base. Veil thin, hairy, buff, having a superior, often nearly absent, hairy zone.

Spores 7–11 x 4–5 μ elliptical, smooth with a small pore at apex, purple-brown spore print. Cystidia (sterile cells) numerous on gills with yellow globules in 3 percent KOH.

Single but more often numerous on conifer debris, sawdust, wood chips mixed with dirt in late S. and F. in PNW, RM, C. A, less commonly in CS. I have encountered massive fruitings along newly constructed roads during wet weather in Alaska.

The non-cespitose habit and the long slender stalks separate it from the other species treated. *Naematoloma udum* (Pers. ex Fr.) Karst. is very similar but has larger spores (14–20 μ long) and is found in eastern North America. There are a group of closely related species with slender stalks described by Smith (1951), and I would not recommend any of these as edibles.

207 NAEMATOLOMA CAPNOIDES (Fr.) Karst.   Edible/Common

Cap orange, moist; gills gray; cespitose clusters on conifer wood

Cap 2–7 cm broad, convex, soon broadly convex, nearly flat in age, sometimes with a low knob, moist but not viscid, smooth, without hairs, or with scattered buff fibrils, reddish orange, orange to cinnamon, margin inrolled at first and pale yellow with buff patches of adhering veil remnants. Flesh thick, white to pallid. Gills attached, close, *white to gray* finally purple-brown. Stalk 5–10 cm long, 4–10 mm thick, equal, dry, scattered hairs to hairy at base, yellowish above a faint ring, tan to rusty brown below. Veil hairy, white to buff, leaving a thin to obscure, superior ring zone. Taste mild.

Spores 6–7.5 x 3.5–4.5 μ elliptical, smooth with a small pore at apex, *purple-brown* spore print. Cystidia (sterile cells) numerous on gills with yellow globules in 3 percent KOH.

Large cespitose clusters, often abundant on conifer wood in F.; widely distributed.

It is often seen in large clusters with many caps, and the purple-brown spore print can usually be seen on one of the shorter caps in the cluster. The gills do not have the olive or greenish coloration of *N. fasciculare,* and the cap is not the brick-red color of *N. sublateritium,* which occurs on hardwood logs and stumps. Both species were once placed in the genus *Hypholoma.*

208 NAEMATOLOMA SUBLATERITIUM   Edible, good/Common

(Fr.) Karst. ["Brick Cap"]

Cap brick-red, moist; gills sometimes tinted olive; cespitose clusters on hardwoods

Cap 2–10 cm broad, convex to broadly convex with a low obtuse knob at times, moist, covered with scattered patches of yellowish hairs, the rest is smooth, without hairs, brick-red over the center, margin buff to buff-pink, inrolled at first. Flesh thick, pale brownish, sometimes staining yellow when bruised. Gills attached, close, white sometimes yellowish rarely with a faint olive tint, soon gray to dingy purple from the spores. Stalk 5–9 cm long, 5–15 mm thick, equal, smooth and white above annular zone to gray, eventually dull brown and flattened hairs below, often staining yellowish over the base. Veil hairy, buff, leaving a thin, superior ring zone. Taste mild to bitter.

Spores 6–7.5 x 3.5–4 μ elliptical, smooth with a small pore at apex, purple-gray spore print. Cystidia (sterile cells) numerous on gills with yellow globules in 3 percent KOH.

In densely cespitose clusters on hardwood stumps and logs in late S. and F., common in eastern North America, rare in western North America. See *N. capnoides.*

(Huds. ex Fr.) Karst. ["Sulphur Tuft"]

Caps yellow-orange, moist; gills greenish yellow; cespitose clusters on hard-woods; taste bitter

Cap 1–8 cm broad, conic, soon broadly convex to nearly flat in age, broadly knobbed at times, moist, hairless or silky over the margin, citron-yellow to orange-yellow over center, sometimes tinted olive, margin incurved at first with hanging veil remains. Flesh up to 4 mm thick, yellowish bruising dingy brown. Gills attached, crowded, narrow, *sulphur-yellow to greenish yellow.* Stalk 5–12 cm long, 3–10 mm thick, narrow, *tapering toward base,* moist to dry, hairless above ring, flattened hairs below, white to light buff above, brown to rusty-brown below, usually darkening in age. Veil hairy, light buff to cream color leaving a superior, hairy ring zone. *Taste very bitter.*

Spores 6.5–8 x 3.5–4 μ elliptical, smooth with a small pore at apex, purple-brown spore print. Cystidia (sterile cells) on gills with yellow globules in 3 percent KOH.

In large or small cespitose clusters on logs and stumps of hardwoods and conifers in Sp., F., and during mild weather in W.; widely distributed. I have seen this commonly in the PNW and SE but less so elsewhere. The green gills, bitter taste, and thin stalk separate this *Naematoloma* from *N. capnoides* and *N. sublateritium.* The bitter taste is retained after cooking, making this abundant species totally inedible.

AGARICACEAE
The Meadow Mushrooms

These are fleshy mushrooms with a white to brown or gray-brown cap which is often covered with appressed squamules (flattened scales), especially in age. The surface is moist to dry but not viscid. The flesh stains yellow, red, or not at all. The gills range from white to pale gray or deep pink and are free or nearly so. In age the maturing spores *turn the gills chocolate-brown,* the color of the spore print. The spores most often have a pore at the apex (Fig. 64) and are elliptical and smooth. The stipe always has a su-perior annulus (ring) which is usually single or sometimes double. It is persistent and easily seen. The cuticle of the cap is filamentous (Fig. 82).

The various species are characteristically found in humus, grassy areas or well-manured ground, with several also in hardwood or conifer duff. Well-fertilized lawns during moist, cool periods often produce abundant crops of several species. The caps will often be partially buried in the grass. The collector should dig the entire specimen out of the ground.

Edibility. This is an excellent edible group. The only reports of any illness after eating the species of *Agaricus* center around species related to *A. silvaticus.* Some people experience gastrointestinal disturbance, but this seems to be a highly individual reac-tion. Any new edible should be tested in the usual way.

The mushroom of commerce, *A. bisporus,* was in reality derived from several different species. The continual propagation of it over several centuries has resulted in several commercial strains which have no exact counterpart in nature (see *A. campestris*).

Agaricus campestris, the "Pink Bottom," is probably the most closely related of any of the common species encountered in North America.

Almost all of the mushrooms in this group can be eaten. The stipe base should be cut off, but there is no need to remove the cap cuticle or the gills since they are good eating. The firm flesh is good for the traditional mushrooms on steak. *Agaricus* is excellent in any dish calling for mushrooms. It is especially good in scrambled eggs, soup, creamed on toast, stuffed with crabmeat, in "Mushroom Loaf," and raw in salads. (See Recipes.)

Key to Agaricus

1 a. Flesh changing to yellow or reddish when bruised, **2**
   b. Flesh not changing when bruised, **8**
2 a. Cap white at first, sometimes yellowish in age, *210. A. arvensis*
   b. Cap colored, **3**
3 a. Cap very large (10–30 cm broad), **4**
   b. Cap smaller 4–12 cm broad, **5**
4 a. Cap yellow-brown; stalk long and slender, *211. A. augustus*
   b. Cap with brown scales; stalk short and broad (see 211. *A. augustus*),
     A. crocodilinus
5 a. Cap gray; staining yellowish then slowly reddish brown, *212. A. placomyces*
   b. Cap pinkish to reddish brown; staining yellow or reddish when bruised, **6**
6 a. Flesh bruising slowly yellowish (see 213. *A. silvaticus*), *A. hondensis*
   b. Flesh bruising reddish, sometimes slowly, **7**
7 a. Stalk glabrous, or nearly so, below the ring, *213. A. silvaticus*
   b. Stalk with fibrillose scales below ring (see 213. *A. silvaticus*), *A. subrutilescens*
8 a. Veil single; gills bright pink, *214. A. campestris*
   b. Veil double; gills pale pink, *215. A. rodmani*

210 AGARICUS ARVENSIS   Edible, choice/Infrequent

Fr. ex Schaeff. ["Horse Mushroom"]

Cap white; flesh bruising yellow

Cap 8–20 cm broad, oval, convex to flat in age, dry, smooth, without hairs with small scales over the center in age, white, creamy white to yellowish brown, margin sometimes covered with hanging veil remnants, *bruises yellow,* darkening to dingy yellowish brown. Flesh thick, firm, white slowly tinged yellowish. Gills free, crowded, whitish but soon grayish pink, blackish brown in age. Stalk 5–20 cm long, 10–30 mm thick, equal sometimes with a small bulb at base, dry, shining and smooth above the ring, cottony white, small scales below, bruising yellow. Veil membranous, white, thin, double before breaking, the lower layer with a characteristic tooth pattern around the stalk, leaving a superior skirtlike annulus. Odor of anise.

Spores 7–11 x 4.5–5.5 μ elliptical, smooth, purplish brown to blackish brown.

Several to numerous in meadows and fields in S. and F.; widely distributed.

Look for the grayish pink gills, anise odor, yellow staining, and cottony white scales on the stalk. It is indeed a delicious edible fungus. *A. silvicola* (Vitt.) Saccardo is very similar but found in woodlands and has smaller spores (5–6.5 x 3–4 μ) but in other respects is indistinguishable. I have recorded it under conifers in the Yukon, Montana, Idaho, Virginia, and in many other locations. *A. xanthrodermus* Genevier also stains yellow, but the cap remains white to dull white at maturity. In addition, the tissue in the stalk base *turns yellow instantly when cut.* This latter species has sometimes caused gastrointestinal disturbance (Group VI toxins).

211 AGARICUS AUGUSTUS Fr.   Edible, choice/Infrequent

Cap yellow-brown; flesh bruising yellow

Cap 10–25 cm broad, large, egg-shaped with a flattened top, convex to flat with a low knob, dry, yellow-brown, covered with dense hazel-brown scales, bruising yellowish, margin only slightly recurved, soon straight. Flesh thick, white or becoming very slowly yellowish in age. Gills free, close, pink to chocolate-brown at maturity. Stalk 8–20 cm long, 15–30 mm thick, equal or enlarged near base, dry, white, nearly hairless, above ring densely hairy, scaly below, white to brownish in age, bruising yellow. Veil membranous, white above, with raised, cottony patches underneath, leaving a superior, soft, skirtlike ring. Odor of anise.

Spores 8–11 x 5–6 μ elliptical, smooth, chocolate-brown spore print.

Scattered to numerous under conifers or in grassy areas in or close to forest cover, in Sp., S. or F.; widely distributed.

The large, yellow cap, stalk base inserted into the ground, and large spores are the chief characters of this Agaricus. A giant closely related, West Coast species is *A. crocodilinus* Murr. It has a short, stout stalk and a white cap covered with scales which darken to brown in age. It also has spores up to 16 μ long (8–16 x 6–8 μ).

212 AGARICUS PLACOMYCES Pk.   Edible (?)/Common

Cap gray; flesh bruises yellow and then reddish

Cap 4–8 (up to 15) cm broad, oval, convex, nearly flat in age, sometimes has a low knob in age, dry, covered with densely arranged *gray-brown* to blackish brown hairy scales, margin incurved and wrinkled at first but soon straight. Flesh firm, white, or dull pinkish in age, staining slowly yellowish then changing to dull reddish brown. Gills free, close, pink, dark chocolate-brown in age. Stalk 3–9 (up to 15) cm long, 7–20 mm thick, enlarged toward base, slender, dry, dull white changing to pinkish brown or dingy brown in age, base slightly enlarged or forming a small bulb. Veil membranous, white, cottony, with soft cottony patches underneath, leaving a persistent, superior ring. Odor disagreeable.

Spores 4.5–6 x 3.5–4.5 μ egg-shaped or broadly elliptical, smooth, chocolate-brown spore print. Cystidia (sterile cells) on gill edge abundant, long (up to 120 μ), club-shaped.

Several to numerous under hardwoods or conifers in late S. or F.; widely distributed.

Some persons have become ill (Group VI toxins) from eating this fungus, but others do not seem to be disturbed by it. I would proceed with caution or avoid this species altogether. The cap has gray hairs with no sign of the red-colored fibrils typically seen in *A. silvaticus.*

213 AGARICUS SILVATICUS Vitt. ex Fr.   Edible/Numerous

Cap reddish brown; stalk hairless; flesh staining reddish

Cap 4–12 cm broad, egg-shaped, convex to broadly convex in age, dry, covered with densely arranged pinkish brown, hairy scales, often bruising red, margin only slightly turned in, straight in age. Flesh thick, firm, white, slowly bruising reddish brown when young or remaining unchanged. Gills free, crowded, grayish pink to dark reddish brown in age. Stalk 6–11 cm long, 10–20 mm thick, enlarging toward base, dry, scattered, flattened white hairs above and below ring, turning dingy pink to pinkish brown in age, often with a bulbous base. Veil membranous, cottony beneath, with raised buff patches, leaving a superior, persistent ring.

Spores 5–7 x 3–4 μ broadly elliptical, smooth, chocolate-brown spore print. Cystidia (sterile cells) on gill edge, club-shaped, numerous, 14–20 x 5–7 μ.

Scattered to numerous under conifers and occasionally hardwoods in Sp. and S., but most often F.; widely distributed. I have recorded it under numerous conifers across North America.

A. placomyces has a gray-brown cap. *A. subrutilescens* (Kauff.) Hotson and Stuntz has hairy scales below the ring. *A. hondensis* Murr. is poisonous to some people. It has a white cap at first (Smith, 1949), but it soon turns dull reddish brown. In addition, the flesh stains yellowish, not red, and it is known only from PSW, PNW, and adjacent C. *A. haemorrhoidarius* Fries instantly stains blood-red when the cap or flesh is bruised, but it is a rare species.

214 AGARICUS CAMPESTRIS Fr.   Edible, choice/Common

["Meadow Mushroom" or "Pink Bottom"]

Cap white at first; gills pink young; flesh does not bruise

Cap 2–10 cm broad, convex, nearly flat in age, dry, smooth, without hairs to hairy, sometimes arranged in flattened scales, usually white at first often becoming light brown, cinnamon-brown, or even dingy brown in age, occasionally remaining nearly white. Flesh thick, firm, white or tinted reddish brown. Gills *free,* crowded, narrow, *pink* at first but clouded purplish brown from the maturing spores. Stalk 2–6 cm long, 10–20 mm thick, equal, dry, white and smooth, without hairs above the ring, white and hairy below, slowly discoloring dingy reddish brown. Veil membranous, white, leaving a torn superior ring.

Spores 5.5–7.5 x 3.5–5 μ elliptical, smooth with an obscure pore at the apex, dark chocolate-brown spore print.

Several to abundant on lawns, grassy areas, and meadows, in Sp. and F. or during long periods of cool moist weather at other times.

This is the common, delicious *Agaricus* found on lawns throughout North America. One must note the characteristic pink, free gills which soon turn chocolate-brown from the maturing spores. The spore print is brown, unlike the white spores of the white Amanitas. *A. hortensis* Cooke is a closely related mushroom (according to Pilat, 1951) which is cultivated commercially. He finds no difference between it and *A. bisporous* (Lange) Moeller and Schaeff. It has basidia with two spores (Figure 74) while *A. campestris* has four spores on each basidium, the spores are larger (7–8.5 x 5–5.5 μ), and it is found on well-manured ground in Europe. Like *A. campestris, A. hortensis* has varieties with white or brown caps which may or may not have flattened scales. We probably do not have *A. hortensis* growing wild in North America as it does in Europe.

215 AGARICUS RODMANI Pk.   Edible, choice/Numerous

Cap white to yellowish; gills pale pinkish; flesh not bruising

Cap 5–12 cm broad, broadly convex, nearly flat in age, dry, flattened hairs but often appearing smooth and without hairs, dull white with dingy-yellow to yellow-brown or even grayish stains, surface finely cracked in age, margin inrolled. Flesh firm, white, does not change color when bruised. Gills nearly free, close, pale pink, pinkish brown to deep blackish brown in age. Stalk 2–8 cm long, 15–30 mm thick, equal or narrowed at base, white and hairless above and below ring. Veil membranous, woolly, white, *double,* leaving a *double ring* which flares outward at first at the center of the stalk.

Spores 4–6.5 x 4–5 μ egg-shaped, smooth, chocolate-brown spore print.

Single to numerous in lawns, on hard-packed soil, frequently around barnyards, in Sp. or early S. and again in late S. and early F.; widely distributed but most common in eastern North America.

The double veil separates it from *A. campestris* which it closely resembles; in fact it will also be found under *A. edulis* Vittadini, or *Psalliota rodmani* Peck. It is thick, fleshy, and extremely good eating.

PAXILLACEAE

There is only one genus, *Paxillus,* in this family and only three species of consequence in North America. One species is sessile (stalk absent) and the other two have long decurrent gills and clay-brown to yellow-brown spores. Two species occur on wood and one on the soil. *Phylloporus rhodoxanthus* resembles this group, and so it is keyed out below. Its gills are also decurrent but they are bright yellow and stain blue when bruised, a reaction commonly encountered in the boletes but never in the Paxillaceae. Many mycologists regard *Phylloporus* as a gilled bolete.

Edibility. The information on *Paxillus* is conflicting, but they are generally rated as inedible. *Phylloporus* is listed in Europe as edible, but I have no information on it from North America.

Key to Paxillus and Phylloporus

1 a. Stalk absent; imbricate (overlapping) on conifer logs, *216. Paxillus panuoides*
   b. Stalk present, **2**
2 a. Stalk eccentric, densely covered with blackish brown short hairs,
      217. Paxillus atrotomentosus
   b. Stalk central, glabrous (hairless), **3**
3 a. Gills dull yellowish olive; flesh bruises brown but never blue,
      218. Paxillus involutus
   b. Gills yellow to lemon-yellow; flesh bruises blue to brown,
      219, Phylloporus rhodoxanthus

216 PAXILLUS PANUOIDES Fr.   Edibility unknown/Numerous

Cap lacking stalk, yellow-brown; gills forked and veined; on wood

Cap 3–8 cm broad, sessile (lacking stalk), petal or fan-shaped, dry, minutely downy, soon glabrous, brownish yellow or olive tinted, margin lobed or wavy. Flesh thin, soft, white. Gills radiate from point of attachment, close, often forked with veins between, pale yellow or pale yellow-orange. Stalk absent. Veil absent.

Spores 4–6 x 3–4 μ short elliptical, smooth, many deep red in Melzer's solution, yellow to buff spore print. Cystidia (sterile cells) absent.

Usually many to overlapping on limbs, logs, and stumps of conifers, in Sp., S., and F.; widely distributed. It is also on wood products in service and sometimes on hardwoods.

This is the only *Paxillus* without a stalk. *Crepidotus,* with which it has been confused, has a dull brown spore print, non-forked gills, and spores which do not turn deep red in Melzer's solution. *Phyllotopsis nidulans* is also similar but has a foul smell, a densely hairy cap, and pink, allantoid (sausage-shaped) spores which are hyaline (transparent) in Melzer's solution.

217 PAXILLUS ATROTOMENTOSUS   Edibility unknown/Numerous
(Batsch ex Fr.) Fries

Cap brown, dry; stalk off-center, dark brown dense hairs; on conifer wood

Cap 4–12 cm broad, convex, flat and depressed somewhat in center at maturity, dry, flattened hairs, tan, rusty brown to dark brown, margin rolled in. Flesh firm, thick, white. Gills short to long decurrent (extending down stalk) in age, close, narrow, forked to sometimes nearly poroid near stalk, mustard-yellow to yellow-brown. Stalk 2.5–10 cm long, 8–25 mm thick, equal, usually off center, robust, *densely hairy, dark brown to blackish brown covering,* sometimes with a rootlike base. Veil absent.

Spores 5–7 x 3–4 µ egg-shaped, smooth, turns yellow to red in Melzer's solution, clay color to mustard-yellow spore print.

Single to several or in clusters on well-decayed conifer sticks, logs, and stumps, sometimes partially buried, in S. and F.; widely distributed.

This robust species has a very distinctive, densely hairy, blackish brown stalk and a dry, off-center brown cap.

218 PAXILLUS INVOLUTUS (Batsch ex Fr.) Fries   Edibility unknown/Numerous

Cap fibrillose, dry, red-brown, inrolled; stalk central, hairless; on ground

Cap 4–12 cm broad, convex, flat with a central depression in age, dry, covered with matted, soft hairs, light reddish brown tinged with olive, sometimes obscurely zoned or spotted, margin inrolled until late age. Flesh thick, dingy yellowish bruising brownish. Gills extend down stalk, crowded, broad, forked sometimes nearly having pores, yellowish olive bruising brown. Stalk 4–10 cm long, 14–20 mm thick, equal or enlarged toward the base, never off-center, dry, hairless, yellowish brown streaked or stained darker brown. Veil absent.

Spores 7–9 x 4–6 µ elliptical, smooth, clay-brown to yellowish brown.

Single or several, near or sometimes on wood, under mixed hardwood-conifer stands in S. and F.; widely distributed. Usually found where conifers are present or at least in a mixed stand.

There are conflicting stories about the edibility of this dull brown plant which has a central, hairless stalk.

219 PHYLLOPORUS RHODOXANTHUS   Edible/Numerous
(Schw.) Bres.

Cap red-brown, dry, feltlike; gills extend down stalk, yellow, bruising blue

Cap 2–6 cm broad, convex, dry, felty to touch, often developing cracks, brown to red or red-brown, yellow flesh exposed in the cracks. Flesh firm, buff to yellow. Gills extend down stalk, fairly well separated, broad, veinlike sometimes nearly having pores, yellow to lemon-yellow, bruising blue or sordid brown. Stalk 2–6 cm long, 2–20 mm thick, widest at apex, tapering toward base, buff but stained dingy-reddish most often over the lower two-thirds. Veil absent.

Spores 11–15 x 4.5–6 µ narrowly elliptical, smooth, orange-brown to yellowish-brown spore print.

Single or several under conifers or in mixed conifer-hardwood stands, in Sp., S., and F.; widely distributed.

This is a good edible but is not often found in sufficient quantity.

BOLBITIACEAE

This family is closely related to the Coprinaceae and many of the species resemble each other. However, the Bolbitiaceae has spore print colors ranging from yellow-brown to cinnamon-brown or earth-brown, and the smooth or ornamented spores have a pore at the apex. The cuticle of the pileus (cap) is cellular (Fig. 80). The pileus is conic and striate to convex in *Bolbitius* and *Conocybe* which also have long, thin stipes (stalks). *Agrocybe* has larger or even robust species with or without an annulus. The spore print is usually brown to earth brown. Bolbitius is found on dung or manured grass, but many of the species in *Agrocybe* and *Conocybe* are on humus, wood mulch, needles, or leaf litter. All are probably saprophytes living on decaying matter.

Edibility. There are no records of poisons with *Bolbitius* or *Conocybe,* but they are too dainty and fragile to serve as food. *Agrocybe* has several species which are very robust and fleshy and if properly identified could be eaten. However, Agrocybes are not readily distinguished, even by good amateur collectors, from such poisonous genera as *Hebeloma.*

Key to Conocybe, Bolbitius, and Agrocybe

1 a. Stalk tall and very thin (1.5–3 mm thick); cap narrowly conical; very fragile, **2**
   b. Stalk thicker; if thin, cap convex or yellow colored, **3**
2 a. Cap white to pale cinnamon, *220. Conocybe lactea*
   b. Cap yellow-brown to reddish brown, *221. Conocybe tenera*
3 a. Cap yellow, bell-shaped (campanulate) in age, margin striate,
      222. Bolbitius vitellinus
   b. Cap convex, cream to tan or brown, not striate, **4**
4 a. Cap cream color; veil absent; in grass, *223. Agrocybe pediades*
   b. Cap tan to brown; veil present; on rotten logs, buried wood, in grass,
      224. Agrocybe praecox, acericola and allies

220 CONOCYBE LACTEA (Lange) Métrod   Edibility unknown/Numerous

Cap narrowly conical, wrinkled, whitish; stalk tall, very thin, white; in grass

Cap 12–30 mm broad, 15–35 mm tall, *narrowly conical,* bell-shaped in age, moist but soon dry, whitish to pale cinnamon-pink or very light tan, irregular folds or wrinkles often over center, striate from margin one-half distance to center. Flesh thin, water-soaked, dull white, stalk center hollow. Gills nearly free, narrow, uneven, close, reddish cinnamon. Stalk 5–12 cm long, 1.5–3 mm thick, enlarging toward base, dry, white with scattered minute, white hairs, hollow, small white or slightly yellowish basal bulb. Veil absent.

Spores 11–14 x 6.5–9 μ broadly elliptical, thickened wall, smooth, with pore at apex, red-brown spore print. Cystidia (sterile cells) on gill edge with a bulbous top. Cap cuticle of round to club-shaped cells.

Scattered in grass, often in great numbers, in Sp. in eastern North America but exact distribution unknown. Caps expand in early morning in dew-laden grass and in sunny areas, usually wilt and are gone by midday.

Known also as *Galera lateritia* Fries and *Conocybe lateritia* (Fr.) Kühn. The long "dunce cap," which is whitish and often wrinkled, along with the long delicate stalk distinguish it at once.

221 CONOCYBE TENERA (Shaeff. ex Fr.) Fayed   **Edibility unknown**/Common

Cap bell-shaped, striate, red-brown; stalk tall and thin, striate, red-brown; in grass

Cap 8–22 mm broad, 8–16 mm high, narrowly conic, bell-shaped in age, dry, striate almost to center, dark reddish to yellow-brown, becoming lighter in age. Flesh thin, watery, brown. Gills free or nearly so, subdistant, narrow, clay-brown to cinnamon-brown. Stalk 4–8.5 cm long, 1.5–2 mm thick, equal, rigid, straight, fine longitudinal striations, *color of cap,* base not appreciably enlarged. Veil absent.

Spores 7.5–10 x 5.5–7 μ elliptical, thickened wall, smooth, with a pore at apex, red-brown spore print. Cystidia (sterile cells) on gill edge with a bulbous top. Cap cuticle composed of pear-shaped cells.

Scattered to abundant on lawns, grassy areas in open, or under forest stands, in Sp. and S., occasionally in F.; widely distributed.

Also known as *Galera tenera* Fr. The same tall slender growth habit as *C. lactea,* but both cap and stalk are brown and the cap is not as narrowly conical. There are several which closely resemble this species and they are difficult to separate.

222 BOLBITIUS VITELLINUS Fr.   **Nonpoisonous**/Common

Cap bell-shaped, viscid, bright yellow, striate; in well-manured grass

Cap 2–5 cm broad, conic, bell-shaped in age, without hairs, viscid, bright yellow, margin striate, grooved in age. Flesh soft, very thin. Gills attached, free in age, close to fairly well separated in age, narrow, reddish gray with minute hairy edges. Stalk 6–12 cm long, 2–4 mm thick, equal or slightly enlarged toward base, minutely hairy at apex, shiny white often tinted yellow. Veil absent.

Spores 10–13 x 6–7.5 μ elliptical, smooth, with a pore at apex, bright rusty orange spore print. Cap cuticle composed of round or pear-shaped cells.

Several to abundant on dung, most often cow dung or manured grass, in Sp., early S., and F.; widely distributed.

This fungus resembles a small yellow *Coprinus* but, of course, there is *no deliquescence* of the gills and the spore print is *not black.*

223 AGROCYBE PEDIADES Fr.   **Edible**/Common

Cap convex, viscid, cream color; stalk tall and thin; in grass

Cap 1–6 cm broad, convex, bell-shaped, without hairs, viscid, cream color to dull yellowish, light reddish brown in age. Flesh firm, pallid. Gills adnate, fairly well separated, whitish, light yellowish brown, dark brown in age. Stalk 4–7 cm long, 2–4 mm thick, equal or slightly larger at base, dry, scattered minute hairs, whitish to light brown. Veil absent. Odor mealy. Taste mild or disagreeable.

Spores 11–13 x 7–8 μ elliptical, smooth, with a pore at apex, dark brown spore print. Cystidia (sterile cells) on gill edge club-shaped with bulbous apex. Cap cuticle of pear-shaped cells.

Several to numerous in lawns and grassy areas in Sp., S., and F.; widely distributed. I have seen large fruitings during cool moist Sp. weather in Maryland.

Also known as *Naucoria semiorbicularis* (Bull.) Fries. This species could be confused with the poisonous *Hebelomas* and is not a recommended edible for the novice collector. *Hebeloma* spores do not have a pore at apex and the cuticle is not composed of pear-shaped cells, but these features must be examined with the microscope.

Cap convex, dry, brown; stalk long and thick; in grass or mulch

Cap 1–7 cm broad, convex, nearly flat in age, without hairs, dry, light tan to light brown. Flesh soft, thick, white. Gills adnate (attached) to adnexed (notched), close, whitish at first to light brown in age. Stalk 4–10 cm long, 4–8 mm thick, equal, without hairs, dry, white above and below annulus, streaked brown from the spores. Veil membranous, white, leaving a superior, persistent, hanging ring.

Spores 8–13 x 5.5–7 μ elliptical, smooth, with a pore at apex, dark rich brown spore print. Cystidia (sterile cells) on face and edge of gills inflated. Cap cuticle of pear-shaped to round cells (Fig. 80).

Solitary to scattered in lawns and grassy areas in Sp. and early S.; widely distributed.

A very similar species, *A. acericola* Peck, is found on rotten logs and stumps or on buried wood. It is common in areas which have been logged in western North America. Usually numerous white strands (rhizomorphs) connect the fungus to its host.

CORTINARIACEAE

This is a very large family with many genera and species. The eight common genera treated here include mushrooms with and without stalks. They are found in a wide variety of habitats but are all related by spores which are some shade of brown and are entire. The gills are always attached to the stalk. Many of the species have the characteristic spiderweb-like (cortinous), fibrous veil which gives the family its name. None are found on dung but many are saprophytes on wood, humus, or soil. A large number of them form mycorrhiza with both hardwoods and conifers.

Edibility. There are several groups of poisonous species in this family. *Galerina* is particularly poisonous, containing Group I toxins which are often fatal and are found only in one other mushroom genus, *Amanita. Inocybe* also has many Group II toxins and should be avoided. *Hebeloma* has Group VI toxins which have not as yet been identified but have resulted in severe poisonings. Much work remains to be done in the study of North American species of *Cortinarius,* and so I would not recommend trying new species of this group. The species of *Gymnopilus* and *Crepidotus* are either bitter or have no flavor.

This leaves the large genus *Pholiota* and *Rozites caperata* as potential sources of edible species. (See the manual on North American species of *Pholiota* by Alexander H. Smith and L. R. Hesler.) Many of the *Pholiota* species are large, fleshy, and good to eat. They, along with *Rozites caperata,* can serve as a major source of food particularly in the late summer and fall.

Key to the Genera of the Cortinariaceae

    1 a. On wood, sometimes buried wood, **2**
       b. On the ground, not from buried wood, **4**
    2 a. Stalk absent, *Crepidotus*
       b. Stalk present, off center or central, **3**
    3 a. Spore print bright rusty orange to bright rusty brown, *Gymnopilus*
       b. Spore print dull brown, *Pholiota* (also see *Galerina autumnalis*)

4 a. Veil membranous; cap orange, wrinkled; never in cespitose clusters, 237. *Rozites caperata*
   b. Veil (if present) hairy and often like a spider web, **5**
5 a. Stalk thin, brittle and fragile; cap thin, fragile, often striate, usually glabrous, (hairless) conic to convex, *Galerina*
   b. Stalk fleshy to pliant, not brittle and fragile; cap not fragile, seldom striate, variously shaped, **6**
6 a. Cap dry, conic to bell-shaped (Figs. 8 and 13), often hairy; spores warted (Fig. 63) to smooth; gill edge covered with clavate (club-shaped) (Fig. 76), often thick-walled cystidia, *Inocybe*
   b. Cap viscid or dry, usually convex; spores never tuberculate (strongly warted); not with above combination of characteristics; stalk always dry, **7**
7 a. Cap viscid; veil present or absent; stipe fleshy, always dry; cylindric cystidia only on gill edge, *Hebeloma*
   b. Cap dry or viscid; veil like a spider web; if cap is viscid cystidia not on gill edge, *Cortinarius*

CREPIDOTUS

This is the only genus in the family without a stalk. The caps are small, convex to fan-shape, with thin flesh, and always on wood. The brown spores are smooth or minutely spiny and are entire. There are 125 species in North America (Hesler & Smith, 1965).

Edibility.   These are bitter or bland and are not edible.

Key to Crepidotus

1 a. Cap covered with hairy squamules (small scales), *225. C. mollis*
   b. Cap hairless to hairy but not covered with small scales, **2**
2 a. On hardwood limbs and logs in S. and F., *226. C. applanatus*
   b. On conifer logs near melting snow (see 226. *C. applanatus*), *C. lanuginosus*

225 CREPIDOTUS MOLLIS (Fr.) Staude   Edibility unknown/Common

Cap convex; stalk absent, dry, covered with small brown scales; usually on hardwoods

Cap 1.5–4 (up to 8 cm) broad, stalk absent, 2–2.5 cm wide, convex, dry, buff-ground color covered with dark brown, fibrous small scales. Flesh soft, thin, white. Gills close to subdistant, narrow, brown to red-brown. Stipe absent, merely a short basal plug covered with minute hairs. Veil absent.

Spores 7–10 x 5–7 μ short elliptical, smooth, yellow-brown spore print. Cystidia (sterile cells) on sides and edges of gills.

Grows in groups to nearly imbricate (overlapping) on the bark of dead hardwoods and sometimes conifers, in Sp., S., and F.; widely distributed.

The dense small scales covering the surface of the cap make this one of the easier species to identify. There is too little flesh to consider these edible.

226 **CREPIDOTUS APPLANATUS** (Pers.) Kummer **Edibility unknown**/Common

Cap fan-shaped, dry, white to cinnamon; stalk absent; gill edges fuzzy; on hardwoods

Cap 1–4 cm broad, petal-like to fan-shaped, moist, hairless to minutely downy, dull white to dull cinnamon or brownish, margin inturned, faintly striate. Flesh soft, thin, white. Gills close, narrow, white but soon brown, edges fuzzy (minutely hairy). Stalk absent, replaced by a short white plug. Veil absent.

Spores 4–5.5 μ round, minutely spiny, brown to cinnamon-brown. Cystidia (sterile cells) present on the gill edge, thin-walled, narrowly club-shaped.

Abundant, sometimes appearing overlapping, usually on hardwood limbs, logs, and stumps, in S. and F.; widely distributed.

There are many look-alike small species of *Crepidotus* which resemble *C. applanatus. C. lanuginosus* Hesler and Smith is a minute (3–15 mm broad) western species which fruits as the snow melts on conifer logs in the Sp.

GYMNOPILUS

This is a group of orange-brown to red-brown species with central stipes on wood or on the ground from buried wood. The spores are bright rusty-orange, smooth or roughened but entire. L. R. Hesler's North American manual (1969) includes 73 species.

Edibility.   These are bitter or bland and not edible.

Key to Gymopilus

1 a. Cap large (5–18 cm broad), yellow-orange; usually in cespitose clusters on ground from buried wood, *227. G. spectabilis*
   b. Cap smaller, golden yellow to orange; single to sometimes cespitose on logs, sticks, and stumps, **2**
2 a. Cap hairy; veil yellowish, *228. G. sapineus*
   b. Cap hairless; veil white (see 228. *G. sapineus*), *G. penetrans*

227 **GYMNOPILUS SPECTABILIS** (Fr.) A. H. Smith   **Nonpoisonous**/Common

Cap large, convex, yellow-orange, dry; stalk thick, orange; taste bitter; cespitose on wood or ground

Cap 5–18 cm broad, convex, nearly flat in age, dry, hairless to hairy in age, buff, yellow to yellow-orange. Flesh firm, pale yellow, red-brown in 3 percent KOH. Gills adnate (attached) to short decurrent (extending down stalk), crowded, mustard-yellow to orange buff, edges minutely hairy. Stalk 3–20 cm long, 8–10 (up to 30) mm thick, equal or club-shaped, dry, scattered hairs, same color as cap above and below ring, brownish near base. Veil membranous, persistent, superior, yellowish. Taste very bitter.

Spores 7–10 x 4.5–6 μ elliptical, roughened, thin-walled, entire, orange or rusty orange spore print. Cystidia (sterile cells) on gill edge, long, narrow, thin-walled.

Single but most often in cespitose clusters on ground from buried wood, stumps, and logs of hardwoods and conifers, in Sp., S., F., and early W.; widely distributed.

Cap convex, golden-yellow to orange; taste bitter; single or in cespitose clusters on wood

Cap 1–9 cm broad, convex, sometimes with a low knob, dry, golden-yellow to bright salmon color or orange with scattered patches of hairs. Flesh firm, yellow. Gills attached, close, yellow to rusty yellow in age, minutely hairy edges. Stalk 3–7 cm long, 4–12 mm thick, equal, dry, buff darkening to yellow-brown below the often missing ring. Veil hairy, fragile, superior, yellow. Taste often very bitter.

Spores 7–10 x 4–5.5 μ elliptical, minutely roughened, thin-walled, bright rusty brown spore print. Cystidia (sterile cells) on gill edge abundant, narrow with rounded heads.

Single to cespitose clusters on hardwoods and conifers in Sp., S., and F.; widely distributed.

G. penetrans (Fr. ex Fr.) Murr. is closely related but has a hairless cap and white veil.

PHOLIOTA

This is the third genus in the Cortinariaceae which is found almost exclusively on wood or from buried wood. *Crepidotus* has no stipe, *Gymnopilus* has bright rusty-orange spores, and *Pholiota* has dark brown to gray-brown spores. Some Pholiotas have caps covered with erect or flattened dense small scales with a dry or viscid layer beneath while others are smooth and viscid. The gills are attached to the stalk, which sometimes has a ring. The smooth spores are entire or sometimes have an obscure pore at the apex found nowhere else in the family. The species are typically found in large cespitose clusters as saprophytes on decaying matter or occasionally as parasites on wood. A. H. Smith and L. R. Hesler (1968) have published a comprehensive treatment of the North American species.

Edibility. I have no reports of specific toxins and most species are unqualified edibles. However, some people appear to be allergic to certain species such as *Pholiota aurea*. Since Pholiotas are large, fleshy species, one can obtain a good quantity for the table. The flavor is quite good and they can be simply sautéed in butter or used in most mushroom recipes.

Key to Pholiota

1 a. Cap granular or covered with scales, **2**
   b. Cap glabrous (hairless), **7**
2 a. Cap brilliant yellow to golden-yellow, **3**
   b. Cap some other color, **4**
3 a. Cap granular, dry, golden-yellow; on ground under alder and Douglas fir; annulus (ring) large and persistent, *229. P. aurea*
   b. Cap brilliant yellow, viscid, with yellow scales; on conifer logs, *230. P. flammans*
4 a. Cap creamy white, viscid, with whitish small scales, *231. P. destruens*
   b. Cap another color, **5**
5 a. Cap dry, yellow-brown; in cespitose clusters, *232. P. squarrosa* and related species
   b. Cap viscid, **6**

6 a. Cap with many erect, pointed scales; usually cespitose, *233. P. squarrosoides*
   b. Cap with flat or some slightly raised scales; single or in cespitose clusters,
      234. P. aurivella and related species
7 a. Cap streaked olive-brown with a cinnamon margin; stalk yellow-green,
      235. P. spumosa
   b. Cap orange-brown to red-brown; frequent on living hardwoods, especially elm,
      236. P. albocrenulata

229 PHOLIOTA AUREA (Fr.) Kummer   Edible (?)/Infrequent

Cap striking golden-yellow, granular, dry; ring large, buff

Cap 5–20 (up to 30) cm broad, convex, convex-knobbed, dry, granulose, golden-yellow to golden orange-brown, margin with ragged, hanging pieces of the veil. Flesh firm, thick, yellowish. Gills adnate (attached) to adnexed (notched), close, yellowish orange to cinnamon. Stalk 5–15 (up to 25) cm long, 1–3 (up to 6) cm thick, enlarged toward the base, hairless above ring, granular surface below, buff overall. Veil membranous, dark buff, leaving a large, superior, persistent ring.

Spores 9–13 x 4–6 μ elliptical, smooth to slightly roughened, pale orange-buff spore print.

Single to numerous, often under or near alder and vine maple or under old growth Douglas fir in PNW and A, in F.

Known also as *Phaeolepiota aurea* (Matt. ex Fr.) Maire ex Konr. and Maubl. and *Togaria aurea* (Fr.) W. G. Smith. The cap is reported to be edible but there are also reports of gastrointestinal disturbance from eating it, (Smith, 1963). One should proceed with caution. The large size combined with the granular, strikingly colored cap make this one of the most distinctive of the gilled fungi. The granular cap surface rubs off easily when dry and is a distinctive field characteristic. *Pholiota erinaceella* (Pk.) Peck also has a granular cap surface but is brown, smaller (1–4 cm broad), and has smooth spores (6–8 x 4–5 μ). It is found primarily in the NE and adjacent C.

230 PHOLIOTA FLAMMANS (Fr.) Kummer   Nonpoisonous/Numerous

Cap brilliant yellow, viscid, hairy scales; on conifer wood

Cap 3–8 cm broad, conic, convex in age with a low umbo, viscid and brilliant yellow beneath a dense covering of lemon-yellow hairy scales, margin fringed with yellow pieces of the veil. Flesh firm, yellow. Gills attached, close, yellow staining dingy brown in age. Stalk 5–10 cm long, 5–10 mm thick, equal or enlarged at base, silky and yellow above ring, covered with dense yellow scales below. Veil membranous, leaving a superior, hairy ring which may disappear in age.

Spores 4–5 x 2.5–3 μ elliptical, smooth, brown spore print. Cystidia present on sides and edges of gills.

Single or in small cespitose clusters on conifer logs (especially western hemlock) and stumps in late S. and F.; widely distributed. It makes a striking contrast to the dark forests of hemlock and Sitka spruce along the southeastern Alaskan coast and may be seen at a distance.

This brilliant yellow fungus is distinguished at once by its color.

231 PHOLIOTA DESTRUENS (Brond.) Gillet   Edible/Numerous

Cap large, viscid, whitish, minutely scaly; on cottonwood and aspen

Cap 8–20 cm broad, convex, broadly convex in age, viscid, whitish, creamy, light brown in age, scattered whitish to buff scales, margin with hanging pieces of veil. Flesh firm, white. Gills adnate to adnexed (notched), close, white to cinnamon in age. Stalk

138

5–18 cm long, 1–3 cm thick, equal or enlarged at base, silky white above ring, covered with scales below, white to brownish in age. Veil membranous, superior, leaving a loose ring which is very near the gills, sometimes absent in age.

Spores 7–9.5 x 4–5.5 μ elliptical, smooth with a pore at apex, brown spore print. Cystidia (sterile cells) present on gill edge only.

Single or more often in cespitose clusters on logs, limbs and dead wood of *Populus* sp., aspens, cottonwood, and balsam poplar, in late F.; widely distributed.

This is a large species with a robust stalk and generally light coloration. This, combined with its occurrence on poplars in late F., creates a distinctive *Pholiota.* It is reported to be edible, but I have not tried it.

232 PHOLIOTA SQUARROSA (Fr.) Kummer   **Edible**/Common

Cap large, dry, dense yellow-brown scales; often in cespitose clusters around aspens

Cap 3–12 cm broad, convex, broadly convex in age, *dry,* covered with dry scales, yellow-brown, margin incurved and covered, at first, with hanging pieces of veil. Flesh firm, pale yellowish. Gills adnate to adnexed (notched) with a fine line on stalk, close to crowded, yellowish sometimes with a greenish tinge to dingy brown in age. Stalk 4–12 cm long, 4–12 mm thick, equal, dry, with yellow-brown scales. Veil membranous, superior, sometimes absent in age. Odor mild or faintly onion-like.

Spores 5–8 x 3.5–4.5 μ elliptical, smooth with a pore at apex, spore print brown.

In cespitose clusters at the base or on the wood of hardwoods and conifers in F.; widely distributed in North America. I have frequently observed this fungus on or at the base of living aspens and birch in northern New York, Montana and the Yukon Territory of Canada.

P. squarrosoides (Pk.) Sacc. has a greenish tint to the gills and a viscid cap. *Pholiota terrestris* Overholts has a dry, brown to cinnamon-brown cap and is usually found on the ground in large, cespitose clusters, most commonly in the PNW. It resembles both of the above species. It is edible but not outstandingly so.

233 PHOLIOTA SQUARROSOIDES (Pk.) Sacc.   **Edible**/Common

Cap cinnamon-buff, viscid with pointed erect scales; in cespitose clusters on hardwoods

Cap 2.5–11 cm broad, conic, convex in age, whitish at first soon cinnamon-buff, viscid with dry, yellow-brown scales which are often pointed and erect. Flesh firm, white. Gills adnate to adnexed (notched), close, white to rusty brown. Stalk 5–10 cm long, 5–15 mm thick, equal, dry, white at apex with recurved buff to rusty-brown scales below. Veil hairy, ring superior, often absent.

Spores 4–6 x 2.5–3.5 μ elliptical, smooth, brown spore print. Cystidia (sterile cells) abundant on face and edge of gills.

Rarely single, usually in large cespitose clusters on hardwoods in late S. and F.; widely distributed. This fungus is particularly common on birch, beech, and maple.

P. squarrosa looks similar but has a dry cap, a green tint to the gills, and more dense cap scales that have no yellow coloration.

234 PHOLIOTA AURIVELLA (Fr.) Kummer   **Edible**/Numerous

Cap large, orange, viscid, flattened wine-red scales; single or several on living trees

Cap 4–16 cm broad, convex to broadly convex in age, sometimes broadly knobbed, viscid, orange with large appressed (flattened) triangular, wine-red scales. Flesh firm, yellow. Gills adnate to adnexed, close, light yellow to brown in age. Stalk 5–8 cm long,

5–15 mm thick, equal, dry, hairy above to scaly below, yellowish brown. Veil hairy, which is sometimes absent, leaving a thin, superior ring.

Spores 7–11 x 4.5–6 μ elliptical, smooth, with a pore at apex, brown spore print.

Solitary or several on living trunks and logs of hardwoods and conifers; widely distributed, in Sp. and F.

P. abietis Smith and Hesler is similar but has smaller spores (5.5–7.5 x 3.5–4.5 μ). It is also edible and is found on conifer wood in the PNW and elsewhere. *Pholiota squarrosa-adiposa* Lange is also a member of this complex. It is edible. According to Smith and Hesler (1968), it is found only in large cespitose clusters on alder and maple logs along the Pacific Coast.

235 PHOLIOTA SPUMOSA (Fr.) Singer   Edibility unknown/Numerous

Cap viscid, hairless, olive-brown; stalk yellow-green; on soil or wood

Cap 2–6 cm broad, conic, nearly flat often with a knob in age, viscid, hairless but appears streaked, olive-brown, margin cinnamon. Flesh soft, yellow to yellow-green. Gills adnate to adnexed, close, yellow, sulphur-yellow to mustard-yellow in age. Stalk 3–5 cm long, 4–8 mm thick, equal, dry, yellow to *yellow-green,* covered with a thin layer of yellow hairs from the veil, dingy brown over the base in age. Veil hairy, thin, leaving an obscure, superior, hairy ring zone.

Spores 7–9 x 4–4.5 μ elliptical or somewhat bean-shaped, smooth, with a pore at apex, brown spore print. Cystidia (sterile cells) on edge and face of gills, conspicuous, long, and thin-walled.

Numerous in cespitose clusters on soil from buried wood or on stumps and logs, usually on conifer wood but reported on hardwoods, in Sp., S., and F.; widely distributed.

This fungus was often also known as *Flammula spumosa* (Fr.) Kummer. It is one of the most common of a complex of small closely related Pholiotas that will be encountered by almost anyone who consistently collects fungi. The green coloration and yellow gills combined with the conspicuous cystidia are the distinguishing characteristics.

236 PHOLIOTA ALBOCRENULATA (Pk.) Sacc.   Nonpoisonous/Numerous

Cap glutinous, orange-brown; stalk dark brown; on live or dead hardwoods

Cap 2.5–8 cm broad, conic to convex, glutinous, orange-brown to red-brown, light brown hairy scalelike veil sometimes remains hanging from margin. Flesh firm, pallid. Gills adnate to adnexed (notched), close, white to gray in age, with white uneven edges, often covered with drops of liquid. Stalk 3–10 cm long, 5–15 mm thick, equal, pallid above, dark brown below ring, often curved from host. Veil fragile, forming a superior ring or absent.

Spores 10–18 x 6–8 μ elliptical, not equilateral, smooth with a pore at apex, thickened wall, brown spore print.

Single or several on stumps, logs, and living hardwoods, rarely on conifers in S. and F., in SE, SO, SW, NE, CS, and adjacent C.

The white gill edges and red-brown cap, combined with its preference for maple and elm. are distinctive characteristics.

237 ROZITES CAPERATA   Edible, choice/Common
(Fr.) Mich. ["Chicken of the Woods"]

Cap orange, frosted at first, wrinkled; veil white, persistent

Cap 5–10 cm broad, nearly oval to bell-shaped in age, dry, *covered at first with minute white hairs that give a frosted appearance,* straw-colored to orange-brown, *margin wrinkled* and turned in at first. Flesh thick, firm, white. Gills adnate, adnexed (notched) in age, close, dull whitish to rusty-colored in age. Stalk 6–12 cm long, 10–20 mm thick, enlarging toward base, without hairs, dingy white. Veil *membranous,* white, leaving a superior, persistent, membranous ring. Taste pleasant.

Spores 11–14 x 7–9 μ elliptical, *finely warted,* dingy-rusty to yellowish spore print.

Several to numerous in mixed conifer-hardwood or conifer stands in late S. and early F.; widely distributed. I have collected this in quantity in many places along the east coast.

This resembles a *Cortinarius* with membranous instead of fibrillose veil. The orange frosted caps which are characteristically egg-shaped at first, combined with the membranous veil and warted spores, are the distinguishing characters of this fungus. The flavor is excellent but not, as the common name suggests, exactly that of chicken.

GALERINA

These are small mushrooms with convex to conic, striate, glabrous (hairless) caps; attached gills; and long, thin, brittle stipes (stalks) not unlike the white-spored genus *Mycena.* The spores are brown with a roughened to warted surface which is not as distinct as the warts in Fig. 63. Near the apiculus a smooth, flat depression called a *plage* may be visible. Some species are on wood but the vast majority are found in deep moss. A book on the North American species by A. H. Smith and R. Singer (1964) includes 199 species.

Edibility. Some species of *Galerina* contain the *deadly poisonous toxins* described in Group I. They should be totally avoided. In fact, it is wise to avoid all small mushrooms that occur in grass and moss and have thin, brittle stalks.

Key to Galerina

1 a. Cap cuticle viscid, dark brown drying light tan, *238. G. autumnalis*
   b. Cap cuticle moist to dry, **2**
2 a. Cap conic; stalk very thin (1–3 mm thick); in moss, *239. G. cerina*
   b. Cap convex; stalk thicker (3–5 mm thick); in grassy areas, *240. G. venenata*

238 GALERINA AUTUMNALIS   Poisonous (deadly)/Common

(Pk.) Smith and Singer

Cap viscid, brown; hairy ring; abundant on wood

Cap 2.5–5 cm broad, convex to convex with a small discreet knob, viscid, hairless, dark brown when moist, light tan in dry weather, margin faintly striate. Flesh thick, light brown, watery. Gills attached, close, broad, rusty-brown. Stalk 1.5–6 cm long, 3–7 mm thick, equal, dry, brown streaked with white fibrils. Veil hairy, leaving a thin, superior, white, hairy ring. Odor and taste mealy.

Spores 8.5–10.5 x 5–6.5 μ elliptical, wrinkled, rust-brown spore print. Cystidia (sterile cells) on gill edge bottle-shaped to fusiform.

Scattered to abundant in dense groups on well-decayed hardwood and conifer logs in late F. and sometimes in early S.; widely distributed.

The deadly poisonous phallotoxins and amatoxins (see Group I toxins) have been isolated and identified in this and other Galerinas.

239 GALERINA CERINA Smith and Singer   Edibility unknown/Common

Cap conic, moist, striate; stalk thin; in deep, moist moss

Cap 5–15 mm broad, conic to convex, hairless, moist, rich dark yellow-brown fading somewhat in age, margin striate. Flesh very thin, watery brown. Gills attached, fairly separated, orange-buff to light brown in age. Stalk 20–40 mm long, 1–3 mm thick, equal, smooth but with a few veil hairs that soon disappear, moist, fragile, a few hairs at the base. Veil hairy, white, leaving a superior ring of hairs which soon disappear.

Spores 9–12 x 5.5–7 μ somewhat bean-shaped, with a loose sac, called *calyptrate* (Fig. 70), over one end, brown spore print. Cystidia (sterile cells) on gill edge fusiform, often with rounded ends.

Grows in groups on hair-cap moss *(Polytrichum),* sphagnum moss, or other mosses in Sp. and early S., rarely F.; widely distributed.

Singer and Smith (1964) have described a number of varieties of this common widespread *Galerina.* The veil is sometimes yellow, the gills sometimes extend down the stalk or one species differs in its microscopic characteristics from the others. I have found this fungus in May in Maryland. The unusual spores, along with the spring fruiting on mosses, are distinctive features. This species is too fragile to be considered edible, but *no Galerina* should be eaten under any condition.

240 GALERINA VENENATA A. H. Smith   Poisonous (deadly)/Rare

Cap dry, brown; hairy ring; in grass

Cap 10–35 mm broad, broadly convex, moist, glabrous, cinnamon-brown, margin waxy and split in age. Flesh thick. Gills attached, fairly well separated, broad, golden-brown to dull cinnamon. Stalk 30–40 mm long, 3–5 mm thick, equal, dry, without hair, brownish, white mycelium just at base. Veil hairy, leaving a superior, hairy ring. Odor and taste mealy.

Spores 8–11 x 6–6.5 μ elliptical, roughened, rusty-brown spore print. Cystidia (sterile cells) on gill edge fusiform, thin-walled.

Scattered in grass in late F. and W.; known only from the PNW.

G. autumnalis has a viscid cuticle but in other respects both species resemble each other closely. *Galerina venenata* has caused severe poisoning (see Group I toxins), and is included here even though it is infrequently encountered.

INOCYBE

These are small brown mushrooms with conic, pointed to convex dry caps which are typically radially fibrillose. The brown gills are adnexed (notched) to adnate. The stalk is fleshy to pliant, smooth to hairy, and sometimes a fibrillose ring is present. The brown, entire spores are smooth to angular (Fig. 69) or warted (Fig. 63) and usually elliptical (Fig. 64). Cystidia (sterile cells) line the gill edge and are usually rounded to pear-shaped, sometimes with thick walls. These are woodland mushrooms; they almost always occur on the ground and most likely form mycorrhiza with trees and shrubs. They are seldom in grassy areas but may be found in tundra where they are probably associated with dwarf willows and birches.

Edibility. Many species of *Inocybe* have been studied and almost without exception found to have toxins in Group II (Robbers, Brady and Tyler 1964). All species of *Inocybe* are *poisonous* and should be strictly avoided.

Key to Inocybe

1 a. Cap sharply conic, yellow-brown; veil absent; odor of green corn,
   241. I. fastigiata
   b. Cap convex, if conic then white; veil present; odor not of green corn, **2**
2 a. Cap white or tinted lilac with a sharply pointed umbo (knob), **3**
   b. Cap rich brown, covered with dense matted hairs, convex, *242. I. lanuginosa*
3 a. Cap white, *243. I. geophylla*
   b. Cap, gills, and stalk tinted lilac (see 243. *I. geophylla*), *I. geophylla* var. *lilacina*

241 INOCYBE FASTIGIATA (Schaeff. ex Fr.) Quel. Poisonous/Common

Cap conic, brown, hairy; veil absent; odor of green corn

Cap 2–6 cm broad, sharply conic, bell-shaped, dry, radially hairy, yellow-brown, margin splitting in age. Flesh thin, white but soon dingy yellowish. Gills notched, close together, white soon tinged olive. Stalk 4–10 cm long, 2–8 mm thick, equal or enlarged downward, dry, finely hairy, whitish to very light brownish, twisted striate. Veil absent. Odor of green corn or spermatic.

Spores 9–13 x 5–7 μ elliptical, smooth not ornamented, dull brown spore print. Cystidia (sterile cells) on gill edge, abundant, club-shaped, thin-walled.

Several to numerous under hardwoods and conifers in S. and F.; widely distributed.

Relatively large concentrations of muscarin (Group II toxins) could cause rather serious poisoning if any quantity of this fungus was eaten. There are many Inocybes that resemble this species and like all other members of the genus, they should be avoided.

242 INOCYBE LANUGINOSA (Bull. ex Fr.) Quel. Poisonous/Common

Cap dark-brown, raised dense hairs, dry; veil present; on wood

Cap 15–35 mm broad, conic to convex with a blunt knob, dry, densely tufted hairs that are matted, raised, or erect, rich dark brown, margin flat or slightly raised in age. Flesh watery, light brown. Gills adnate to notched, fairly well separated, white at first, soon dull brown. Stalk 2.5–8 cm long, 2–7 mm thick, equal, dry, pallid with a covering of dark brown hairs. Veil buff, hairy, soon gone leaving no ring.

Spores 6.5–10 x 4.5–7 μ elliptical with distinctive nodules (tuberculate), yellow-brown spore print. Cystidia (sterile cells) on gill edge broadly club-shaped, thickened walls.

Several together in small groups in wood debris, on the ground, or on well-decayed wood and stumps, in both conifer and hardwood forests, in S. and F.; widely distributed.

This species also has several similar looking, closely related Inocybes with different spore sizes and other microscopic differences. They should all be strictly avoided. *Inocybe casimirii* Velenowsky and *I. leptophylla* Atk. are either the same or closely related, and both resemble *I. lanuginosa*.

243 INOCYBE GEOPHYLLA (Sow. ex Fr.) Kummer   **Poisonous**/Common

Cap white to lilac, dry; veil white; odor disagreeable

Cap 15–30 mm broad, conic to bell-shaped with a pointed knob, dry, covered with flattened hairs, white, margin upturned in age and often split. Flesh thin except in center, white. Gills adnate to adnexed (notched), close, broad, white to grayish white to clay color in age. Stalk 2–6 cm long, 2–4 mm thick, equal, minutely hairy with hairs forming an annular zone, white to grayish white. Veil hairy, white, leaving a faint, hairy ring. Odor somewhat disagreeable.

Spores 7–10 x 4.5–6 μ elliptical, smooth, not ornamented, dull clay-brown spore print. Cystidia (sterile cells) common on sides and edge of gills, club-shaped, thick-walled.

Several to abundant under hardwoods and conifers, sometimes on lawns, in S. and F.; widely distributed.

I. geophylla var. *lilacina* Peck has a lilac coloration to the cap, gills, and stalk. I have seen it growing with the white variety. Toxins have been reported in a large number of Inocybes and they should all be avoided.

HEBELOMA

The caps are convex, cream color to red-brown and always viscid. The gills are attached and sometimes covered with minute water drops. The stipe is fleshy, dry, and ringed in some species. The spores are smooth to finely warted and more clay-brown in color than in *Cortinarius.* The gill edges always have long cylindric to club-shaped cystidia (sterile cells) similar to those in Fig. 75. Hebelomas are on the ground and probably form mycorrhiza with various shrubs and forest trees.

Edibility.   They are *very poisonous* with many records of poisonings, probably because some species are so fleshy and good looking. The toxins are not known and are therefore placed in Group VI.

Key to Hebeloma

1 a. Veil absent; cap brown to red-brown, **2**
   b. Veil present, consists of fine hairs (check young specimens),
      244. H. mesophaeum
2 a. Cap 4–6 cm broad; gills white to gray, covered at first with fine water drops,
      245. H. crustuliniforme
   b. Cap smaller or larger; gills lack water drops, **3**
3 a. Cap 2–4 cm broad (see 245. *H. crustuliniforme*), *H. hiemale*
   b. Cap 7–12 cm broad (see 245. *H. crustuliniforme*), *H. sinapizans*

244 HEBELOMA MESOPHAEUM (Pers. ex Fr.) Quel.   **Poisonous**/Common

Cap convex, viscid, brown; hairy veil; taste radish-like to bitter

Cap 1–4 cm broad, convex to broadly convex in age, with a broad knob, viscid, brownish over the center, gray-brown to cream color over the margin, which is incurved at first. Flesh firm, dull white, brown in the stalk. Gills adnate (attached) to adnexed (notched), close, broad, white to brownish or gray-brown in age, margins with fine hairs. Stalk 3.5–8.5 cm long, 3–5 mm thick, slender, equal, dry, hairy, white changing gradually to brown toward the base. Veil of fine hairs (like spider web) leaving a superior, hairy ring and sometimes hairs on cap margin. Its taste is radish-like but soon bitter; don't swallow it.

Spores 8–11 x 4–6 μ broadly elliptical, minutely wrinkled, surface entire, yellow to olive-brown spore print. Cystidia (sterile cells) only on gill edge, long and narrow.

Several to numerous, sometimes cespitose, two or three together, on hard bare ground or grassy areas, sometimes in sphagnum moss, in F., occasionally in Sp. and S.; widely distributed.

The poisons in *Hebeloma* are unknown and are included in Group VI. The species in this genus should be learned and totally avoided.

245 HEBELOMA CRUSTULINIFORME   Poisonous/Infrequent

(Bull. ex Saint-Amans) Quel.

Cap viscid, cream color; water drops on gills; radish odor; bitter taste

Cap 4–6 cm broad, convex, broadly convex with a low knob in age, hairless, viscid, cream color over the incurved margin to brown or red-brown at the center. Flesh firm, white, pale brown in age. Gills notched, crowded, narrow, white to gray, covered with fine water drops, edges minutely hairy. Stalk 3–8 cm long, 8–23 mm thick, robust, equal but often enlarging to form a bulbous base, covered with fine hairs, dry, nearly smooth toward base, dull white, sometimes light yellowish tan over the base. Veil absent. Odor of radishes. Taste very bitter; don't swallow it.

Spores 9.5–12.5 x 5.5–7.5 μ broadly elliptical to almond-shape, minutely wrinkled surface, entire, yellow-brown to brown spore print. Cystidia (sterile cells) only on gill edge, long and narrow.

Single or several in open areas under hardwoods or mixed woods, in F.; widely distributed but exact range unknown.

The poison is unknown and included in Group VI. *Hebeloma sinapizans* (Paul. ex Fr.) Gillet is closely related but is larger and has a cap up to 11 cm broad, spores 10–14 μ long, and cystildia (sterile cells) with broad bases. *Hebeloma hiemale* Bres. is a small species (cap 2–4 cm broad) without the radish-like odor.

CORTINARIUS

In numbers, this is the largest genus of gilled mushrooms in North America and is estimated to have from 300 to 600 species. They are fleshy, have convex caps and a rusty-brown to cinnamon-brown spore color. They possess a typical spider-web-like (cortinous) veil. If a hairy annulus (ring) colored rusty-brown from the spores is not present on the stalk, the long rusty-brown veil hairs can often be seen on the stipe. The spores have a slightly wrinkled surface and are entire. The viscid cap species of *Cortinarius* have no cystidia (sterile cells) on the gill edge, which distinguishes them from *Hebeloma*.

Five subgenera have been recognized. Subgenus *Myxacium* includes all the species with viscid stipes. Subgenus *Bulbopodium* has a viscid cap and a large basal bulb. A viscid cap is also present in subgenus *Phlegmacium* but the stipe is equal to club-shaped. Subgenus *Telamonia* has a dry to moist cap but the colors fade as it grows older or dries, while subgenus *Cortinarius* differs only in that the cap colors do not fade.

The species are generally found on the ground where at least some and probably all form mycorrhiza with trees, shrubs, and other higher plants.

Edibility. Some of the very distinctive identifiable species can be eaten safely. However, many of the species are not well known or even described in the literature. I have no records of toxins in this huge genus, but I would not recommend trying any but the well-known species until more studies of the genus are published.

Key to Cortinarius

1 a. Cap viscid, **2**
   b. Cap dry, **4**
2 a. Cap wrinkled to corrugated, *246. C. corrugatus*
   b. Cap smooth, **3**
3 a. Cap and stalk slimy, viscid, *247. C. collinitus*
   b. Cap viscid soon dry; stalk dry with a basal bulb, *248. C. glaucopus*
4 a. Cap lilac to dark violet, **5**
   b. Cap another color, **7**
5 a. Cap, gills, and stalk dark violet, *249. C. violaceus*
   b. Cap, gills, and stalk lilac to pale lilac, **6**
6 a. Cap silvery white with silky hairs, tinted lilac, *250. C. alboviolaceous*
   b. Cap lilac with dull, thick hairs, *251. C. traganus*
7 a. Stalk circled by several red zones of veil tissue, *252. C. armillatus*
   b. Stalk not as described above, **8**
8 a. Gills reddish cinnamon to blood-red, **9**
   b. Gills yellowish cinnamon to brown, **10**
9 a. Gills striking blood-red; cap yellowish cinnamon, *253. C. semisanguineus*
   b. Gills reddish cinnamon; cap red-brown, *254. C. cinnabarinus*
10 a. Gills yellowish to yellowish cinnamon; cap yellow-brown, *255. C. cinnamomeus*
   b. Gills dark cinnamon-brown, very broad; cap purplish brown, *256. C. torvus*

246 CORTINARIUS CORRUGATUS Pk.   Edibility unknown/Numerous

Cap viscid, yellow-brown, corrugated; gills purplish; under hardwoods

Cap 5–10 cm broad, broadly convex to bell-shaped, viscid, hairless, roughly corrugated, yellow-brown to reddish brown. Flesh thin, firm, white. Gills attached, close, broad, purplish to rusty cinnamon. Stalk 6–10 cm long, 5–15 mm thick, enlarging somewhat toward base and terminating in a basal bulb, moist with a viscid annular zone, scattered hairs; yellowish with a tint of brown. Veil hairy with a thin glutinous layer leaving a viscid ring zone.
  Spores 10–13 x 7–9 μ elliptical, roughened, rusty brown spore print.
  Single to several, often in clumps under moist hardwoods, in S. and F.; in eastern North America.
  The distinctive corrugated cap sets this apart from other species of *Cortinarius*.

247 CORTINARIUS COLLINITUS Fr.   Edible/Common

Cap rich brown, glutinous; gills pinkish at first; stalk with glutinous pallid or bluish bands

Cap 3–9 cm broad, convex to nearly flat in age, glutinous, rich brown to orange-brown, margin faintly striate, whitish or light brown at first. Flesh firm, white. Gills attached, nearly distant, pinkish buff to rusty-brown in age. Stalk 6–12 cm long, 6–15 mm thick, equal or narrowing toward the base, plates or bands of thick glutin which are pale bluish or violet to light yellowish brown in age, dull white flattened hairs beneath. Veil pure white hairs with a thin glutinous layer over them.

Spores 10–15 x 6–8 μ almond-shaped, wrinkled wall, rusty brown spore print.

In troops under conifers and hardwoods, in F.; widely distributed. I have often found it in the southeast in F. in great quantity.

The plates or bands of nearly colorless glutin tinted blue to violet make this a distinctive species. Cap colors are quite variable, and so are the stalk colors in age. *C. muscigenus* Peck is similar but lacks the glutinous bands. *C. trivialis* Lange is also closely related but the bands on the stalk are composed of yellow-brown hairs, the lower stalk is also yellow-brown, and the spores are also smaller, 10–13 x 6–7 μ.

248 CORTINARIUS GLAUCOPUS Schaeff. ex Fr.   Edibility unknown/Common

Cap viscid, cinnamon; gills violet-gray; stalk dry, lilac

Cap 3.5–8 cm broad, convex, broadly convex in age, viscid with flattened hairs, cinnamon to red-brown tinged sometimes with lilac-gray. Flesh thick, white to buff. Gills attached, close together, lavender to violet-gray, rusty brown in age. Stalk 3–8 cm long, 15–25 mm thick, enlarging downward to a basal bulb, dry, streaked with flattened hairs, pale lilac to pale lavender, obscure annular zone turns brown from the spores. Veil hairy, leaving an obscure, superior ring of scattered hairs.

Spores 7–9 x 4–5 μ elliptical, roughened; rusty brown spore print.

Several to numerous under hardwoods and conifers in Sp., S., and F.; widely distributed. Sp. fruitings have only been observed near snowbanks in the PNW. This common species has violet gills, a viscid cap, and a bulbous, lavender stalk.

249 CORTINARIUS VIOLACEUS (Fr.) S. F. Gray   Edible/Numerous

Cap dry, dark violet; gills and stalk dark violet; under conifers

Cap 5–12 cm broad, convex, often somewhat knobbed, dry, *dark shining violet* covered with hairs in erect tufts over the center. Flesh thick, firm, grayish violet, does not bruise. Gills adnate, fairly well separated, broad, *dark violet* color of the cap. Stalk 7–16 cm long, 10–25 mm thick, enlarged downward, dry, flattened hairs, also *dark violet*. Veil hairy, violet, leaving a superior, fibrillose ring.

Spores 12–17 x 8–10 μ elliptical, minutely roughened to warted, rusty brown spore print. Cystidia (sterile cells) present on face of gills.

Single or scattered on ground in northern evergreen (conifer) forests in F., in the northern United States and adjacent Canada. It is not uncommon in the PNW where I have seen it in late September.

250 CORTINARIUS ALBOVIOLACEUS (Fr.) Kummer   Edible/Common

Cap dry, lilac tinted white; gills and stalk color of cap; flesh tinted violet

Cap 3–6 cm broad, bell-shaped, convex to nearly flat with a knob, dry, flattened silky hairs, silvery white, pale lilac tinted white, margin incurved until mature. Flesh thin, light violaceous. Gill adnate, close, broad, pale violet to grayish violet, cinnamon-brown in age. Stalk 4–8 cm long, 5–20 mm thick, enlarging to a club-shaped basal bulb, dry, silky fibrillose, white tinted color of cap. Veil fibrillose white, leaving a superior, fibrillose ring.

Spores 7–10 x 4–6 μ elliptical, roughened, rusty-brown spore print.

Single to several under mixed conifers and hardwoods in late S. and F., in eastern North America.

251 CORTINARIUS TRAGANUS Weinm. ex Fr.   **Edibility unknown**/Numerous

Cap dry, lilac with thick flattened hairs; flesh tinted yellowish

Cap 4–10 cm broad, convex, nearly flat sometimes with a low broad knob in age, dry, flattened hairs, pale lilac overall, sometimes cracked in age. Flesh thick, firm, tinted yellowish. Gills attached, fairly well separated, broad, pale lilac but soon yellow-brown. Stalk 4–6 cm long, 1–3 cm thick, enlarging to a club-shaped bulb, dry, fibrillose, pale lilac-like cap above and below ring. Veil hairy, white tinted lilac soon streaked brown from the spores, leaving a superior, hairy ring which may disappear in time. Odor pungent and disagreeable.

Spores 7.5–10 x 5–6 μ elliptical to almond-shaped, coarsely wrinkled, rusty-brown spore prints.

Several to numerous in old-growth conifer woods, often in deep moss, in F.; widely distributed but most common in PNW.

252 CORTINARIUS ARMILLATUS (Fr.) Kummer   **Edibility unknown**/Common

Cap moist, red-brown; gills cinnamon; stalk with rings of red veil tissue

Cap 5–12 cm broad, convex, bell-shaped in age, moist fresh, minutely hairy, reddish brown, veil pieces adhering to the margin. Flesh thick, soft, pallid. Gills attached, well separated, broad, cinnamon to rusty brown in age. Stalk 7–15 cm long, 10–25 mm thick, enlarging toward base, dry, dull brownish, circled by several red zones of veil tissue. Veil hairy to nearly membranous, collapsing on expanding to form the concentric rings. Odor, if present, of radishes.

Spores 10–12 x 5–6.5 μ elliptical, nearly tuberculate (strongly warted), rusty-brown spore print.

Single to several under mixed conifer hardwoods in S. and F., eastern North America. It is uncommon in western North America, but I have seen it. This stately fungus is easy to spot through its concentric red rings.

253 CORTINARIUS SEMISANGUINEUS Fr.   **Edibility unknown**/Common

Cap yellow-brown, dry; gills blood-red; yellowish stalk

Cap 2–5 cm broad, convex, nearly flat with a low knob in age, dry, minutely hairy, yellowish cinnamon. Flesh firm, whitish buff. Gills adnate, crowded, *striking blood-red.* Stalk 3–6 cm long, 3–6 mm thick, equal, hairy, yellow. Veil hairy, leaving an indistinct superior, hairy ring.

Spores 5–8 x 3–5 μ elliptical, coarsely wrinkled, rusty brown spore print.

Several to scattered in low moist woods, often in deep moss under hardwoods and conifers in S. and F.; widely distributed. It is common in southeastern Alaska in late July. Look for the very striking blood-red gills and yellowish stalk.

254 CORTINARIUS CINNABARINUS Fr.   **Edibility unknown**/Common

Cap bright red-brown, dry; gills cinnamon; stalk red-brown

Cap 25–40 mm broad, convex, convex with a low knob, dry, felty to touch, bright red-brown. Flesh firm, water-soaked brown. Gills attached, fairly well separated, cinnamon to reddish cinnamon. Stalk 2–6 cm long, 3–8 mm thick, equal or enlarged toward base, often curved, streaked with hairs, color of cap or slightly lighter, shining, hollow. Veil fibrillose (cortinous) also reddish, sometimes leaving an obscure, superior, ring zone. Odor of radish or mild.

Spores 7–9 x 4.5–5.5 μ elliptical, roughened, rusty brown spore print.

Several to numerous at times under hardwoods and conifers, in S. and F.; widely distributed. Most often observed in oak or beech woods in eastern North America.

255 CORTINARIUS CINNAMOMEUS Fr.   Edibility unknown/Numerous

Cap yellow-brown, dry; gills yellowish cinnamon; stalk cinnamon-buff

Cap 2–5 cm broad, convex to nearly flat in age, dry, yellow-brown, olive-brown to cinnamon-buff, flattened shiny hairs. Flesh thin, straw-yellow to olive-yellow. Gills adnate, close, broad, olive-yellow to yellowish cinnamon. Stalk 3–6 cm long, 3–7 mm thick, equal, dry, often curved, finely hairy, yellow to cinnamon-buff. Veil fibrillose (cortinous), leaving an inconspicuous, superior, hairy ring.

Spores 6–8 x 4–5 μ elliptical, coarsely wrinkled, rusty brown spore print.

Single or several in needle duff under conifers, in Sp., S., and F.; widely distributed. I have found it under lodgepole pines in Idaho in the early Sp. and under Virginia pine in Maryland in F., to cite two extremes of the fruiting range.

Some mycologists call this *Dermocybe cinnamomea* (L. ex Fr.) Ricken.

256 CORTINARIUS TORVUS Fr.   Edibility unknown/Infrequent

Cap convex-knobbed, moist, purple-brown; gills very broad, brown

Cap 4–8 cm broad, convex, often convex-knobbed in age, moist but not viscid, frosted with very fine hairs, purplish brown to copper-brown, margin sometimes wrinkled in age. Flesh firm, grayish purple to brownish in age. Gills attached, fairly well separated, very broad, dark cinnamon-brown. Stalk 4–7 cm long, 6–8 mm thick, enlarging toward base, pallid, whitish violet, silky above ring. Veil hairy, leaving a superior, persistent ring.

Spores 8–12 x 5–6 μ elliptical, roughened, rusty brown spore print.

Scattered in deep debris under hardwoods or mixed woods in late S. or F.; widely distributed.

The very broad, deep brown gills and the convex-knobbed, purplish brown cap are the chief field characteristics of this fungus.

CANTHARELLACEAE
The Chanterelles

These highly prized edibles range in shape from vaselike and deeply depressed in the center (Fig. 86) to that of a regular mushroom. A pileus (cap) is present, even though it may be deeply depressed, whereas the coral fungi (Clavariaceae) have a complete hymenium over the ends of the upraised branches (Figs. 100–101), and form no pileus. The spore-bearing surface (hymenium) is used to differentiate between genera. *Craterellus* has a smooth to slightly wrinkled hymenium and white spores. *Cantharellus* is nearest to a gilled mushroom with gill-like blunt ridges and white to buff spores (Fig. 86). *Gomphus* differs from both by its wrinkled and irregularly ridged hymenium (Fig. 87) which is almost porous at times. It has minutely ornamented light orange-red spores. *Polyozellus* has only one species, which is deep violet-black with rough tuberculate spores (Fig. 63). Chanterelles tend to reappear in the same locality, if not the same area, each year.

Edibility. This is one of the truly delicious groups of edible mushrooms. Most species have an excellent flavor and only a few in the genus *Cantharellus* do not taste good. *Cantharellus cibarius,* the Golden Chanterelle of the French, *pfifferling* in Germany, or *kantarellas* in Scandinavia, has long been a highly regarded edible in Europe. It is widespread in North America from such diverse areas as Rock Creek Park in Washington, D.C., to the coastal forests of Washington State. A large western species, *C. subalbidus,* the White Chanterelle, has the same flavor but is larger and has thicker flesh than *C. cibarius.* Chanterelles are prepared and served as a side dish, in cream, braised with meat, or in biscuits. Regardless of the recipe chosen, a general rule is to cook them slowly over a low fire.

Key to Chanterelles

1 a. Fruiting body with yellow, orange to red coloration, at least over the gills, **2**
    b. Fruiting body purple, lilac, gray to blackish brown but not as above, **12**
2 a. Fruiting body colored the same overall; cap glabrous (hairless) or nearly so, **3**
    b. Fruiting body not colored the same overall; cap with squamules (small scales) or recurved scales, **7**
3 a. Fruiting body red to reddish orange, *257. Cantharellus cinnabarinus*
    b. Fruiting body orange, yellowish to nearly white, **4**
4 a. Fruiting body very small (8–25 mm) (see 261. *C. tubaeformis*), *C. minor*
    b. Fruiting body medium to large, **5**
5 a. Fruiting body thin with short discontinuous gills or without gills and smooth over large areas, yellow to orange, *258. Craterellus cantharellus*
    b. Fruiting body fleshy with well-formed gills, **6**
6 a. Fruiting body nearly white with light orange tints (see 259. *C. cibarius*), *Cantharellus subalbidus*
    b. Fruiting body orange, *259. Cantharellus cibarius*
7 a. Cap funnel to trumpet-shaped with large recurved scales, **8**
    b. Cap convex to somewhat depressed in center with squamules, **10**
8 a. Fruiting body small (3–7 cm broad); fruiting in spring (see *G. floccosus*), *Gomphus bonarii*
    b. Fruiting body large (5–15 cm broad); fruiting in fall, **9**
9 a. Gills low blunt ridges, buff to yellow, *260. Gomphus floccosus*
    b. Gills nearly having pores, deep yellow (see 260. *G. floccosus*), *G. kauffmanii*
10 a. Gills contorted and wrinkled, nearly absent, orange (see 261. *C. tubaeformis*), *C. lutescens*
    b. Gills blunt but well formed, **11**
11 a. Cap dark brown; stalk yellow (see 261. *C. tubaeformis*), *C. infundibuliformis*
    b. Cap yellowish brown, stipe grayish orange, *261. C. tubaeformis*
12 a. Never cespitose; in deep moss; small cap (12–45 mm broad), *262. C. umbonatus*
    b. In cespitose clusters; on ground; large caps, **13**
13 a. Gills absent; hymenium smooth; very thin flesh; ashy gray to blackish, *263. Craterellus cornucopioides*
    b. Gills present; thick fleshy, purple, fruiting body, **14**
14 a. Deep purple; tuberculate (strongly warted), nearly round spores, *264. Polyozellus multiplex*
    b. Light purplish to purplish brown; wrinkled, elliptical spores, *265. Gomphus clavatus*

257 CANTHARELLUS CINNABARINUS   Edible/Numerous

Schw. ("Red Chanterelle")

Cap funnel-shaped, red; under hardwoods

Cap 12–40 mm broad, convex-depressed, lobed and irregular margin, dry, lacking hairs, *cinnabar-red* to *reddish orange*. Flesh very thin, white. Gills extend down stalk, well separated, forked, veined, narrow, blunt, pinkish to nearly color of cap. Stalk 20–40 mm long, 4–9 mm thick, equal or slightly smaller toward base, color of cap.

Spores 7–11 x 4–5.5 μ elliptical, thin-walled, pink spore print.

Solitary to scattered under hardwoods in S. and early F. in eastern North America.

258 CRATERELLUS CANTHARELLUS (Schw.) Fries   Edible, choice/Infrequent

Cap funnel-shaped, yellowish; flesh thin; in clusters under hardwoods

Cap 3–14 cm broad, deeply depressed to funnel-shaped, hairless to minutely tomentose (woolly), yellow to orange, margin thin, inrolled at first. Flesh thin (5–9 mm), white to color of cap. Gills nearly absent or as short, low, incomplete veins, light yellowish, tinted pink. Stalk 2.5–10 cm long, 5–25 mm thick, tapering toward base, dry, hairless, cream color or color of gills, often hollow.

Spores 5.5–10.5 x 3.5–6.5 μ elliptical, smooth, pale salmon to pinkish yellow spore print.

In small to large clusters under hardwoods in eastern North America but most frequently found in SE and So. in S.

In Europe it is often called *Cantharellus odoratus* (Schw.) Fr. The reduced veins, fragrant odor, thin consistency and pinkish coloration separates *Craterellus cantharellus* from the fleshy, orange-colored, thick-veined *Cantharellus cibarius* which it otherwise resembles. They are both rated as choice edibles.

259 CANTHARELLUS CIBARIUS   Edible, choice/Common

Fr. ("Golden Chanterelle")

Cap funnel-shaped, orange; flesh thick; gills blunt orange

Cap 3–10 cm broad, depressed with recurved margin at first to deeply depressed in center in age, dry, hairy to hairless in age, whitish yellow, yellow to orange-yellow. Flesh firm, thick, light yellowish. Gills extend down stalk, well separated, narrow and blunt, forked, color of cap or paler. Stalk 2–8 cm long, 5–25 mm thick, somewhat smaller at base, dry, solid, hairless, color of cap or near it.

Spores 8–11 x 4–5.5 μ elliptical, thin-walled, pale yellow spore print.

Scattered, numerous to sometimes in cespitose clusters, under hardwoods and mixed woods in S. and F.; widely distributed.

The yellowish orange fruiting bodies with narrow blunt gills are very distinctive. There is an often large, white species, *C. subalbidus* Smith and Morse which is found in the PNW and that I find flavorful. Care must be taken not to mistake these for the "Jack-O-Lantern," 119. *Omphalotus olearius* which is poisonous.

260 GOMPHUS FLOCCOSUS (Schw.) Singer   Edibility unknown/Numerous

Cap trumpet-shaped with raised yellow to orange scales

Cap 5–15 cm broad, depressed to funnel-like or trumpet-shaped when mature, surface dry, clothed with flattened, then erect and even recurved scales (Fig. 87), buff, yellowish to dull orange, scales usually darkest. Flesh fibrous, white. Gills extend down stalk, low, blunt ridges which form or give rise to irregular veins, buff to yellow. Stalk not distinct from cap (whole fruiting body stands 8–20 cm high), buff color of gills, hollow from *deep* trumpet-like cap.

Spores 12–17 x 6–7.5 μ elliptical, wrinkled, ochre spore print.

Solitary but more often several to cespitose clusters under conifers at all elevations in late S. and F.; widely distributed.

G. kauffmanii (A. H. Smith) Corner is a closely related large western (rare in SE) species clay-color to tawny-olive scales on the cap and with a very poroid, yellow hymenium. *G. bonarii* (Morse) A. H. Smith is a similar small, western spring fruiting species. Some people may be sensitive to the 3 species covered here (Group VI toxins).

261 CANTHARELLUS TUBAEFORMIS Fr. Edibility unknown/Numerous

Cap convex, yellow-brown with raised small brown scales; flesh thin; gills blunt, yellow

Cap 10–40 mm broad, convex, soon convex-depressed with an incurved margin, hairless, roughened or with upturned small dark brown scales, yellowish brown. Flesh very thin, dull yellowish. Gills extend down stalk, well separated, narrow, blunt with numerous small veins in between, gray tinted yellowish or orange. Stalk 15–50 mm long, 3–6 mm thick, equal, smooth or furrowed, oval or flattened, moist, hairless, grayish orange, hollow in age.

Spores 8–13 x 5–8 μ broadly elliptical, thin-walled, white spore print.

Scattered to numerous or even cespitose clusters on wet soil, in moss, occasionally on well-decayed wood in S. and F.; widely distributed.

This is not a fleshy type of edible. The closely related species are not well known; some have caused some mild poisonings but most are commonly collected and eaten. *C. infundibuliformis* Fries has a dark brown cap and yellow stalk. *C. lutescens* Fries has a yellow-brown cap and obscurely wrinkled orange hymenium (not gill-like) and is often found in sand on old log roads. *C. minor* Peck is often very small, yellow overall with a salmon tint to the gills. The ranges of these are not truly known, but I have collected them in eastern North America.

262 CANTHARELLUS UMBONATUS Fr. Edible/Common

Cap small, gray; gills gray, forked; in moss

Cap 12–45 mm broad, convex-knobbed, inrolled at first, dry to moist, hairless, *gray* to gray with an olive tinge. Flesh thin, white. Gills extend down stalk, *forked,* white tinted with *gray.* Stalk 6–9 cm long, 4–8 mm thick, moist, equal, thickening just at base, pinkish buff staining reddish brown over the base.

Spores 8–14 x 3–5 μ narrowly elliptical, thin-walled, white spore print.

Single to scattered in moss (*Polytrichium* ssp. most commonly noted) in S. and F.; in eastern North America.

This rather distinctive fungus is always associated with *Polytrichium* or other mosses in Maryland and Virginia where I have found it in quantity every F. It is also called *Geopetalum carbonarium* (Fr.) Pat. and *Cantharellula umbonata* (Gmelin ex Fr.) Singer.

263 CRATERELLUS CORNUCOPIOIDES Edible, choice/Numerous

Fr. ["Horn of Plenty"]

Cap trumpet-shaped, dark gray-brown; flesh thin; gills absent, gray-black

Cap 2–6 cm broad, deeply depressed (trumpet-shaped), dry, stiff hairs and small scales, dark gray-brown. Flesh very thin, brittle, dingy brown. Gills absent, only slightly wrinkled to smooth, ash-gray to blackish with a sheen. Stalk only a dingy brown, base without a sheen.

Spores 11–15 x 7–9 μ elliptical, smooth, white spore print.

In cespitose clusters under hardwoods in eastern North America and under oak along the Pacific coast, in S. and F.

The nearly smooth, gray hymenium (without gills), the overall drab appearance, combined with its growth in numerous cespitose clusters when it fruits, sets this good edible apart from the other chanterelles.

264 POLYOZELLUS MULTIPLEX Murr.   Edible, good/Rare

Cap and gills violet to deep purple; in cespitose clusters under boreal conifers

Cap 1–5 cm broad, flat, smooth, incurved margins, densely grouped together in large cespitose clusters, *deep purple, violet-black or blue-black.* Flesh soft, thick, blue-black. Gills low ridges with many veins, almost like pores at times, pale violet. Stalk 10–40 mm long, 10–25 mm thick, mostly grown together and fused, violet-black.

Spores 5.5–6.5 x 4.5–5.5 μ almost round, *tuberculate* (strongly warted), thin-walled, white spore print.

Cespitose clusters under conifers, often in blueberry patches, northern U.S. and adjacent C. I have found it in the Payette National Forest in Central Idaho under pure Engelmann spruce and as far south as Colorado at high elevations.

265 GOMPHUS CLAVATUS (Fr.) S. F. Gray   Edible, good/Common

Cap club-shaped, fine scales, purplish brown; gills shallow veins, light purplish

Cap 3–10 cm broad, flat to depressed, dry, hairless to minutely scaly, usually cespitose, margin lobed to deeply lobed, light purplish to faded purplish brown. Flesh firm, white or tinged cinnamon. Gills extend down stalk to base, blunt, thick, interconnected by veins, frequently forked or almost poroid, light purple to light purplish brown. Stalk 1–8 cm long, 1–2 cm thick, usually expanding from a compound base, blending at once into the funnel-shaped cap, color of gills to purple drab just at the narrow smooth base which is covered with white mycelium.

Spores 10–13 x 4–6 μ elliptical, wrinkled, yellowish orange spore print.

Cespitose, often many in one cluster, under conifer forests in late S. and F.; widely distributed.

G. pseudoclavatus (A. H. Smith) Corner is a smooth-spored, similar species known from the CS and westward.

CLAVARIACEAE
The Coral Fungi

The coral fungi appear as erect finger-like or club-shaped stalks (Figs. 100–101) which may be intricately branched (Fig. 103) or branched from a large fleshy base (Fig. 102). They do not have a pileus (cap) or depressed top like the chanterelles (Fig. 86). The upper part of the "club" or the branches serve as the spore-bearing surface. If a coral fungus is laid on its side on white paper and covered to preserve moisture, the usual spore print can be obtained. The spores range from elliptical to subglobose (nearly round) and are white to cream or even brown. They have either one or more oil drops in each spore or they have none. The spore surface is smooth to ornamented and a few species have amyloid (blue) reactions in Melzer's solution. The flesh ranges from soft and brittle to leathery and tough. Ferric sulphate ($FeSO_4$) mixed to 10 percent in water will turn the flesh of the brown-spored genus *Ramaria* green.

There are many coral fungi in North America, and we have included only some of the most common edibles. Regional studies have been published such as those by Coker (1923) and Doty (1944), but there is no comprehensive treatment of the North American group.

Edibility. Only one, *Ramaria formosa,* has been reported to cause severe poisoning. The toxins are unknown and they are placed in Group VI. Other species can cause gastrointestinal disturbance to individuals, the effect being that of a strong purgative.

The Ramarias also contain some of the best edible species. *Ramaria botrytis* and the yellow *Ramaria* in the western mountains in the spring is a favorite food (especially in a spaghetti sauce) for most of my family. They are difficult to clean but worth the effort; a soft brush helps in cleaning.

Key to Coral Fungi

1 a. Single, cylindrical to clavate (club-shaped), **2**
   b. Multiple-branched fruiting bodies, **5**
2 a. Cylindric, deep violet, brittle clubs; taste disagreeable, *266. Clavaria zollingeri*
   b. Not violet, shaped differently, **3**
3 a. Expanded at the top to a flattened club; orange-yellow; taste sweet,
      267. Clavariadelphus truncatus
   b. Club-shaped but not flattened; color different; taste bitter, **4**
4 a. Wrinkled over the top; orange-red, *268. Clavariadelphus pistillaris*
   b. Narrowly club-shaped; salmon colored, *269. Clavariadelphus ligula*
5 a. Delicately branched, without a fleshy base; spores white, **7**
   b. Coarse branches, blunt or flattened, **6**
6 a. Coarse branches with flattened broad ends, *275. Sparassis radicata*
   b. Coarse, blunt branches from fleshy base; spores orange-brown to red-brown, **9**
7 a. Branches irregular; smoky gray, *270. Clavulina cinerea*
   b. Branches irregular to equal branching; not colored as above, **8**
8 a. Branches irregular, with fine tips; white tinted yellow, *271. Clavulina cristata*
   b. Branches equal with crownlike tips; pale pink to tan, *272. Clavicorona pyxidata*
9 a. Fleshy base with orange-red coloration; fleshy bruises red-brown to black,
      273. Ramaria formosa
   b. Robust, fleshy base; flesh does not bruise as above, **10**
10 a. Yellow branches with white over the base (see 274. *R. botrytis*), *Ramaria* (yellow
       species)
   b. Pink to red branches, **10**
11 a. Branches a delicate pink, *274. Ramaria botrytis*
   b. Branches bright pink to reddish overall (see 274. *R. botrytis*), *Ramaria subbotrytis*

266 CLAVARIA ZOLLINGERI Lev.   Edibility unknown/Common

Single, cylindric, deep violet, brittle; taste disagreeable

Fruiting body 2–8 cm high, 2–5 mm thick, round to somewhat flattened, rarely branched, cylindric, clustered or cespitose, smooth, nearly pointed at top and abruptly narrowed at base, *deep violet to amethyst.* Flesh *brittle,* pale violet. Taste somewhat disagreeable.

Spores 4–7 x 3–5 μ nearly round, thin-walled, without oil drop, white spore print. Basidia mostly 4-spored.

Sometimes under the name *C. amethystina* Fr., but this lilac species is actually a *Clavulina* with two-spored basidia; it is also branched. *Clavaria vermicularis* Fries is a closely related, pure white, brittle species which is usually 6–12 cm high with the same growth habit and white, elliptical spores 5–7 x 3–4 μ.

267 CLAVARIADELPHUS TRUNCATUS (Quel.) Donk   Edible/Numerous

Club-shaped, often flattened at the top, orange-yellow; taste sweet

Fruiting body 6–15 cm high, 2–9 cm thick, club-shaped, often with a broad or even flattened apex which may rupture, revealing the hollow interior, smooth at base to wrinkled or veined near the top, bright yellow to golden yellow to orange-yellow to pinkish brown below with a white, hairy covering over the base. Flesh white, thin, hollow at top, turns green in ferric sulphate solution. Taste sweet.

Spores 9–13 x 5–7 μ elliptical, smooth, thin-walled, pale orange-red spore print.

Scattered to several under conifers in needle litter, in S. and F.; widely distributed. It is closely related to *Clavariadelphus pistillaris* (Fr.) Donk but is more yellow and has a broader apex.

268 CLAVARIADELPHUS PISTILLARIS   Edible/Common
(Fr.) Donk

Club-shaped, wrinkled, orange-red; taste bitter

Fruiting body 7–30 cm high, 2–6 cm thick, club-shaped, longitudinally rugose (coarsely wrinkled) to wrinkled, orange-red, brownish red tinged yellowish, white and finely downy just over the base. Flesh firm, whitish, hollow in age. Taste bitter.

Spores 10–16 x 5–10 μ elliptical, smooth, thin-walled, white to buff spore print.

Single to many in hardwoods in eastern North America and mixed hardwoods in western North America in Sp. and F.

269 CLAVARIADELPHUS LIGULA (Fr.) Donk   Edibility unknown/Numerous

Clavate to flattened, salmon-colored clubs; taste bitter

Fruiting body 2–6 cm high, 3–12 mm thick, cylindric, narrowly club-shaped to flattened, surface roughened, salmon color to orange-buff, fine white downy base. Flesh firm, not brittle, white, green in ferric sulphate solution. Taste slightly bitter.

Spores 12–15 x 3–4.5 μ elliptical, thin-walled, white spore print. Basidia usually 4-spored.

Gregarious in clusters or even cespitose in conifer duff under many different conifers in S. and F.; widely distributed.

Clavariadelphus sachalinensis (Imai) Corner is very similar but has an ochre spore print and spores 16–24 μ long. It is common in Alaska and probably elsewhere according to Wells and Kempton (1968).

270 CLAVULINA CINEREA (Fr.) Schroet.   **Edible**/Common

Multiple short, irregular, smoke-gray branches

Fruiting body 2.5–10 cm high, 2–6 cm thick, medium size, densely covered with short, often irregular, thin branches, smoke gray, sometimes bluish gray or even brownish in age, whitish just at the base. Flesh firm, grayish white.

Spores 6.5–11 x 7–10 μ nearly round, one oil drop, smooth, white spore print. Basidia club-shaped, 2-spored. Clamp connections present.

Scattered to abundant on the ground in moss or needle duff under conifers or mixed woods in late S. and F.; widely distributed.

The gray coloration separates this from the very similar *Clavulina cristata* which also has the delicate, fine branches.

271 CLAVULINA CRISTATA (Fr.) Schroet.   **Edible (?)**/Common

Multiple branched, dainty with fine tips, white to yellowish

Fruiting body 2.5–8.5 cm high, 2–4 cm thick, dainty fine branches with fine tips, white tinged yellowish. Flesh somewhat tough, sometimes hollow.

Spores 7–11 x 6.5–10 μ nearly round, slightly thick-walled, smooth, one oil drop per cell, white spore print. Basidia club-shaped, 2-spored (Fig. 74).

Single to several in fields and under hardwoods and conifers, in late S., F., and W.; widely distributed.

The fine, thin, delicate branches with small pointed projections (cristate) are characteristic of this species. The basidia are 2-spored.

Ramaria stricta (Fr.) Quel. is somewhat similar but tends to be pale brown to yellow-brown in age. It occurs on wood and has consistently parallel straight branches. Neither species is worth eating but neither is poisonous.

272 CLAVICORONA PYXIDATA (Fr.) Doty   **Edible, good**/Common

Multiple, slender, equal branches with crownlike tips; taste mild

Fruiting body 8–13 cm high, 6–10 cm thick, tough, pliable, tan to pale pinkish with *repeated slender, equal, branching and clear yellow crownlike tips,* pubescent whitish to pinkish brown over the base. Flesh tough, white. Taste mild.

Spores 4–5 x 2–3 μ elliptical, smooth, white spore print.

Gregarious or in cespitose clusters on dead wood (especially poplar and willow) in Sp. and S.; widely distributed.

273 RAMARIA FORMOSA (Fr.) Quel.   **Poisonous**/Numerous

Robust fleshy base with orange-red branches; flesh bruises black; taste bitter

Fruiting body 7–25 cm high, 3–15 cm thick, pinkish buff to light orange-red with lemon-yellow, usually blunt, branched tips, often branching almost from the base. which is whitish. Flesh white *turning red-brown to black when bruised.* Taste faintly bitter.

Spores 8–15 x 4–6 μ oblong-elliptical, minutely roughened, red-brown spore print.

On humus or in soil under hardwoods or occasionally conifers in late F., W., and early Sp.; widely distributed.

Contains Group VI toxins and is reported to cause digestive disturbance. The white flesh which bruises brown to black is the principal distinguishing characteristic.

274 RAMARIA BOTRYTIS (Fr.) Ricken Edible, good /Common

Robust fleshy base with blunt, pink branches; taste mild

Fruiting body 7–20 cm high, 10–30 cm thick, cauliflower-like, robust with thick numerous branches arising from a large, fleshy base, whitish to tan with deep pink to occasionally red or purple branch tips. Flesh somewhat brittle, white. Odor pleasant.

Spores 12–30 x 4–6 μ smooth *with fine, longitudinal striations,* light orange-brown spore print.

Scattered to abundant under hardwoods and conifers in S. and F.; widely distributed. I have seen large fruitings of this fungus in the PNW in early F., but it is also found in late Sp. and early S.

The pink to red branch tips and massive cauliflower-like appearance are characteristic of this species. Sometimes only the dense head may appear above the duff with the rest of the fruiting body partially buried. We have eaten this fungus and consider it a good edible. *Ramaria subbotrytis* (Coker) Corner is bright pink to reddish overall, with spores 7–9 x 3–3.5 μ; usually rare but the most colorful of our coral fungi. A yellow *Ramaria* is the common edible spring coral in the RM. It is often found in great quantity, especially under or near firs *(Abies* and *Pseudotsuga).* Some people experience a purgative effect following a meal of it, but four out of five members of my family, including the author, enjoy it every spring. It is often found in sufficient quantity to be preserved by canning.

275 SPARASSIS RADICATA Edible, good /Infrequent

Weir ["Cauliflower Fungus"]

Multiple flattened branches, large, rounded, white to yellowish; under conifers

Fruiting body 25–50 cm high, 15–35 cm thick, fleshy, white to yellowish, consisting of a series of cauliflower-like branches with flattened tips which arise from a deeply rooted large central stalk, 2–4 cm thick. Flesh somewhat tough, white.

Spores 5–6.5 x 3–3.5 μ egg-shaped, smooth, white spore print.

Single on ground under conifers in F., widely distributed but most common in the PSW, RM, PNW, and adjacent C.

Sparassis crispa Wulfen is an eastern species which is much more flattened and has short, flat branches and somewhat larger spores. It is common under Virginia pine in sand. Both species are edible and good, but they must be cooked slowly. Young specimens are more fleshy and are preferred. The conspicuous cauliflower-like appearance makes *Sparassis* a distinctive fungus.

BOLETACEAE
The Boletes

Boletes are often mistaken for a member of the gilled mushrooms or the two groups are simply lumped together. The general appearance and fleshy nature of the fruiting body certainly makes them seem similar. However, the spore-bearing surface beneath the cap must always be examined. In the boletes it will consist of minute holes (pores) which are the mouths of short or often long tubes. The spores are produced in these tubes and following forcible discharge into the tube they fall out into the air, where

157

they are carried off by the wind. If the cap is severed from the stipe and placed on a white paper, the discharged spores will fall in heaps on the paper. In this way spore prints can be obtained and the true color of the spores determined.

To distinguish between bolete genera the arrangement of the pores must be examined. In *Suillus* and *Fuscoboletinus,* they are arranged in loose radial rows (*boletinoid,* Fig. 22) extending from the stipe to the pileus (cap) margin. *Suillus* has olive-brown spores and *Fuscoboletinus* has pinkish to reddish brown spores. *Gyrodon,* with a single species associated with ash trees, has very shallow, bright yellow pores (3–5 mm thick). *Leccinum* is characterized by black tufts of hairs on the stipe called scabers and like *Boletus* has pores which have no radial arrangement. *Boletus* has a moist to dry pileus and a stipe which is usually smooth or covered with a fine netlike (reticulate) ornamentation. *Tylopilus* has pink spores. The tubes are also pink, not radially arranged, and most species have a very bitter taste.

Boletes characteristically have a fleshy central stipe and pileus. The polypores with which they might be confused also have united tubes (pores), but they are woody, leathery, or corky. If a polypore is fleshy, it does not have a central stalk with the typical mushroom-like pileus as shown in Figures 21–23. At this time two comprehensive regional books have appeared: one covering the northeastern boletes by Snell and Dick (1970), and the other covering the boletes of Michigan by Smith and Thiers (1971).

Edibility. Several members of the genus *Boletus* are poisonous. They either bruise blue or have red tube mouths or both. These striking field characteristics clearly separate the poisonous species from the others. The toxins are not known, so they are included in Group VI. Considered in total, this is one of the really fine and safe edible groups. Most boletes are fleshy and provide a good quantity of food, and many have rather distinctive flavors. If the cap is covered with glutin (slime), the cuticle should be removed. The pores should be removed in older specimens since they become slimy when cooked. Boletes are excellent in most of the general recipes, but we recommend serving them with "German White Sauce" (see Recipes).

Key to the Genera of the Boletaceae

1 a. Cap and stalk covered with long, shaggy, gray-black hairs, *276. Strobilomyces floccópus*
   b. Cap and stalk may be hairy but not as above, **2**
2 a. Tubes contorted, not consistently arranged vertically, mouths not open; known only from western America, *277. Gastroboletus turbinatus*
   b. Tubes straight, vertically arranged, mouths open; widely distributed, **3**
3 a. Tubes radially arranged (boletinoid) in rows, sometimes irregular, (Fig. 22) from the stalk to the margin but always most noticeable near the stalk, **4**
   b. Tubes not radial, mouths often round or angular (Fig. 21), **6**
4 a. Mouths of tubes uneven, 2–5 mm broad, tubes very shallow (1–4 mm long), light yellow with tints of green at first; cap deep rich brown, dry; always near ash trees, *278. Gyrodon merulioides*
   b. Mouths of tubes even, tubes longer and lacking the above combination of characteristics, **5**
5 a. Cap usually dry to moist; spore print wine-red to reddish brown, *Fuscoboletinus*
   b. Cap usually viscid to slimy; if dry, pores radially arranged in rows from the stipe to the margin; spore print cinnamon, olive-brown to brown, *Suillus*

6 a. Mouths of tubes white then pink in age, bruises brown; spore print pink; stipe reticulate (with netlike ridges); taste extremely bitter, *279. Tylopilus felleus*

  b. Mouths of tubes maybe white but *not p̄ink* in age and bruises brown; spore print not pink; stalk reticulate or not; taste mild to bitter, **7**

7 a. Stalk with conspicuous dark brown to black scabers (hairy tufts) at maturity, Leccinum

  b. Stalk never as above, **8**

8 a. Spore print yellow; pores white to pale yellow in age; stalk not reticulated, *Gyroporus*

  b. Spore print another color; and not with above combination of characteristics, *Boletus*

276 STROBILOMYCES FLOCCOPUS   Edible/Numerous

(Vahl. ex Fr.) Karst. ["Old Man of the Woods"]

Cap shaggy from gray-black scales; stalk also shaggy

Cap 4–15 cm broad, convex, dry, covered with erect blackish brown to gray-black scales, scales and veil pieces hanging from margin. Flesh white, staining red and finally black when bruised. Tubes with large mouths, white, then gray, stain like flesh. Stalk 5–14 cm long, 5–20 mm thick, nearly equal, shaggy, same color as cap. Veil white to gray, cottony. Taste mild.

Spores 9–16 x 8–12 μ almost round, strongly warted, black spore print.

Occurs alone or in small numbers, often along streams, on the ground, or on well-rotted wood under hardwoods or mixed conifer hardwoods, mostly in NE, So., and SE in S.

The "Pine Cone Fungus" or "Old Man of the Woods" is very distinctive (Fig. 23), but its taste is not outstanding. It does not rot easily and may be found still standing in the fall and covered with a green mold. It is also listed as *Strobilomyces strobilaceus* (Scop. ex Fr.) Berk. A closely related, smaller species with wrinkled spore ornamentation, *S. confusus* Singer, is found in NE and SE.

277 GASTROBOLETUS TURBINATUS   Edibility unknown/Infrequent

(Snell) Smith and Singer

Cap blackish brown, dry; tubes twisted and contorted, yellow to red, stains blue

Cap 2–8 cm broad, convex, dry, felty, blackish brown, golden brown to red with orange tints. Flesh firm, yellow to pinkish, bruising blue. Tubes 1–2 cm long, mouths round when visible, up to 1 mm broad, *twisted and contorted,* yellow to reddish in some, instantly blue when bruised. Stalk 1–7 cm long, 8–20 mm thick, short, dry, marbled yellow, orange-yellow to red, bruising blue. Veil none.

Spores 10–18 x 5.5–9.5 μ elliptical, smooth, *no spore deposit,* spores are not discharged.

Scattered under western conifers (often hemlock and fir), often deep in the duff with only cap showing, in S. and early F. in RM and PNW.

This Gastromycete is a close relative of the bolete, distinguished by contorted pores and a stalk that barely protrudes beyond the cap. The spores are not forcibly discharged, so no spore print can be obtained. Ten species have now been described in *Gastroboletus* (Thiers and Trappe, 1969), but the one described here is encountered most frequently.

278 GYRODON MERULIOIDES (Schw.) Singer   **Edible**/Numerous

Cap yellow-brown; tubes shallow, radial, yellow; flesh ± bruises blue; under ash

Cap 5–25 cm broad, nearly flat to slightly depressed, felty dry, dull yellow-brown to red-brown. Flesh yellow, sometimes slowly becoming bluish green. Tubes *very shallow,* 1–2 mm deep, mouths long and 2–3 mm wide, radial and nearly lamellate (in thin layers), uneven crosswalls, bright yellow to dull, dingy yellow in age, bruises slightly blue. Stalk 1–6 cm long, 10–30 mm thick, *lateral,* dry, partially reticulate at apex, yellow at first, stained reddish over lower part in age. Veil absent.
Spores 7–11 x 5–7.5 µ broadly elliptical, smooth, brownish red to yellowish ochre spore print.
Single to scattered under or near ash, follows distribution of ash in SE, CS, NE, and adjacent C. in S. and F.
Lateral stalk, very thin gill-like to wrinkled, yellow spore-bearing surface and constant association with ash are the distinctive characteristics.

279 TYLOPILUS FELLEUS (Bull. ex Fr.) Karst   **Nonpoisonous**/Frequent

Cap dry, tan; tubes pink; stalk reticulate (with netlike ridges); taste bitter

Cap 8–30 cm broad, broadly convex, dry but feels slimy in wet weather, nearly hair-less, tan to crust-brown. Flesh soft, white, slightly pink where bruised. Tubes 1–2 cm deep, mouths small 1–2 per mm, often depressed at stalk, white at first but soon *deep pink.* Stalk 4–12 cm long, 15–25 mm thick, enlarged below or sometimes bulbous, *reticulate,* dry, pallid to pale brown. Veil absent. Taste *very bitter.*
Spores 9–15 x 3–5 µ long elliptical, smooth, rose or pink spore print.
Single or scattered on or near rotten evergreen (conifer) stumps or buried wood in CS, SE, and NE, in S. and F.
The pink spore print, tubes, and brown stalk with netlike ridges are distinguishing characteristics. Closely related *Tylopilus plumbeoviolaceus* (Snell) Snell and Dick has a smooth stalk with netlike ridges only at the apex and violet coloration on the cap and stalk. *Tylopilus alboater* (Schw.) Murr. is an eastern bolete growing under oak; it has a black to grayish brown cap, flesh which stains dingy pink to black, and a bluish purple to nearly black partially reticulate (with netlike ridges) stalk.

FUSCOBOLETINUS

All of the species in this genus have elliptical spores that are red-brown or purplish brown with a reddish tint. In all other respects they are similar to *Suillus* including the radially arranged tubes referred to as boletinoid.

Edibility. They are all either edible or nonpoisonous and bland. Their flavor is not exceptional and one can use them best in casseroles with other ingredients.

Key to Fuscoboletinus

1 a. Cap dry, with pink, dense fibrils; flesh pink in narrow zone beneath cuticle, *280.*
   F. *ochraceoroseus*
   b. Cap viscid to glutinous, not as described above, **2**

280 FUSCOBOLETINUS OCHRACEOROSEUS   Edible/Common

(Snell) Pomerleau and Smith

Cap dry, dense pink fibrils; tubes radial, yellow; flesh pink in zone beneath cap cuticle

Cap 8–25 cm broad, broadly convex, flat in age, dry, pink hairs with white beneath, darkening in age, margin sometimes yellow with the white remains of the veil tissue adhering. Flesh yellow, pink under cap hairs in a narrow zone, turns slightly bluish green when bruised or cut. Tubes thin, mouths 1–2 mm wide, radial, extend down stalk, straw-yellow to dingy-brown in age. Stalk 3–5 cm long, 1–3 cm thick, netlike ridges, base often enlarged, dry, strawy-yellow, white near base. Veil thin, whitish, sometimes leaves temporary zone over stalk.

Spores 7.5–9.5 x 2.5–3.5 μ long cylindrical, smooth, dark reddish brown spore print.

Occurs alone or in groups in S. and early F. usually near western larch in PNW and adjacent C. It is confused with *Suillus lakei* which has more orange coloration on the cap and an olive-brown spore print. *Suilleus pictus* is a similar eastern species found under white pine and produces an olive-brown spore print.

281 FUSCOBOLETINUS SPECTABILIS   Edible/Infrequent

(Pk.) Pomerleau and Smith

Cap viscid, red with gray veil patches; in eastern larch bogs

Cap 4–10 cm broad, convex, flat in age, viscid, red between the grayish, flattened patches of the veil. Flesh light yellow, stains pink slowly to brown. Tube mouths 1 mm or more deep, yellow, radial, same stain as flesh. Stalk 4–10 cm long, 10–15 mm thick, yellow above, dingy, red, gelatinous ring, streaked red and gray below, dry. Veil red, cottony hairs with gelatinous layer beneath. Taste astringent. Odor somewhat aromatic.

Spores 9–13 x 5–6.5 μ elliptical, smooth, purplish brown spore print.

Scattered under eastern larch (tamarack) in bogs in NE and adjacent C, in S. and F. A distinctive red cap with gray patches sets this apart from other boletes. It is also known as *Boletinus spectabilis* (Pk.) Murr.

282 FUSCOBOLETINUS AERUGINASCENS   Edible, good/Common

(Secr.) Pomerleau and Smith

Cap glutinous, whitish to smoke-gray; tubes radial; flesh bruises blue-green

Cap 3–12 cm broad, convex to flat in age, sticky to glutinous in wet weather, nearly white, smoky gray to wood brown in dry weather. Flesh white, staining blue-green when exposed or bruised. Tube mouths 1–3 mm deep when mature, white, gray to brown in age, radial, with flesh-colored stain and colors paper blue-green. Stalk 4–6 cm long, 8–12 mm thick, nearly equal, with netlike ridges (reticulate), white above ring, viscid, smoke-gray to brownish gray below. Veil membranous, cottony hairs, white, leaves a flattened ring. Taste mild.

Spores 8–12 x 4–5 μ almost tapering at ends, smooth, reddish brown spore print.

Occurs scattered to numerous under larch (tamarack) trees; widely distributed late S. and F. but in PNW often in Sp.

Also known as *Suillus aeruginascens* (Secr.) Snell. It often appears in great numbers in the PNW during cool wet Sp. or F. weather.

283 FUSCOBOLETINUS SINUSPAULIANUS   Edibility unknown/Rare

Pomerleau and Smith

Cap glutinous, yellow-brown to red-brown, hairless; flesh orange-buff, does not show bruises

Cap 3–13 cm broad, convex, hairy but sticky, brownish red, margin incurved at first, with hanging pieces of the veil. Flesh yellowish to orange-yellow. Tubes up to 1 cm deep, mouths angular (1–2.5 mm broad), radial, yellow-brown to dark brown. Stalk 4–12 cm long, 1–3 cm thick, equal, dry, reticulate (netlike ridges) and yellow-brown above ring, gray-brown spotted with red below. Veil whitish leaving a grayish brown annular zone and torn pieces on the cap margin.

Spores 8–12 x 4–5 μ elliptical, smooth, deep purple-brown spore print.

Occurs scattered in boreal conifer forests. Known only from eastern Canada where it fruits in autumn.

SUILLUS

The species have elliptical spores which are olive-brown to brown but not red-brown. The pores are radially aligned (boletinoid) but if they are not obviously so, then the pileus is viscid to glutinous. There are about 50 species in North America and most of them are mycorrhizal with various conifers.

Edibility. There are no poisonous species in this genus. Many of the species are rated as excellent edibles. *Suillus cavipes* is known as the "Mock Oyster" and is especially good in "Mushroom and Sour Cream Casserole" and "Mushroom and Cheese Casserole" (see Recipes).

Key to Suillus

1 a. Veil and/or annulus (ring) present, **2**
b. Veil absent, **8**
2 a. Veil does not leave a ring but remains as ragged pieces on cap margin, *284. S. americanus* and *S. sibericus*
b. Veil leaves a ring, **3**
3 a. Cap viscid to glutinous (in dry weather, test it by wetting it), **4**
b. Cap dry, **7**
4 a. Cap with red to orange-red hairs over the viscid surface, *290. S. lakei*
b. Cap with a smooth, viscid to glutinous surface or not as above, **5**

5 a. Cap glabrous (hairless), bright yellow to red; veil yellow; under or near larch, *285. S. grevillei*
   b. Cap and veil another color; under or near various conifers, **6**
6 a. Cap convex-knobbed, viscid, streaked olive-buff; in or near lodgepole pine and jack pine forests, in swamps or wet areas, *286. S. umbonatus*
   b. Cap convex but not knobbed, viscid to glutinous, variously colored but not olive-buff; widely distributed, *287. S. luteus* and related species
7 a. Stalk base hollow; cap with brown to orange-brown squamules (small scales); widely distributed, *288. S. cavipes*
   b. Stalk base solid; cap with red fibrils; under eastern white pine, *289. S. pictus*
8 a. Stalk without hairs, smooth, lacking glandular dots, *291. S. brevipes*
   b. Stalk covered with cinnamon to reddish brown glandular dots, **9**
9 a. Flesh and tubes staining blue (sometimes slowly) when bruised; cap orange-yellow; pores deep brown and minute at first, *292. S. tomentosus*
   b. Flesh and tubes not staining; cap differently colored; pores not deep brown at first, **10**
10 a. Cap yellow, covered with flattened grayish squamules (small scales) and stained red, *293. S. hirtellus*
   b. Cap cinnamon to pinkish gray or variously colored, but without squamules and red stains, *294. S. granulatus* and closely related species

284 SUILLUS AMERICANUS (Pk.) Snell   **Edible**/Common

Cap buff with veil patches on margin; flesh yellow, stains brown

Cap 3–10 cm broad, convex with a low knob, viscid, pale buff to cinnamon streaked with reddish hairs, margin with a white to buff, soft, cottony, hanging veil, disappearing only at maturity. Flesh very narrow, yellow to orange-yellow, staining brown when bruised. Tubes about 4–6 mm deep, mouths large (1–2 mm broad), short decurrent (extending down stalk), mustard-yellow to yellow-brown in age. Stalk 3–9 cm long, 3–10 mm thick, nearly equal, often crooked, dry, lemon-yellow, covered densely with cinnamon, red to red-brown glandular dots. Veil cottony, thick, usually not leaving ring but hanging pieces on cap margin.
   Spores 8–11 x 3–4 μ subfusiform (Fig. 65), smooth, dull cinnamon spore print.
   Occurs under eastern white pine in late S. and early F.; distributed throughout NE and adjacent C.
   The very narrow crooked stalk combined with the cottony, marginal veil remnants are distinguishing characteristics. I have encountered large fruitings under white pine in southern New Hampshire and in Massachusetts. *Suillus sibiricus* (Sing.) Singer has brown spots on the cap, is darker or more dingy yellow, has a thicker stalk, and is found under western white pine but is a closely related species. See also 286. *S. umbonatus,* an associate of the lodgepole pine which could be confused with *S. sibiricus.*

285 SUILLUS GREVILLEI (Kl.) Singer   **Edible**/Common

Cap glutinous, bright yellow to red; veil yellow, sticky; under larch

Cap 5–15 cm broad, egg-shaped at first to broadly convex, hairless, glutinous, bright yellow overall to a yellow margin with varied red coloration over the center to deep, rich chestnut-red, especially in western material. Flesh pale lemon-yellow, dingy pinkish or not changing when bruised, sometimes yellow-green just at base of stalk. Tubes about 6–15 mm deep, mouths 1–2 per mm, attached or slightly depressed, radial,

163

yellow, dingy in age, bruising brown. Stalk 4–10 cm long, 10–30 mm thick, nearly equal, somewhat enlarged at base, yellow and reticulate (netlike) above ring which is cottony white to buff and may have a sticky surface, streaked reddish brown over yellow below ring, dry but without glandular dots. Veil pale lemon-yellow, somewhat sticky in wet weather.

Spores 8–11 x 3–4 μ elliptical, smooth, olive-brown spore print.

Occurs cespitose to abundant under or near larch (tamarack) throughout PNW, RM, CS, NE, and C, in F. In the PNW often seen in early S.

The proportion of bright yellow and red color in the cap varies considerably. Generally there is much more red in western collections whereas eastern variants are often completely yellow. *Suillus luteus* is closely related but has glandular dots on the stalk and duller cap coloration.

286 SUILLUS UMBONATUS Snell and Dick   Edible/Infrequent

Cap convex-knobbed, viscid, olive-buff; under lodgepole and jack pine

Cap 3–9 cm broad, egg-shaped soon *convex-knobbed,* viscid appearing streaked, olive-buff. Flesh pale yellow, bruising dingy cinnamon. Tubes 4–6 mm deep, mouths 1–2 mm broad, radial, yellow to yellow-green, flesh colored stain. Stalk 3–9 cm long, 4–8 mm thick, equal, pale yellow above the gelatinous ring, whitish below, covered with pallid to yellow glandular dots. Veil sticky to gelatinous, whitish at first to dingy pinkish cinnamon.

Spores 7–10 x 3.5–4.5 μ elliptical, smooth, dull cinnamon-brown spore print.

Scattered, often in moist areas in sphagnum or various mosses, under lodge-pole pine in PNW and RM. Also under jack pine in eastern C.

The distinctive features of this species are the *convex-knobbed* cap and narrow stalk along with the streaked cap and the habitat. We have found it in Glacier Park, Montana close to the bog mat surrounding the pond at McGee Meadow.

287 SUILLUS LUTEUS (Fr.) S. F. Gray   Edible, good/Numerous

Cap glutinous, brown; tubes yellow; ring, glutinous, purple-brown

Cap 4–17 cm broad, egg-shaped at first, convex to broadly convex in age, hairless, viscid to glutinous, forming a layer which can be peeled off, brown shading to yellow-brown or reddish brown. Flesh white to pale yellow. Tubes about 3–7 mm deep, mouths 3 per mm, adnate, radial in age, white at first to pale or deep yellow with a brownish tint in age. Stalk 4–11 cm long, 10–25 mm thick, equal, yellow, reticulate (netlike ridges), with pink to pinkish brown glandular dots, a sticky to glutinous purplish ring zone, pale yellow to nearly white and glandular dotted below. Veil white, membranous with a purplish zone on the underside, gelatinous in humid weather.

Spores 6–11 x 2.5–4 μ elliptical, smooth, dull cinnamon spore print.

Occurs scattered to numerous under pines and spruce in late S. or F. in SE, NE, and adjacent C.

There are several closely related species which are all edible. These include *S. acidus* (Pk.) Singer with a yellow to buff cap which has an acid-tasting glutin and brown pores. *Suillus cothurnatus* Singer, a southeastern species, has a baggy ring in which the lower edge flares and is very close to *S. pinorigidus* Snell and Dick, which has orange tubes. *Suillus subluteus* (Pk.) Snell, a jack pine associate in the Lake States, cannot be easily distinguished by field characteristics. *Suillus subolivaceus* Smith and Thiers has olive-brown tubes and is found in the PNW under mixed conifers. I have collected and eaten most of these closely related species and find them all quite good.

288 SUILLUS CAVIPES (Opat.) Smith and Thiers   Edible, choice/Common

["Mock Oyster"]

Cap dry, red-brown hairs; stalk yellow-brown with hollow base; under larch

Cap 3–12 cm broad, convex to flat in age with a low knob, dry, densely hairy, fibrils tawny brown, reddish brown to yellow-brown especially at margin, tips of fibrils pallid. Flesh soft, white, unchanged when bruised. Tubes 3–5 mm long, mouth angular (up to 1 mm broad), extend down stalk, radial, pale yellow to greenish yellow in age. Stalk 3–9 cm long, 8–20 mm thick, large at apex, narrowing toward base, dry, yellow and sometimes with netlike ridges above a thin annulus, thinly fibrillose below, tawny-brown to yellow-brown, when cut lengthwise, the solid upper part gives way to a lower hollow base. Veil fragile, fibrillose, white to light orange-red, soon gone.

Spores 7–10 x 3.5–4 μ short elliptical, smooth, dark olive-brown spore print.

Numerous under western larch (tamarack) and eastern larch in F. in PNW, RM, CS, NE, and adjacent C; also in other regions where plantations of larch have been established.

The brown, dry, hairy caps, hollow stalk, and association with larch are used to de-limit this from other species of *Suillus*. It also appears under the name *Boletinus cavipes* (Opat.) Kalchbr.

289 SUILLUS PICTUS (Pk.) Smith and Thiers   Edible, choice/Common

Cap dry, red hairs; flesh yellow bruising red; under eastern white pine

Cap 3–12 cm broad, conic, convex to nearly flat in age, dry, covered with thick red hairs, yellow flesh beneath, margin at first thickly covered with red hairs. Flesh firm, yellow, slowly changing to reddish when cut or bruised. Tubes 4–8 mm deep, large and angular, 0.5–5 mm wide, radially arranged, often extending down stalk, yellow to brownish in age, staining reddish brown where injured. Stalk 4–12 cm long, 8–20 mm thick, equal or somewhat larger toward the base, yellow above ring, whitish below with zones and patches of dull red hairs. Veil dry, hairy, whitish, leaving a superior, hairy annular zone.

Spores 8–12 x 3.5–5 μ elliptical, smooth, olive-brown spore print.

Solitary to scattered under eastern white pine in eastern North America. It is very common in central and northern New York; I have seen large fruitings near Saranac Lake. Fruiting in S. and F.

This species is also known as *Boletinus pictus* Peck.

290 SUILLUS LAKEI (Murr.) Smith and Thiers   Edible/Common

Cap viscid, reddish orange small scales; flesh yellow not bruising; under western conifers

Cap 4–20 cm broad, convex, nearly flat in age, tacky to viscid beneath a layer of reddish to orange-brown small scales, surface appears streaked. Flesh firm, yellowish, unchanged when cut or bruised. Tubes 5–10 mm deep, mouths small but up to 2 mm wide in age, radially arranged, orange-brown to nearly salmon color, in age dingy brown and staining brown when bruised. Stalk 4–12 cm long, 1–4 cm thick, equal or narrowed somewhat at the base, yellow and smooth above ring, yellow but staining brown below, with or without glandular dots. Veil dry, hairy, leaving a superior ring consisting of a ring of loose hairs sometimes hard to detect in age.

Spores 7–10 x 3–4 μ elliptical, smooth, dull cinnamon-brown spore print.

Solitary to scattered in Sp. and S. in the PNW, in F. on the Pacific Coast. Abundant during cool moist periods under mixed conifers, especially western hemlock, Douglas fir, and western red cedar. It is one of the most common boletes encountered in Sp. in northwest Montana and Idaho.

291 SUILLUS BREVIPES (Pk.) Kuntze   Edible, good/Common

Cap glutinous, red-brown; stalk white to buff, smooth; veil absent

Cap 5–10 cm broad, convex, nearly flat in age, glutinous, dark reddish brown to yellow-brown, but quite variable; lighter in age. Flesh white, yellowish in age, unchanged when cut or bruised. Tubes about 4–10 mm deep, mouths small (1–2 per mm), radial near stalk, yellow. Stalk 2–5 cm long, 1–3 cm thick, equal, dry, white to pale yellow, without glandular dots. Veil absent.

Spores 7–10 x 2.5–3.5 μ elliptical, smooth, cinnamon spore print.

Occurs scattered to numerous or even in cespitose clusters under pines throughout most of North America in S., F., or rarely W.

S. albivelatus Smith, Thiers, and O. K. Miller, has a veil and sometimes an annular zone. Other closely related species are discussed following *S. granulatus.*

292 SUILLUS TOMENTOSUS   Edible, good/Common
(Kauf.) Snell, Singer and Dick

Cap viscid, wooly, orange; orange glandular dots on stalk; tubes brown, bruises blue

Cap 5–15 cm broad, convex to broadly convex in age, coarsely tomentose (woolly) at first, nearly hairless in age, sticky to viscid, orange-yellow. Flesh pallid to yellow *bruising blue or greenish blue.* Tubes 1–2 cm deep, *minute at first,* 2 per mm in age, attached or short decurrent (extending down stalk), somewhat reddish in age, dark brown to dingy yellow in age, stain blue when bruised. Stalk 3–12 cm long, 1–3 cm thick, equal or club-shaped, dry, yellow-orange to orange and covered with glandular dots. Veil absent.

Spores 7–12 x 3–4 μ elliptical, smooth, dark olive-brown spore print.

Occurs alone or in groups, frequently under lodgepole pine or other two-needle pines in late S. and F. in the PNW and adjacent C north to A. It is found less commonly in the Lake States under jack pine.

The yellow-orange cap, orange glandular dotted stalk, combined with the dark brown young tubes and blue bruising of the tissue are the distinguishing characteristics. I have seen extensive fruiting of this fungus in Alberta in July and in Montana in early August.

293 SUILLUS HIRTELLUS (Pk.) Kuntze   Edible/Infrequent

Cap viscid, yellow stained red with gray scales; red glandular dots on stalk; veil absent

Cap 4–15 cm broad, convex, viscid and yellow under flattened grayish small scales and smaller reddish stained areas, in age nearly hairless, staining brown. Flesh pale yellow, not staining. Tubes narrow 4–8 mm deep, mouths elongated, 1–2 mm broad, pale yellow, bruising brown or not at all. Stalk 3–8 cm long, 5–20 mm thick, near equal, base somewhat larger, dry, covered with reddish brown or purple-brown dots over a yellow ground color. Veil absent.

Spores 7–11 x 2.5–3.5 μ nearly oblong, smooth, dull cinnamon spore print.

Occurs alone, scattered to numerous under pine, spruce, or aspen in late S. and F.; in NE and SE.

Suillus hirtellus var. *thermophilus* Singer is a deep southern variety with red coloration on the stalk, fruiting in S. and F., while *S. hirtellus* var. *cheimonophilus* Singer, also from the South, has yellow cap hairs and fruits in W. Both varieties are found under southern pines.

294 SUILLUS GRANULATUS (Fr.) Kuntze — Edible, good/Common

Cap glutinous, cinnamon; red glandular dots on stalk; veil absent

Cap 5–15 cm broad, convex, broadly convex in age, viscid to glutinous, mottled cinnamon to pinkish gray, darker when young. Flesh white to yellowish, a watery green line above the tubes. Tubes about 1 cm deep, mouths small (2 per mm), subdecurrent (almost extending down stalk), radial, pale honey-yellow, brownish in age. Stalk 4–8 cm long, 1–2.5 cm thick, equal, dry, yellow at apex, white with cinnamon over the base, covered with tan to reddish brown glandular dots. Veil absent.

Spores 7–10 x 2.5–3.5 μ elliptical, smooth, dingy cinnamon spore print.

Occurs alone to abundant under pines throughout northern U.S. and C, in S. and F., occasionally Sp. in PNW.

A closely related species with a yellow to tawny cap is *S. punctatipes* (Snell and Dick) Smith and Thiers. *Suillus placidus* (Bonorden) Singer is similar but with an ivory-white cap. *Suillus albidipes* (Pk.) Singer is not glandular-dotted over the stalk and the veil forms a roll of cottony tissue around the cap margin but leaves no ring.

S. punctipes (Pk.) Singer has more orange in the cap, an orange-brown stalk, and is found in NE and SE, has a dry cottony veil which adheres to the margin of the cap in a roll. *Suillus brevipes* (Pk.) Kuntze is similar but has a somewhat darker pileus and *lacks glandular dots*. *S. subaureus* (Pk.) Snell has a yellow, viscid cap beneath finely flattened woolly hairs, has glandular dots on the stalk, and is found under aspen and oak in central North America. All the closely related species are edible.

LECCINUM

The conspicuous *brown to blackish brown scabers* (Fig. 21) on the stalk separate *Leccinum* from the other bolete genera. The cap is usually dry or viscid when wet, and a narrow sterile margin without pores is often present around the outer edge of the cap. The pore mouths are various shades of yellow to yellow-brown but never red and not radially arranged. The flesh ranges from unchanged when bruised to bruising blue and/or red.

Edibility. All of the species of *Leccinum* are edible. Once the scabrous stalk can be recognized and the genus identified, then the rule cautioning beginners to avoid blue-staining boletes can be ignored. For further information see the edibility section under the Boletaceae.

Key to Leccinum

1 a. Cap matted fibrillose (hairy), dry; flesh bruising red then dingy blue; under mixed conifers in western North America, *295. L. fibrillosum*

   b. Cap glabrous (hairless) or nearly so, viscid when wet; flesh stained or unstained; widely distributed, **2**

2 a. Cap white to gray-brown, sticky to viscid in wet weather; flesh unchanged when bruised, **3**
   b. Cap rusty-red to red-brown, almost viscid or dry in wet weather; flesh staining red with gray overtones, **4**
3 a. Cap white or nearly so, (see 296. *L. scabrum*), *L. albellum*
   b. Cap gray-brown, *296. L. scabrum*
4 a. Cap bright rusty-red to orange-red; widely distributed, *297. L. aurantiacum*
   b. Cap more often orange-cinnamon with olive tones; under aspen in the northern boreal forest, *298. L. insigne*

295 LECCINUM FIBRILLOSUM Smith, Thiers, and Watling   Edible/Numerous

Cap dry, hairy, red-brown; flesh bruising red then purplish

Cap 8–25 cm broad, convex, broadly convex in age, dry matted hairs concentrated into coarse, small scales, brown to red-brown, sterile, narrow, marginal flap of hanging veil tissue. Flesh firm, pallid staining reddish then purplish drab. Tubes 1–2 cm deep, mouths 1–2 per mm, depressed to nearly free at stalk. Stalk 4–12 cm long, 2–5 cm thick, sometimes even but very often enlarged at center and tapering toward both ends, dry, white but covered with dense long black scabers and coarse scales.
Spores 14–20 x 4–5 μ narrowly fusiform smooth, olive-brown spore print.
Scattered to locally common under mixed western conifers, often in dense huckleberry in PNW in late S. and early F.
Long black scales on cap and stalk and short, thick, often centrally swollen stalk combined with the conifer habitat are its distinctive characters.

296 LECCINUM SCABRUM (Bull. ex Fr.) S. F. Gray   Edible, good/Common

Cap viscid, gray-brown, hairless; flesh white not staining; under aspen and birch

Cap 4–20 cm broad, convex to broadly convex, hairless, viscid, whitish at first, gray-brown to dull yellow-brown, margin free of hanging veil remains. Flesh firm, white, *not staining* when bruised. Tubes 8–15 mm long, mouths small, (2–3 per mm), deeply depressed at stalk, a very narrow sterile margin, white then tan with a pinkish tint in age, does not stain when bruised or becomes faintly yellow. Stalk 7–15 cm long, 7–15 mm thick, enlarging toward the base, dry, pallid to whitish with *dark brown to black* scabers, which, when numerous, appear almost reticulate (netlike). Veil inconspicuous, no ring or marginal patches.
Spores 15–19 x 5–7 μ subfusiform (Fig. 65), smooth, olive-buff spore print.
Several to numerous under aspen and, in particular, birches; widely distributed in S. and F. after wet cool periods.
The comparatively graceful thin stalk, brown cap, clean margin, and nonstaining flesh, along with the presence of birch, are distinguishing characteristics. Research by A. H. Smith, H. Thiers, and R. Watling (1966, 1967) has revealed many closely related species. My wife and I have eaten many of these (they are all edible) and found them good to choice. *Leccinum albellum* (Pk.) Singer looks like these but has a white cap.

297 LECCINUM AURANTIACUM (Fr.) S. F. Gray   Edible, good/Common

Cap dry, rusty-red, large; flesh white, bruises red and then blue

Cap 5–20 cm broad, robust, convex to broadly convex, flattened hairs to minutely woolly, dry to sticky in wet weather, bright rusty red to orange-red, red to reddish brown, at first with marginal veil tissue which remains in patches. Flesh firm, thick, white, bruises *wine-red with gray or faded blue-gray overtones.* Tubes 1–2 cm deep, mouths round, small (3–5 per mm), depressed at stalk, white, olive-gray to olive-brown in age or when bruised, narrow sterile margin. Stalk 10–20 cm long, 20–50 mm thick, robust, enlarging to a wide base or expanding noticeably in the center, dry, pallid to white and covered with brown (in age black) scabers (tufts) almost to the base. Veil remains only as marginal tissue on cap.

Spores 13–18 x 3.5–5 μ subfusiform, smooth, yellow-brown spore print.

Scattered to numerous usually under aspen and pines; widely distributed in S. and F.

This fungus is typically more robust than *L. scabrum* and has a larger, chunky stalk, a red cap, and red-staining flesh when bruised. There are many closely related species (see notes following *L. scabrum*).

298 LECCINUM INSIGNE Smith, Thiers, and Watling   **Edible, good**/Common

Cap dry, orange-cinnamon with olive tints; flesh white staining red then blue

Cap 4–15 cm broad, convex to broadly convex, dry to sticky when wet, orange-cinnamon sometimes with olive tones, margin exceeds tubes and results in hanging pieces of tissue. Flesh firm, white *staining to reddish gray* when bruised, then to bluish gray, resulting in a livid stain. Tubes 15–20 mm deep, mouths 2–4 per mm, depressed at stalk, whitish at first, soon pale olive to sometimes wood-brown. Stalk 8–12 cm long, 10–30 mm thick, robust, enlarging toward base, dry, whitish with dense brown scabers (tufts) which blacken at maturity.

Spores 13–16 x 4–5.5 μ subfusiform (Fig. 65), smooth, dark yellow-brown spore print.

Common and often abundant under pure stands of aspen, often mixed with conifers (jack pine, lodgepole pine, and white spruce) in Sp. and early S. in CS, but in S. in PNW and A.

This species is included here because it is the large, abundant *Leccinum* of the North; I have seen massive fruitings of it in the southwest Yukon Territory of Canada, Glacier Park in Montana, and elsewhere.

GYROPORUS

The genus is distinguished from other boletes by the yellow spore print. The cap is always dry and the stipe lacks reticulations (netlike ridges), glandular dots, and scabers (tufts).

Edibility. I have not found a sufficient quantity to make a meal of either of these boletes.

Key to Gyroporus

  1 a. Cap dry, white; stalk white; flesh stains blue, *299. G. cyanescens*
    b. Cap dry, brown; stalk carrot color; flesh white, bruises brown, *300. G. castaneus*

299 GYROPORUS CYANESCENS (Bull. ex Fr.) Quel.   **Edible, choice**/Infrequent

Cap dry, white; tubes white; stalk uneven, white; flesh stains blue

Cap 4.5–12 cm broad, broadly convex, cottony with small scales or hairy patches, dry, margin incurved at first, whitish to yellowish. Flesh white but instantly blue when cut. Tubes 3–7 mm deep, mouths small, depressed at stalk, white at first to yellow in age, turns blue instantly when bruised. Stalk 5–10 cm long, 15–35 mm thick, irregularly swollen, crooked, same color as cap, deep blue when bruised, felted, hollow with a soft pith. Veil absent.

Spores 7–11 x 4–6 μ oblong–elliptical, smooth, pale yellow spore print.

Single or scattered on gravel or sandy soil under hardwoods, sometimes under mixed hardwood-conifers in eastern North America in S. and early F.

The whitish sporocarp (spore-forming body) with its cottony cap, instantly bruising blue everywhere, and the very irregularly swollen stalk make this one of the most distinctive boletes. I have not eaten it but I have seen good fruitings of it in northern New York.

300 GYROPORUS CASTANEUS (Bull. ex Fr.) Quel.   **Edible, good**/Common

Cap dry, brown; tubes white to buff; stalk uneven, orange; flesh bruises brown

Cap 3–8 cm broad, convex to flat in age, minutely woolly, dry, orange-gray to rusty red or chestnut-brown. Flesh of cap white, bruising brown, stalk interior hollow or filled with a cottony, fragile pitch. Tubes 3–7 mm deep, mouths 1–3 per mm, depressed and not attached to stalk, white to light yellow in age. Stalk 3–8 cm long, 6–22 mm thick, enlarged somewhat toward base, dry, uneven with a band of pure white beneath tubes, the rest is carrot color, light orange to reddish brown. Veil absent.

Spores 7–11 x 4.5–6.0 μ oblong-elliptical, smooth, light yellow spore print.

Single to several under hardwoods from S. to F.; widely distributed.

The very uneven, hollow stalk resembles a common carrot root. I have not found enough at one time to make a meal.

BOLETUS

The pileus is dry and not viscid to glutinous. The tubes are not radially arranged. The stipe is smooth or has a netlike (reticulate) pattern of fine ridges but does not have scabers (tufts) as in *Leccinum* or the glandular dots often encountered in *Suillus* and *Fuscoboletinus*. The spores are smooth, elliptical, and olive-brown to yellow-brown. The species are found on the ground and most species, in all likelihood, form mycorrhiza with various trees, both hardwood and confier.

Edibility.   Species with red tube mouths or those that have flesh or pores that bruise blue should be avoided (see discussion under Boletaceae). Some of the truly superb edibles are found in this genus and the species can be safely eaten, even by the beginner, as long as the above rule is observed. *Boletus edulis* (*Steinpiltz* in Germany, cépe in France) is a choice edible. We have collected and enjoyed it in quantity in central Idaho and southeastern Alaska. It is best served with "German White Sauce" (see Recipes). In our field camp we found it delicious with eggs and in soup. It is usually the mushroom illustrated on the packages of imported soup mixes.

170

Key to Boletus

1 a. Fruiting body flesh and/or tubes staining blue when cut or bruised (immediately or after several minutes), **8**

 b. Fruiting body flesh and/or tubes do not stain blue when cut or bruised (brown, red, or dingy yellow stains may be observed), **2**

2 a. Stalk reticulate (netlike) over *one-third of length* or more, **7**

 b. Stalk with coarse hairs to smooth but not reticulate, **3**

3 a. Growing on a puffball (usually *Scleroderma* sp., *301. B. parasiticus*

 b. Not growing on a puffball, **4**

4 a. Tube mouths red-brown; taste very peppery, *302. B. piperatus*

 b. Tube mouths another color, **5**

5 a. Cap and stalk striking pink, chrome-yellow just at stalk base, *303. B. chromapes*

 b. Cap brown, reddish brown, or brown with white spots, **6**

6 a. On conifer stumps and logs in western North America; red-brown tomentose (woolly) pileus, *304. B. mirabilis*

 b. On ground in eastern North America; brown, hairless cap ($\pm$ white spots), *305. B. affinis*

7 a. Stalk tall and thin, nearly equal with deep, shaggy to ornamented reticulations, *306. Boletellus russellii* and related species

 b. Stalk bulbous or clavate (club-shaped) with shallow, clearly formed, initially white reticulations, *307. B. edulis* and related species

8 a. Cap often not expanded fully; tubes contorted, mouths often closed, no spore print possible, yellow, red to orange-red, bruising dark blue; western North America under conifers, *277. Gastroboletus turbinatus*

 b. Cap expanding fully; tubes vertical, mouths open at maturity, yields spore print; flesh bruises blue, **9**

9 a. Tube mouths red to orange-red; ⅓ or more of stalk reticulate, **10**

 b. Tube mouths not red; stalk smooth or reticulate, **13**

10 a. Cap 10–20 cm broad; stalk club-shaped to bulbous; known only on West Coast in North America, **11**

 b. Cap 6–12 cm broad; stalk nearly equal; known only from eastern North America, **12**

11 a. Cap 10–20 cm broad, pale gray with pinkish hue; stalk bulbous; under oaks along West Coast, *308. B. satanus*

 b. Cap 10–17 cm broad, dark brown; stalk club-shaped or equal; under mixed woods of conifer and hardwood along West Coast, *309. B. eastwoodiae*

12 a. Cap blood-red; tube mouths deep red; stalk deep red and lacerate-reticulate, *310. B. frostii*

 b. Cap marbled brick-red to olive-brown; tube mouths orange-red to orange-yellow; stalk nearly smooth with fine reticulations just at apex, *311. B. luridus*

13 a. Stalk reticulate; yellow sometimes flushed red; PNW, RM to A, along coast, *312. B. calopus* var. *frustosus*

 b. Stalk smooth, **14**

14 a. Cap brown with conspicuous cracks revealing pink flesh, *313. B. chrysenteron*

 b. Cap variously colored, without cracks, **15**

15 a. Stalk deep maroon-red with a conspicuous yellow band at the apex; known only from western North America (see 313. *B. chrysenteron*), *B. smithii*

 b. Not as described above, **16**

16 a. Tube mouths bright yellow; flesh and tubes bruise dark blue, *314. B. bicolor*
   b. Tube mouths white to greenish yellow or gray; flesh and tubes slowly bruising light blue to blue-green, **17**

17 a. Tube mouths white to gray; cap gray to buffy gray, *315. B. pallidus*
   b. Tube mouths light greenish yellow, **18**

18 a. Cap red-brown, viscid; only in eastern North America, *316. B. badius*
   b. Cap blackish brown, dry; only in western North America, *317. B. zelleri*

301 BOLETUS PARASITICUS Fr.

Nonpoisonous/Infrequent

Cap dry, yellow-brown, parasitic on a puffball

Cap 4–7 cm broad, convex, dry, felted, sometimes cracked, yellow-brown to olive-gray. Flesh white, yellowish when bruised. Tubes 5–10 mm deep, mouths large, extend down stalk, yellow to olive in age. Stalk 2.5–10 cm long, 5–20 mm thick, equal, brownish yellow, nearly hairless. Veil absent. Taste bitter.

Spores 12–18 x 4–5.5 μ nearly elliptical, smooth, dark olive-brown spore print.

Usually one to three fruiting bodies on each *Scleroderma* (a puffball), on which this curious bolete grows. It has been reported from widely separated areas in the Northern Hemisphere, usually in S. and early F. Also known as *Xerocomus parasiticus* (Bull. ex Fr.) Quel.

302 BOLETUS PIPERATUS Fr.

Nonpoisonous/Common

Cap viscid, dull red; tubes deep red; taste very peppery

Cap 2–8 cm broad, convex, hairy or felted on margin, rest hairless, viscid when wet, usually dull red, tawny to cinnamon. Flesh firm, buff, pink near tubes. Tubes 3–5 mm deep, mouths angular 1–3 per mm, attached or short decurrent (extend down stalk), radial near stalk, *deep red* to *rusty red*. Stalk 2–8 cm long, 4–8 mm thick, thickest at apex, narrowing gradually toward base, felted to hairless, dry, yellowish to rusty brown. Veil absent. Taste *very peppery*.

Spores 7–10 x 3–4 μ short elliptical, smooth, cinnamon to brown spore print.

Solitary to several, rarely numerous under various conifers in S. and F.; widely distributed, common under Sitka spruce in Alaska. Its distinctive peppery taste and dull red cap with rusty red pores distinguish it from other boletes. Some authors have included this species in the genus *Suillus*.

303 BOLETUS CHROMAPES Frost

Edible/Numerous

Cap, tubes, and stalk are pink to rose color; stalk base yellow

Cap 4–12 cm broad, convex, dry to sticky, felty, striking *pink to rose* overall, in old age tan. Flesh white, pink below cap cuticle. Tubes 5–8 mm deep, mouths small (2–3 per mm), depressed at stalk, white then pink to pale brown in age. Stalk 6–14 cm long, 10–20 mm thick, often curved nearly equal, covered with fine dots or even coarse short pinkish scabers (dark tufts of hairs) at times, dry, pinkish with a striking chrome-yellow base surrounded by yellow mycelium. Veil absent.

Spores 10–15 x 4–5 μ nearly elliptical, smooth, pinkish brown spore print.

Scattered to numerous under a wide variety of hardwoods and conifers in SE, NE, CS, and adjacent C, in Sp., S., and F.

172

The striking pink color and chrome yellow stalk base surrounded by yellow mycelium make this fungus easily recognized. It has been placed in *Tylopilus* by A. H. Smith and H. Thiers (1971) where it probably belongs, but because of the short scabers (dark, hairy tufts) on the stipe it has also been put in *Leccinum* (Snell and Dick, 1970). I have encountered good fruitings of it in northern Michigan.

304 BOLETUS MIRABILIS Murrill   Edible, choice/Common

Cap large, brown, woolly; tubes yellow; on wood of western conifers

Cap 5–16 cm broad, convex, densely woolly to hairy, dry, dark red-brown. Flesh firm, lemon-yellow, sometimes reddish when bruised. Tubes deep (10–30 mm), mouths angular 0.5–1 mm broad, depressed just at stalk, bright sulphur-yellow, mustard-yellow in age. Stalk 12–22 cm long, 2–8 cm thick, club-shaped to bulbous, widest near base; variable, large, white reticulations which darken in age over apex or entire surface in some, whitish just at base. Veil absent.

Spores 14–24 x 6.5–9 μ nearly elliptical, smooth, olive-brown spore print.

Common on conifer logs and stumps (noted on western hemlock) in PNW, RM, A, and rarely in CS, fruiting in S. and F.

The red-brown, hairy cap and striking yellow pores combined with the unusual occurrence of a bolete on conifer wood are distinctive field characteristics.

305 BOLETUS AFFINIS Pk.   Edible/Numerous

Cap dry, olive-brown; tubes white young; taste of fresh meal

Cap 5–11 cm broad, convex, dry, hairless, but sticky when wet, olive-brown, rich yellowish brown. Flesh soft, white, slightly pinkish in stipe. Tubes 5–15 mm deep, mouths small (2–3 per mm), adnate or depressed at stalk, white to dingy yellowish in age. Stalk 4–8 cm long, 10–25 mm thick, nearly equal or enlarged at apex, no reticulations or a few just at apex of stalk, white to pinkish streaked with brown. Veil absent. Taste of meal.

Spores 8.5–16 x 3–4 μ elliptical, smooth, bright rusty-red spore print.

Scattered to numerous under hardwoods in CS, NE, and adjacent C and the SE, fruiting in S. and F.

A variety, *B. affinis* Peck var. *maculosus* Peck, has conspicuous white spots over the cap.

306 BOLETELLUS RUSSELLII (Frost) Gilbert   Edible/Rare

Cap dry, brown, scaly; stalk glutinous, reticulate-lacerate

Cap 2.5–9 cm broad, convex, dry, felted, in raised tufts or scales, brown to yellowish brown. Flesh thin, pale yellow, unchanging when bruised. Tubes 6–8 mm deep, mouths large, angular, buff to greenish yellow in age. Stalk 5–20 cm long, 10–25 mm thick, slender with coarse, shaggy reticulations, *glutinous and swelling* in wet weather, flesh color to reddish brown. Veil absent.

Spores 13–17 x 7–10 μ elliptical, strongly longitudinally ridged, olive-brown spore print.

Alone or in small groups under hardwoods in Sp., S., and early F.; in the CS, NE, and SE. I have collected it in northern Michigan under aspen. This long slender bolete resembles *Boletellus betula* (Schw.) Gilbert, but the latter has a viscid, yellow to red cap, bright yellow upper stalk and warted spores. It is common in North Carolina north to Maryland. However, both are reported to be edible and good. A third tall, slender, infrequently found, eastern bolete is *Boletus projectellus* Murr. It has long, ornamented reticulations (netlike ridges), grayish brown to yellow-brown cap and stalk, and large, thick-walled, *smooth* spores (20–32 x 6–12 μ). I have found it under Virginia pine in Maryland.

307 BOLETUS EDULIS Bull. ex Fr. Steinpilz   Edible, choice/Common

Cap large, dry, brown; stalk robust with white reticulations

Cap 8–25 cm broad, broadly convex, dry, moist to slippery feeling or viscid in wet weather, hairless but often uneven crust, brown, cinnamon to red-brown. Flesh firm, white sometimes reddish just under the cap, unchanging when bruised. Tubes 10–40 mm deep, mouths small (2–3 per mm), depressed just at stalk, white at first, slowly yellow, yellowish olive to olive-brown in age. Stalk 10–20 cm long, 2–5 cm thick, equal, enlarged at the base to bulbous, dry, white reticulations overall or just at apex of stalk, whitish to yellow or yellow-brown. Veil absent.

Spores 12–21 x 4–6 μ nearly elliptical, thin-walled, olive-brown spore print.

Solitary to scattered under conifers, or mixed conifer-hardwoods, or hardwoods (see below), fruiting from Sp. to S. and F. in RM, to typically S. and F. on Alaskan coast and elsewhere in N.A. In A. under pure western hemlock and other conifers.

The shape of the stalk and color of the cap vary a great deal throughout North America. Generally the forms with a very bulbous stalk are found in the RM, PNW, and A. The equal or club-shaped stalk is most often encountered in areas east of the RM. I have found this form in late S. and early F. in New York and New England. A form with a very brown or even light brown cap is often seen in the RM. Regardless of shape and color there is no blue staining of any tissue and the tubes and reticulations are white at first. The following species with a nonviscid cap have been described mostly based on cap color and different tree associations: *B. nobilis* Pk., *B. variipes* Pk., *B. atkinsonii* Pk., and *B. gertrudiae* Peck. *Boletus separans* Peck has lilac or purplish coloration. These closely related species are edible. Sometimes reticulations are found in the upper stalk of *B. mirabilis,* and this species should be checked out in the key.

308 BOLETUS SATANUS Lena   Poisonous/Rare

Cap large, dry; tubes greenish yellow to deep red mature, bruising blue

Cap 10–20 cm broad, convex, robust, dry, pale olive to olive-buff. Flesh very thick (2–5 cm), pale olive, bruising blue at once when young. Tubes 1–2 cm deep, mouths small (2 per mm or less), attached to stalk or shallow depressed, *greenish yellow at first,* deep blood-red at maturity, blue when bruised. Stalk 6–12 cm long, 3.5–7 cm thick, conspicuously bulbous, dry, reticulate, pink over bulbous portion, rest colored like cap. Veil absent.

Spores 12–15 x 4–6 μ elliptical, smooth, olive-brown spore print.

Known only from the West Coast under oak, but widely reported in Europe.

309 BOLETUS EASTWOODIAE (Murr.) Sacc. and Trott.   Poisonous/Infrequent

Cap large, dry, red-brown; tubes dark red young and old, bruising blue

Cap 10–17 cm broad, convex, robust, dry, minutely woolly, dark brown to dark red-brown, sometimes aerolate (finely cracked), margin incurved and light buff at first. Flesh very thick (up to 4 cm), yellow, instantly bruising blue. Tubes 5–10 mm deep, mouths small (less than 1 mm), adnate to depressed at stalk, *deep blood-red at all ages,* instantly blue when bruised. Stalk 8–15 cm long, 2–5 cm thick, equal, club-shaped to bulbous, dry, yellow to yellow-orange with deep red reticulations, bruises blue. Veil absent.

Spores 13–16 x 5–6.5 μ elliptical, smooth, olive-brown spore print.

Widely distributed under mixed conifer-hardwood forests only along the West Coast where it fruits in F.

The dark brown cap, blood-red tube mouths when young, and the stalk of different shape and color separate *B. eastwoodiae* from *B. satanus.*

310 BOLETUS FROSTII Russell   Nonpoisonous/Infrequent

Cap, tubes, and stalk blood-red; bruising blue in all parts

Cap 6–15 cm broad, convex to flat in age, smooth, viscid, entirely *blood-red.* Flesh firm yellowish, bruising blue. Tubes 7–10 mm deep, mouths round, 2–3 per mm, adnate or depressed at stalk, deep red, *exuding yellow drops when young,* bruising dingy blue. Stalk 4–11 cm long, 10–30 mm thick, enlarging downward, dry, *blood-red* with *lacerate reticulations,* occasionally yellow just at base. Veil absent.

Spores 12–17 x 4–6 μ oblong-elliptical, smooth, yellow-brown spore print with olive tint.

Scattered to several in oak woods in CS, NE, SE, and So. Fruiting in S. and early F. The deep red color overall and deep stalk reticulations combined with the curious yellow drops on the young pore mouths are distinctive field characteristics. This is one of the distinctive species in a group of confusing red-tubed species which bruise blue. Some are definitely not good to eat, and a thorough knowledge of the species is recommended before they are collected for the table.

311 BOLETUS LURIDUS Schaeff. ex Fr.   Poisonous/Infrequent

Cap multicolored; tubes yellow to red; stalk orange-red, reticulate, bruises blue

Cap 8–20 cm broad, convex, moist to viscid when wet, felty, color extremely variable with *mixtures of olive, yellow, red, and brown,* slowly blue when handled or bruised. Flesh whitish, but blue when bruised. Tubes 8–15 mm deep, mouths 2–3 per mm, depressed at stalk and nearly free, *yellow to olive with shadings of orange and red,* blue where bruised. Stalk 4–15 cm long, 2–3 cm thick, nearly equal, with netlike ridges, *yellow infused with orange and with red,* mostly over lower part, turning blue where touched. Veil absent.

Spores 9.5–17 x 4–7 μ elliptical, smooth, olive-brown spore print.

Infrequent; in mixed hardwood-conifer forests in CS, NE, and SE. Fruiting in S. and F.

The reticulate-stalked *B. speciosus* Frost is closely related to the above, reddish overall, and does not have the display of different tints of color on the cap and stalk which is so typical of *B. luridus.* There are reports of poisoning attributed to this species in North America, so it should be avoided. *B. subvelutipes* Peck is a confusing species with the stalk sometimes described as reticulate (Snell and Dick, 1970) and sometimes not. The tube mouths are deep red and stain blue when bruised. It should be considered in toxins Group V.

312 BOLETUS CALOPUS var. *frustosus*   Poisonous/Infrequent
(Snell and Dick) O. K. Miller and R. Watling

Cap large, dry, brown; tubes yellow; stalk reticulate; flesh bruising blue; western N.A.

Cap 5–25 cm broad, convex, robust, dry, brown to dark brown, finely cracked (aerolate) in age, bruising blue. Flesh firm, pale yellow, bruising instantly blue. Tubes 5–20

175

mm deep, mouths small (2–3 per mm), round, depressed at stalk, lemon-yellow with a greenish cast, instantly bruising blue *when mature.* Stalk 7–9 cm long, 3.5–5 cm thick, nearly equal or swollen in center, reticulate at maturity, straw-yellow with red stains at the base. Veil absent. Taste bitter.

Spores 10–14 x 4–5 μ nearly elliptical, smooth, ochre-brown spore print.

Alone or scattered under western conifers (esp. fir and hemlock) in PNW and A. Fruiting in late S. or early F.

313 BOLETUS CHRYSENTERON Fr.   Edible/Numerous

Cap dry, brown with deep cracks showing pink flesh; flesh bruising slowly blue

Cap 3–11 cm broad, convex to flat in age, dry, woolly, finely cracked (aerolate) to deeply cracked (rimose) in age, olive-brown to red just along margin with pink flesh showing in the cracks. Flesh white to red stained near cuticle of cap, slowly blue or green when bruised. Tubes 4–8 mm deep, mouths angular, 1–2 per mm, attached or slightly extending down stalk, pale yellow, dingy brown with greenish blue when bruised. Stalk 3–9 cm long, 3–16 mm thick, nearly equal, often curved, dry, not reticulate, yellow sometimes with red streaks below, sometimes blue when bruised. Veil absent.

Spores 9–15 x 4–5 μ nearly elliptical, smooth, olive-brown spore print.

Occurs alone, scattered to several, under hardwoods in eastern N.A. in Sp., S., and F.

The pink flesh in the cracks, red stains on the flesh, and blue color on bruised tubes distinguish it from *B. subtomentosus* Fries. Because there are conflicting reports of the edibility of *B. chrysenteron* and closely related species, it is not recommended. A green mold is often seen on the fruiting body, and this is further reason to avoid it. *Xerocomus chrysenteron* (Bull. ex St. Amans) Quel. is another name for this fungus. *Boletus smithii* Thiers has a maroon-red stalk with a conspicuous yellow apex and is found only in western North America.

314 BOLETUS BICOLOR Pk.   Edible, good/Numerous

Cap dry, red; red smooth stalk; flesh yellow staining blue

Cap 8–28 cm broad, convex to flat in age, dry, perhaps sticky in wet weather, hairless to finely woolly, red to purple-red sometimes stained yellow. Flesh yellow staining blue when bruised but fading again to yellow. Tubes 5–14 mm deep, mouths seldom up to 1 mm broad, usually adnate, bright yellow to olive in age, bruising blue. Stalk 4–15 cm long, 7–30 (sometimes 60) mm thick, nearly equal, even, hairless, if reticulate it is so only at apex which is yellow, the rest is deep red, slowly bruising blue. Veil absent. Slight radish odor in some.

Spores 8.5–12 x 3.5–4.5 μ narrowly elliptical, smooth; olive spore print.

In small numbers under hardwood (especially oaks), late S. or early F., in CS, NE, SE, and So.

The red cap and nonreticulate red stalk combined with the blue staining when bruised are the identifying characteristics. *Boletus speciosus* Frost is similar but has a reticulate, largely yellow stalk and is also edible. *B. miniatoolivaceus* Frost has more yellow and even olive tints in the cap, a yellow nonreticulate, smooth stalk with red stains, and a much faster change to blue when bruised. It is also considered poisonous (see Group V) and should be avoided. Two small species (*B. fraternus* Peck and *B. rubeus* Frost) are often found in lawns and grass in eastern North America. Both have dark red caps, yellow pores, and flesh rapidly staining blue (see Snell and Dick, 1970).

315 BOLETUS PALLIDUS Frost   Edible/Common

Cap dry, gray; tubes white tinted gray; flesh white staining slowly blue

Cap 5–15 cm broad, convex, smooth, felted, dry, gray to buffy gray. Flesh white, light blue when bruised in some. Tubes 10–15 mm deep, mouths angular 1–2 per mm, adnate or depressed at stalk, white to olive-gray in age, staining blue when bruised. Stalk 8–15 cm long, 7–25 mm thick, enlarged slightly toward base, dry, white to light grayish brown in age, lightly reticulated over upper half. Veil absent. Taste faintly bitter.

Spores 9–12 x 3.5–5 μ elliptical, narrowed at apex, smooth, olive spore print.

Scattered under oak and other hardwoods in S. and F. throughout CS, NE, C, SE and So. The pallid coloration and slow to weak blue staining are good field characters.

316 BOLETUS BADIUS Fr.   Edible/Numerous to infrequent

Cap red-brown, viscid; tubes greenish yellow bruising blue; stalk reticulate

Cap 3.5–11 cm broad, convex to flat at maturity, minutely felted, viscid in wet weather, red-brown, dark brown sometimes olive. Flesh white to light pink or yellowish, sometimes bruising light blue. Tubes 7–15 mm deep, mouths 0.5–1 mm broad, attached or somewhat depressed at stalk, light greenish yellow, dull blue-green when bruised. Stalk 4–10 cm long, 8–25 mm thick, equal or enlarged toward base, dry, finely powdered, reddish brown over a yellow ground color, reticulate and whitish just at apex Veil absent.

Spores 10–18 x 3.5–5 μ narrowly elliptical, smooth, olive to olive-brown spore print.

Single to scattered under conifers, especially pine, in S. and F. Common in NE and CS but uncommon in NE, C, SE, So., and unknown elsewhere.

The red-brown cap and stalk, viscid cap in wet weather combined with the greenish yellow tubes and very slight blue color when bruised are a distinctive combination of characters. Also known as *Xerocomus badius* (Fr.) Kuehn. ex Gilbert.

317 BOLETUS ZELLERI Murr.   Edibility unknown/Numerous

Cap black-brown, dry; stalk red; flesh slowly blue

Cap 4–12 cm broad, convex to flat in age, dry, finely powdered at first, deep brown, blackish brown to dark chestnut-brown. Flesh firm, buff to yellow, slow change to blue when bruised or not at all. Tubes 5–15 mm deep, mouths large (1–2 mm broad), depressed at stalk, yellow, olive-yellow in age. Stalk 5–10 cm long, 1–4 cm broad, equal, dry, hairless, red to brownish red may be yellow just at base. Veil absent.

Spores 9–15 x 4–6 μ elliptical, smooth, olive-brown spore print.

Scattered under western conifers, in PNW in F.

The deep brownish black caps and deep red stalk combined with the flesh very slowly bruising blue are distinctive characteristics. Another name is *Xercomus zelleri* (Murr.) Snell.

POLYPORACEAE
The Polypores

The polypores are the most obvious group of fungi in the woods of North America during all seasons of the year. The perennial fruiting bodies, or "conks" (Fig. 1) as they are commonly called, may be seen season after season on old or dying trees. The abundant imbricate "shelflike" (Fig. 2) fruiting bodies of some species can literally cover stumps, old logs, and snags. These annual species usually grow during the summer months and reach maturity in the fall. All of the polypores have minute to large tubes. The spore-bearing surface (pore surface) is located on the underside of the pileus (Figs. 1–2). When examined with a hand lens, only the exterior opening or tube mouth is visible. In each species these tube mouths, or pores, are a definite size and shape (Overholts, 1953).

Only one other group has united tubes and these are the boletes. It is important to remember that the boletes are fleshy, with central stipes and resemble typical gilled mushrooms. Polypores, on the other hand, often have lateral stipes or no stipe at all. They range from woody to leathery or fleshy. If they are fleshy, they seldom have a central stipe and therefore do not resemble a gilled mushroom. Polypores are very important as saprophytes on dead logs, limbs, and stumps. Some, such as *Fomes pini,* are important saprophytes on the dead vascular tissue in the center of living trees and are called heart-rotting fungi. A few, such as *Fomes annosus,* can become deadly parasites and attack living root cells of forest trees. In some areas widespread mortality of forest and shade trees has resulted from these parasitic root rots. Very few polypores are not actually found in wood, and the role of these species has not been determined as yet. There are many species of polypores; only the most common or edible are included here.

Edibility.   Most polypores are too leathery or woody to be edible and others are tasteless or bitter. I know of no poisonous species.

Several species are excellent edibles. Perhaps the two outstanding edibles are *Polyporus frondosus* and *Polyporus umbellatus;* I consider them excellent. *Polyporus sulphureus* and *P. squamosus* are edible when young, but I would rate them below the former two species. *Fistulina hepatica* is also good but not a true polypore. I know of no other polypores rated as edible and commonly eaten in North America.

Polyporus frondosus is usually very large. We use the caps and upper branches of older fruiting bodies, but almost all of the younger ones can be eaten. It is very good in any recipe, but we like it best in "Mushroom Soup" and "Mushroom Loaf" (see Recipes).

Key to Fistulina and the Polypores

1 a. Spore-bearing surface composed of free tubes, *318. Fistulina hepatica*
   b. Spore-bearing surface composed of pores (united tubes with flesh between, **2**
2 a. Fruiting body forms large, woody, perennial conks (Fig. 1), hoof-shaped to shelflike; to check age, cut in two and count the number of layers of tubes, **3**
   b. Not as above, **11**
3 a. Cap surface smooth, appearing shellacked; shelflike fruiting body, **4**
   b. Cap surface, rough, uneven or not appearing shellacked; hooflike to shelflike fruiting body, **6**
4 a. Cap surface red to orange with white to brown flesh, **5**
   b. Cap surface gray-brown; pores bruising brown when scratched, *319. Ganoderma applanatum*

178

5 a. Cap red; flesh white, *320. Ganoderma tsugae*

    b. Cap orange-red; flesh light brown (see 320. *G. tsugae*), *G. lucidum*

6 a. Cap surface smooth, not cracked and rough, **7**

    b. Cap surface rough and cracked, **8**

7 a. Cap surface dull gray-brown to gray, hoof-shaped (see 323. *F. pini*), *F. fomentarius*

    b. Cap surface brown, usually shelflike or nearly flat with a small upturned cap; pores white, *321. F. annosus*

8 a. Cap with a conspicuous red band near the margin, *322. F. pinicola*

    b. Cap without any red band, roughly to deeply cracked, **9**

9 a. Cap surface rough, brown; on conifer wood, *323. F. pini*

    b. Cap surface deeply cracked; on hardwoods, **10**

10 a. Cap surface deeply cracked, blackish brown; on locust (see 323. *F. pini*), *F. rimosus*

    b. Cap surface deeply cracked, grayish black to black on many different hardwoods (see 323. *F. pini*), *F. igniarius*

11 a. Stalked, single, or several caps, sometimes very large, **12**

    b. Convex cap without any stalk; often imbricate (overlapping) and shelflike, **18**

12 a. Short lateral stalk with angular to hexagonal pores *324. Favolus alveolaris* and *Polyporus squamosus*

    b. Central well-developed stalk with one or several caps, **13**

13 a. Single cap from a central stalk, **14**

    b. Multiple caps from a central stalk, **16**

14 a. Cap tan; stalk with a black foot; on hardwood (see 325. *P. arcularius*), *P. elegans*

    b. Cap dark brown; stalk yellow-brown, **15**

15 a. Cap margin lined with hairs; pores large (1–2 mm wide), *325. P. arcularius*

    b. Cap margin smooth; pores small (2–3 per mm) (see 325. *P. arcularius*), *P. brumalis*

16 a. Sulphur-yellow, glabrous (hairless), *326. P. sulphureus*

    b. Another color, **17**

17 a. Many separate, gray-brown caps on white stalks; white pores, *327. P. frondosus*

    b. Fused rusty brown, hairy caps; greenish yellow pores, *328. P. schweinitzii*

18 a. Pores lilac to purplish; usually in densely overlapping clusters, **19**

    b. Pores another color; single to overlapping (if sulphur-yellow, see 326. *P. sulphureus*), **20**

19 a. On hardwood logs and stumps (see 331. *P. versicolor*), *P. pargamenus*

    b. On conifer logs and stumps (see 331. *P. versicolor*), *P. abietinus*

20 a. Pores very large and irregular (1–3 mm wide) or arranged in gill-like fashion, gills, **21**

    b. Pores round, minute, pure white or smoke-gray, **22**

21 a. Cap densely hairy, rusty red to yellow-brown; pores gill-like to irregular, *329. Lenzites saepiaria*

    b. Cap convex, ash-gray, fine short hairs; pores large (1–3 mm wide) and irregular, *330. Daedalea quercina*

22 a. Cap with multicolored red, yellow, and blue zones; pores minute, pure white; in dense overlapping clusters on hardwood, *331. P. versicolor*

    b. Cap without zones or with dull brown zones; pores white to gray, **23**

23 a. Cap hairy, yellow-brown to gray-brown; pores white (see 331. *P. versicolor*), *P. hirsutus*

    b. Cap minutely hairy, smoke-gray; pores gray to smoky gray (see 331. *P. versicolor*), *P. adustus*

318 FISTULINA HEPATICA Schaeff. ex Fr.   Edible, good/Common

["Beefsteak Fungus"]

Lateral stalk; flesh oozing red juice; free tubes; on oaks

Fruiting body large, cap 10–30 cm broad, nearly flat, minutely roughened, moist, red-orange to liver-colored. Flesh 2–6 cm thick, soft to fibrous in age, zones white to reddish oozing a reddish juice. Stalk lateral, 4–8 cm long, 1–3 cm thick, colored same as cap. Tubes separate from each other (see color photo), similar to individual pipes, 10–15 mm long, white to buff, staining reddish when handled.
Spores 4–5.5 x 3–4 μ egg-shaped, smooth, pale rusty brown spore print.
On hardwood stumps or living trees at base, usually oak, in F.; widely distributed.
The texture of beef and the oozing red juice give it its name. It is a good edible and sought by many mushroom hunters.

319 GANODERMA APPLANATUM   Edibility unknown/Common

(Pers. ex Wallr.) Pat. ["Artist's Conk"]

Cap gray-brown, shellacked, shelflike, woody; pores minute, white, bruising brown

Fruiting body assumes a thin shelflike form, called a conk, on logs and stumps with caps up to 30 cm broad, gray to gray-brown appearing shellacked, hard, smooth with some cracks. Flesh soft, corky, brown. Pores minute, 4–6 per mm, bright white but *bruising brown* when scratched.
Spores 6–9 x 4.5–6 μ egg-shaped, truncate at one end, finely spined, brown spore print.
Solitary to several on logs and stumps of hardwoods or occasionally conifers, but also often on wounds of living trees; perennial, widely distributed.
The pore surface has often been used to etch pictures, hence the nickname "Artists' Conk." The fungus produces a white rot in heartwood and sapwood, and the infected trees sometimes blow over at the point of the rot.

320 GANODERMA TSUGAE Murr.   Edibility unknown/Common

Cap red, shellacked, shelflike, woody; pores white, do not bruise

Fruiting body forms a shelflike, stalkless or short-stalked conk on stumps and logs. Caps 5–25 cm broad, hairless, *shellacked, red to reddish orange.* Flesh tough, punky when dry, white. Pores 4–6 per mm, white to brown in age. Stalk lateral when present, 3–15 cm long, 1–4 cm thick, also shellacked and reddish orange.
Spores 9–11 x 6–8 μ egg-shaped, apex truncate, thick-walled with pores through the wall, white spore print.
Single to several on or near stumps and logs of conifers or less frequently on hardwoods in S. and F.; in SE, CS, NE, and adjacent C. In New England it is occasionally found on beech, where it produces a thin stalk.
G. lucidum (Fr.) Karst. is similar but found on hardwoods and has a light brownish context. *Ganoderma oregonense* Murrill is often very large (up to 80 cm broad) with large spores 10–16 x 7.5–9 μ. It is found on conifer logs in the PSW, PNW, and A.

180

321 FOMES ANNOSUS (Fr.) Cooke   **Nonpoisonous**/Common

Cap often reduced or slightly upturned, brown; pores white, 2–4 per mm; on wood in duff

Fruiting body stalkless, nearly flat with a small upturned or shelflike cap, 4–15 cm broad, brown, dry, margin brown and rounded. Flesh tough, white. Pores nearly round, 2–4 per mm, white to yellowish in age.

Spores 3.5–5 x 3–4 μ nearly round, thin-walled, white spore print.

One to several, at or under the duff on stumps, logs or on living conifers and hardwoods. Infection centers often create areas with many dead trees.

This is one of the most destructive root diseases of forest and shade trees. Infection centers often develop in twenty- to forty-year-old plantations of red pine and white pine in the NE. and C. Southern pines are also very susceptible and the fungus is often found on ponderosa pine and other conifers in the western United States. It may exist both as a saprophyte and parasite. The infected roots are reduced to a white spongy pulp.

322 FOMES PINICOLA (Swartz ex Fr.) Cooke   **Edibility unknown**/Common

["Red Belt Fungus"]

Cap hard, convex, with a red marginal zone; no stalk; pores small white

Fruiting body a convex conk 5–40 cm broad with a hard, somewhat zoned surface, typically black-brown with a *red zone or belt near the margin* which has a very narrow white band. Flesh very tough, light wood-color. Pores 3–5 per mm, white, 3–5 mm thick.

Spores 5–7 x 4–5 μ egg-shaped, yellowish spore print.

On logs and stumps of hardwoods and conifers but rarely on living trees except in A. and the Yukon where it is often on living trees, perennial and visible during all seasons; widely distributed.

323 FOMES PINI (Thore ex Fr.) Karst.   **Edibility unknown**/Common

Cap rough and cracked, brown, woody; no stalk; pores irregular, orange-brown

Fruiting body a stalkless, woody conk 5–20 cm broad, surface rough, cracked, brown to brownish black with minute hairs which are rubbed off in age. Flesh tough, rusty reddish. Pores 2–4 per mm, often irregular in shape, ochre-orange. All parts black in 3 percent KOH solution.

Spores 4–6 x 3.5–5 μ round to nearly round, smooth, pale brown spore print. Cystidia 15–30 μ long, pointed, thick-walled, brown (known as setae Fig., 79).

On living trunks or dead logs and stumps of conifers, perennial and visible during all seasons; widely distributed.

Known also as *Trametes pini* Thore ex Fr. This fungus causes serious loss of heartwood through decay in living conifer trees. The decay is visible as a series of small white pockets filled with white mycelium. The fruiting bodies are quite similar to a number of other species of *Fomes* (Lowe, 1957). On locust one is apt to find *F. rimosus* (Berk.) Cooke, which has a dingy blackish brown cracked surface. *Fomes fomentarius* (L. ex Fr.) Kickx. is most often encountered on aspen, birch, and maple (Fig. 1). The hoof-shaped conk has a smooth gray to grayish brown cap with gray pores. *Fomes igniarius* (L. ex Fr.) Kickx is probably the most common hardwood heart rot. The cap is irregularly cracked and grayish black with a gray brown to brown pore surface containing thick-walled, pointed cells 12–20 μ long. It is most often seen on yellow birch and trembling aspen.

324 FAVOLUS ALVEOLARIS (D.C. ex Fr.) Quel.   **Nonpoisonous**/Common

Fruiting body single on a short, lateral stalk; pores hexagonal

Fruiting body 1–8 cm broad, cream, reddish buff to brick-red, dry, convex or fan-shaped. Flesh tough, thin, white. Pores large, 0.5–2 mm wide, angular to hexagonal, white to buff. Stalk short, lateral, white.

Spores 9–11 x 3–3.5 μ nearly elliptical, smooth, white spore print.

Single to several on limbs and twigs of hardwoods, in S. and F.; widely distributed.

The hexagonal, nearly diamond-shaped, large pores give this species its most distinctive field characteristic. *Polyporus squamosus* Micheli ex Fr. has a much larger cap, 6–30 cm broad, with large brown appressed scales. The pores are large (1–2.5 mm wide) and also somewhat angular. It is tough in age but reported to be edible when young and usually fruits in Sp.

325 POLYPORUS ARCULARIUS Batsch ex Fr.   **Nonpoisonous**/Common

Cap brown, dry with margin clothed in hairs; stalk central, brown; on sticks.

Fruiting body convex to convex-depressed just in center, dry, yellow to dark brown with a row of hairs on the margin. Flesh thin, white, tough. Pores large, 1–2 mm wide, white, extend down stalk. Stalk 2–6 cm long, 2–4 mm broad, central, dry, hairless, yellow to dark brown.

Spores 7–11 x 2–3 μ cylindric, smooth, white spore print.

Several, usually close together on hardwood sticks, logs, or stumps, in S. and F.; widely distributed.

P. brumalis Pers. ex Fr. is similar but has smaller pores (2–3 per mm) and marginal cap hairs are rare. *P. elegans* Bull. ex Fr. has a tan cap, minute pores (4–5 per mm), and a distinctive black stalk base. It is very common on hardwoods, especially willow and alder, throughout N.A.

326 POLYPORUS SULPHUREUS (Bull.) Fries   **Edible**/Common

Imbricate clusters with sulphur-yellow cap and pores

Fruiting body either a large cluster (30–60 cm wide) with a central stalk on the ground or an overlapping series of caps on stumps, logs, or standing trees, each cap 5–25 cm broad, sulphur-yellow, hairless. Flesh firm, margin soft to tough near center, white to yellowish. Pores 2–4 per mm, bright sulphur-yellow to cream color.

Spores 5–7 x 3.5–4.5 μ broadly elliptical, smooth, white spore print.

Growing as described above on a wide variety of hardwoods and conifers in early Sp., but most commonly in late S. and F.; widely distributed.

The edible portion of this polypore is the trimmed margin of the young caps. Although edible, the taste is not as appealing to me as it is to others. The older fruiting bodies turn a dull white but when fresh there are no other species which have the bright sulphur-yellow color.

327 POLYPORUS FRONDOSUS Dicks. ex Fr.   **Edible, choice**/Common

Large cespitose cluster of gray-brown fleshy caps with white pores

Fruiting body large, fleshy, up to 60 cm broad composed of many smaller overlapping caps, 2–8 cm broad, gray to gray-brown, dry, hairless to minutely fibrillose. Flesh firm, white. Pores 1–3 per mm, small, white to yellowish in age. Stalk large, compound, white.

Spores 5–7 x 3.5–5 μ egg-shaped, smooth (no amyloid reaction and not ornamented), white spore print.

Usually solitary near but not on stumps and snags of oaks or other hardwoods in F. in eastern North America.

This is one of the truly delicious edibles, and its bulk makes a prize find. *Polyporus umbellatus* Pers. ex Fr. is similar, also edible, but has small (1–4 cm broad), centrally attached round caps; I have encountered it in New Hampshire, but it is rare. In western North America, *P. osseus* Kalch. often develops a large fruiting body with many mouse-gray caps. It usually has 3–5 pores per mm, a bitter taste, and is found on conifer wood which is sometimes buried. I have no information on the edibility of the latter species.

328 POLYPORUS SCHWEINITZII Fr. Edibility unknown/Common

Large cespitose cluster of rusty brown, fused caps; pores mustard-yellow tinted green

Fruiting body a centrally stalked series of fused caps 15–30 cm broad, densely hairy, rusty brown to orange-brown. Flesh spongy, watery, yellow-brown. Pores 1–3 per mm, mustard-yellow to greenish yellow. Cherry-red turning black where touched with KOH (caustic potash).

Spores 5.5–8 x 4–5 μ elliptical, smooth, white spore print.

Solitary to several on the roots, lower trunk near the ground, or on the ground from buried roots of *living and dead conifers* in S. and F.; widely distributed.

This fungus rots the heartwood of the roots of living conifers. The strength of the root system is impaired and the trees are often wind-thrown. *P. tomentosus* is similar, not so spongy, tan to yellow-brown with thick-walled, pointed cells (setae) and smaller spores (4–6 x 3–5 μ). It is a weak parasite on conifer roots and very abundant fruitings are evident at times.

329 LENZITES SAEPIARIA (Walf. ex Fr.) Fries Edibility unknown/Common

Cap densely hairy, zonate; no stalk; pores brown, often gill-like

Fruiting body lacks a stalk, tough, fibrous, annual conk. Cap 1–7 cm broad, covered with dense, short hairs, zoned with color bands, bright rusty-red to yellowish red with a light whitish to yellowish margin. Flesh fibrous, 1–3 mm thick, yellow-brown. Pores usually gill-like, sometimes mazelike, dull brown.

Spores 6–10 x 2–4 μ cylindric, smooth, white spore print.

Single to several on conifer logs, stumps, boards, bridge timbers, and other wood products, in S. and F.; widely distributed.

L. trabea Pers. ex Fr. is similar but not as hairy and dull in color. It is usually on hardwoods and also on boards and wood in service.

330 DAEDALEA QUERCINA L. ex Fr. Edibility unknown/Common

Cap ash-gray, tough; no stalk; pores white, large and irregular

Fruiting body lacks a stalk, tough, leathery, convex, perennial for two to three years. Cap 3–15 cm broad, fine, short hairs, whitish to ash-gray. Flesh thick, tough, pallid to light brown. Pores very large, long, irregular, and narrow, 1–3 mm wide, sometimes almost gill-like, thick-walled, whitish to flesh-colored.

Spores 5–6 x 2–3 μ cylindric, smooth, white spore print.

Single or several on stumps and logs of hardwoods, especially oak; visible in all seasons, eastern North America.

This fungus is rather thick with large, obvious thick-walled pores. *D. confragosa* Bolt. ex Fr. has a thin, convex, zoned cap, which is dingy ashy gray, has smaller pores (0.5–1.5 mm wide), and large spores (7–9 x 2–2.5 μ). *D. unicolor* Bull. ex Fr. is imbricate with a zoned cap which is usually covered with a growth of green algae. Both of the above species are widely distributed.

331 POLYPORUS VERSICOLOR L. ex Fr.   **Edibility unknown**/Common

Overlapping cluster of colorful zoned caps; pores white, minute

Fruiting body forming dense overlapping clusters on sticks, stumps, and logs. Caps 2–5 cm broad, leathery, colorfully zoned white, yellow, red to green and bluish green, velvety from a dense hairy surface. Pores minute, 3–5 per mm, white. Stalk absent.

Spores 4–6 x 1.5–2 μ cylindric, smooth, white spore print.

On dead hardwoods but also conifers in late S. and F.; widely distributed.

This colorful polypore is one of the most common and widely distributed species in North America. It is much too leathery to be eaten but it will most certainly be observed by anyone who is mushroom-hunting. Use a hand lens to observe the delicate pores. There are a number of common polypores which are not so colorfully zoned but which resemble this species in their growth habit and leathery, stalkless, overlapping caps. Several of the more common species include *P. abietinus* Dicks. ex Fr. which has white, gray to ash-colored hairy zones on the cap and white pores tinted violet. It is found on the dead wood of conifers. A very similar polypore, *P. pargamenus* Fries, is larger but has the same appearance and is found on dead hardwoods. *Polyporus hirsutus* Wulf. ex Fr. has a hairy yellow, gray to brownish cap but no zones; one color predominates in a given cap and the pores are white. *Polyporus adustus* Willd. ex Fr. is quite similar but the pores are gray to grayish black. There are many more small polypores with similar growth habits. At least one of these will be found in every wood lot but they are all far too tough to eat.

HYDNACEAE
The Teeth Fungi

This group of fungi has teethlike spines (Figs. 92–94) on which the basidia (Fig. 74) and basidiospores are borne. Most of the group resemble gilled mushrooms or polypores (Fig. 92) when first encountered. When the fungus is viewed from beneath, however, the distinctive teeth are seen. Two species grow on trees and resemble polypores very closely. *Echinodontium tinctorium* is a woody, hoof-shaped conk with a cracked surface and closely resembles a *Fomes*. However, it has very blunt, thick teeth and is found only in the Pacific Northwest on conifer trees. *Steccherinum septentrionale* resembles a large, imbricate, fleshy *Polyporus* on living hardwood trees, but the crowded teeth leave no doubt that it is not a polypore. Hericiums are distinctive and resemble none of the other fleshy fungi. The delicate white teeth, combined with their white color and lack of any form of a pileus, create a very distinctive form (Fig. 93). There are many members of the Hydnaceae in North America and at least 100 known species. The delicious common edibles and some of the other common and unusual members of this group are included here. The work of Kenneth Harrison (1961) will provide additional information on the subject, but a work on the North American species has yet to appear.

Edibility. The nonedible hydnums (teeth fungi) are so bitter that I doubt that anyone could swallow a mouthful. Thelephoric acid has been found in the genus *Hydnellum,* one of the extremely bitter-tasting groups. It would be both unsafe and foolhardy to attempt to eat the bitter species. *Dentinum repandum,* which closely resembles a mushroom, and all of the species of *Hericium* are very good edibles. Large numbers of *D. repandum* and the large size of the Hericiums, especially *H. coralloides,* are prize finds for the collector. *Dentinum repandum* rivals any edible mushroom in flavor and can be used in almost any mushroom dish. *Hericium* is good sautéed in butter and served with vegetables. The Oregon Mycological Society Cookbook, *Wild Mushroom Cookery,* recommends "Hericium Marinade" (see Recipes).

Key to the Teeth Fungi

1 a. Cap absent; fruiting body with long, white, delicate hanging teeth, **2**
   b. Cap present; stalk central, off center, or lacking, **5**
2 a. Oval, fleshy mass with long spines but no branches, **3**
   b. Multiple branches with long spines, **4**
3 a. White; on wounds of living hardwoods (see 332. *H. coralloides*),
     Hericium erinaceus
   b. Yellowish; on wounds of living conifers, (see 332. *H. coralloides*),
     Hericium weirii
4 a. Teeth hanging only from the branch tips, *332. Hericium coralloides*
   b. Teeth hanging from along the branch (see 332. *H. coralloides*),
     Hericium ramosum
5 a. Only on pine cones; cap small; stalk off center, *333. Auriscalpium vulgare*
   b. Never on cones; cap large; stalk central or lacking, **6**
6 a. Stalk absent, sessile on living trees, **7**
   b. Stalk present and central, **8**
7 a. Woody conk on conifer in the PNW, *334. Echinodontium tinctorium*
   b. Shelflike, fleshy, overlapping on hardwood in eastern North America,
     335. Steccherinum septentrionale
8 a. Flesh brittle and easily broken; without color zones, **9**
   b. Flesh very tough, not breaking easily; zonate (color zones), **11**
9 a. Cap orange, hairless, dry; taste mild, *336. Dentinum repandum*
   b. Cap brown to red-brown with scales, dry; taste bitter, **10**
10 a. Cap brown with coarse raised scales; taste somewhat bitter,
     337. Hydnum imbricatum
   b. Cap red-brown with small, mostly flat scales; taste very bitter (see
     337. H. imbricatum), *Hydnum scabrosum*
11 a. Stalk bright violet; fragrant but disagreeable odor (see 338. *H. aurantiacum*),
     Hydnellum suaveolens
   b. Stalk another color; odor not as above, **12**
12 a. Cap whitish at first with wine-red drops of liquid; odor resinous (see
     338. *H. aurantiacum*), *Hydnellum diabolus*
   b. Cap light orange without liquid; stalk orange; odor mild,
     338. Hydnellum aurantiacum

332 HERICIUM CORALLOIDES (Scop.) Pers.   **Edible, good**/Common

Large, many branches, with delicate, white teeth; on logs

Fruiting body large, 10–25 cm broad, many branches with delicate long white teeth which arise at the branch tips and hang downward, pure white to yellowish in age, attached to wood by a stout, thick stalk. Flesh fibrous, white.

Spores 4.5–5.5 x 5–6 μ nearly round, amyloid, white spore print.

On the logs of hardwoods and conifers in the PNW, CS, and NE in late S. and F.; widely distributed.

This beautiful, delicately branched fungus with a very good taste is also known as *Hydnum caput-ursi* Fries. It is easy to identify; all of the Hericiums are edible. *Hericium ramosum* (Bull. ex Merat) Let. (or *H. laciniatum* in some books) is similar but has spines along the branches (Fig. 93), not just at the end. *Hericium erinaceus* (Bull.) Pers. is a large, rounded, fleshy fruiting body with no branches, but with long, coarse spines (10–30 mm long) hanging from it. It occurs on the wounds of living hardwoods. *Hericium weirii* is similar to *H. erinaceus* but is more yellowish and is found in the PNW on living conifers. They are all edible and a choice find.

333 AURISCALPIUM VULGARE S. F. Gray   **Edibility unknown**/Numerous

["Pine Cone Fungus"]

Cap small, brown, hairy; teeth fine; stalk off center; on pine cones

Cap 5–20 mm diameter, broadly convex to flat, off center, covered with dense, dark brown hairs, (Fig. 94). Flesh firm, light brown. Teeth very fine, needle-like, dense, light brown. Stalk 1.5–8 cm long, 1–2 mm broad, *attached at one side,* densely hairy, dark brown, rigid.

Spores 5–6 x 4–5 μ almost round, with minute spines that show an amyloid (blue) reaction, white spore print.

Single or in nearly cespitose clusters *always on pine cones* in S. and F.; widely distributed. This fungus has a unique appearance and a unique habitat on pine cones.

334 ECHINODONTIUM TINCTORIUM   **Edibility unknown**/Common

Ellis and Ever. ["Indian Paint Fungus"]

Woody, hoof-shaped conk; on conifer trees in PNW and RM

Fruiting body woody, large, 10–25 cm broad, hoof-shaped (Fig. 1), surface cracked rough, dingy blackish brown, stalkless. Flesh tough, zoned, bright cinnamon to rusty red. Teeth stout, gray, with blunt ends, brittle and dry.

Spores 5.5–7 x 3.5–4.5 μ elliptical, minutely spiny, white spore print.

Single to several on living western conifers, especially western hemlock and grand fir, during Sp., S., and F.; known in the RM, PNW, and adjacent C but not distributed west of the coast mountains.

This fungus resembles a *Fomes.* The presence of fruiting bodies or conks on a tree indicates that extensive heart rot of the stem below and just above the conk has taken place. The sporophores soon die on downed trees. The common name, Indian Paint Fungus, comes from the belief that it was used as a source of dye by the Indians of the Pacific Northwest.

335 STECCHERINUM SEPTENTRIONALE

Nonpoisonous/Numerous

(Fr.) Banker

Large overlapping, stalkless clusters; on living hardwoods

Fruiting body large, 15–30 cm broad, 10–15 cm wide, imbricate (overlap) one above the other from a common stalkless base, surface dry, densely hairy, dingy buff to yellow-brown in age. Flesh 2–4 cm thick, white, zoned, tough. Teeth 5–20 mm long, narrow, dull white, pliant. Odor of ham when dried.

Spores 4–5.5 x 2.5–3 μ elliptical, smooth, white spore print.

Large single clusters on wounds of living hardwoods in eastern North America; widely distributed except in late S. and F.

It is too tough to eat. It is important as a heart rot of shade and commercial trees. I have seen large clusters on Sugar Maple in New Hampshire. This is also well known as *Hydnum septentrionale* Fr.

336 DENTINUM REPANDUM (L. ex Fr.) S. F. Gray

Edible, choice/Common

Cap dry, orange; teeth cream color; flesh yellowish, brittle

Cap 1.5–10 cm broad, convex, broadly convex in age, hairless, dry, margin wavy, buff to orange or faded orange. Flesh thick, soft, *brittle,* light yellowish. Teeth 4–8 mm long, various lengths intermixed, cream color. Stalk 2–8 cm long, 6–20 mm thick, equal or enlarged at base, hairless, dry, white, solid.

Spores 6.5–9 x 6.5–8 μ nearly round, smooth, white spore print.

Solitary to numerous under hardwoods, conifers, and mixed woods, in S. and F.; widely distributed.

Also known as *Hydnum repandum* Fries. This fungus resembles a gilled mushroom until the teeth are observed beneath the cap. *D. umbilicatum* (Peck) Pouzar is a small species that has a darker cap and is usually found in swamps and bogs. I have collected it in lodgepole pine bogs near Juneau, Alaska. It is edible and delicious.

337 HYDNUM IMBRICATUM Fr.

Edible/Common

Cap dry with raised, brown scales; teeth long, brown; taste bitter

Cap 8–20 cm broad, convex with a depressed center, dry, covered with coarse, raised, brown scales. Flesh breaking but somewhat tough, white to very light brown, does not turn black in 3 percent KOH solution. Teeth extend down stalk, 5–15 mm long, brown. Stalk 4–10 cm long, 15–30 mm thick, enlarging toward base, dry, smooth, light brown. Taste somewhat bitter.

Spores 6–8 x 5–7 μ nearly round, strongly warted, brown spore print.

Single to several under hardwood and conifer forests in S. and F.; widely distributed.

The brown cap and conspicuous, thick, raised scales separate this hydnum from all others. I have encountered luxuriant fruitings of it along the southeastern coast of Alaska as well as in many other habitats across North America. *Hydnum scabrosum* Fries is similar but the cap scales are not as coarse and it has more red coloration in it; the stalk base is black to bluish to bluish black; lastly it is very bitter and the flesh blackens in 3 percent KOH solution.

338 HYDNELLUM AURANTIACUM Edibility unkown/Common

(Batsch ex Fr.) Karst.

Cap dry, orange-brown, tough; teeth fine, brown; flesh tough, has color zones

Cap 5–15 cm broad, flat to depressed in age, white but soon orange to brown, surface velvety, dry, uneven with knobs and lumps. Flesh tough and fibrous, zoned, dull brown. Teeth 3–4 mm long, fine, brown with light tips. Stalk 1.5–5 cm long, 8–15 mm thick, equal, felted, dingy orange-brown.

Spores 5.5–7.5 x 5–6 μ nearly round, strongly warted, brown spore print.

Single to several under conifers in S. and F.; widely distributed. Harrison (1961) finds it common in Nova Scotia and I have found it in both the SE and PNW.

Hydnellum suaveolens (Scop. ex Fr.) Karst. has a grayish violet cap, bright violet coloration on the stalk, and a fragrant but disagreeable odor. *Hydnellum diabolus* Banker is whitish when young and red-brown in age, exudes red drops of liquid on the cap, and has a sweet, resinous odor. *Hydnellum peckii* Banker is similar but has no resinous odor.

188

Gastromycetes

Stinkhorns, Bird's nests, Earthstars and Puffballs

The general name for the puffballs and their allies comes from the word *gastro,* meaning stomach. This refers to the fact that the spore-bearing surface is inside the fruiting body. For example, the powdery inside of a mature puffball is, in fact, the mature spores. There is no forcible discharge of the spores which in such groups as the gilled mushrooms allowed us to obtain spore prints. Other means of dissemination of the spores has been accommodated by the strange variety of fruiting bodies in the following groups.

Edibility.   The edible members of the Gastromycetes, the puffballs, are among the easiest of all mushrooms to identify. I certainly would urge any beginner to learn to recognize the edible puffball.

Key to the Stinkhorns, Bird's Nests, Earthstars and Puffballs

1 a. Erect stalks often covered over the tip with green slime; odor very disagreeable; arising from a tough, fleshy "egg" (Figs. 36–39), Stinkhorns (Phallales)
   b. Not slimy; no disagreeable odor; not shaped as above, **2**
2 a. Fruiting body resembles a miniature bird's nest (Figs. 35 and 98), Bird's-nest fungi (Nidulariales)
   b. Fruiting body not like a bird's nest, **3**
3 a. Fruiting body star-shaped (Fig. 33 and 97), Earthstars (Lycoperdales)
   b. Fruiting body round (Fig. 26), pear-shaped (Fig. 27), or sometimes on a stalk (Figs. 28 and 31), **4**
4 a. Stalk present, **5**
   b. Stalk not present, although a short sterile base may be present, True puffballs (Lycoperdales)
5 a. Spore sac *on* a dry, woody, or gelatinous stalk, Stalked puffballs (Lycoperdales)
   b. Stalk runs up into or through the spore sac, often resembles an unopened gilled mushroom (Fig. 29), False puffballs (Hymenogastrales)

PHALLALES
The "Stinkhorns"

The sight and smell of a stinkhorn is something most collectors never forget. The young stinkhorn develops within an "egg." When completely mature, the stalk begins to elongate, rapidly rupturing the "egg." Full elongation of the stalk may take place in one-half to three hours and the "egg" then remains around the base, forming a volva (Figs. 36–39, and 99). The slimy green to gray-green layer over the top of the fully developed fruiting body is composed of basidia and basidiospores (Fig. 74). The disagreeable odor attracts insects which walk over the sticky spores. The spores stick to the insects' feet and become distributed as the insects move about. This is certainly very different from the forcible spore discharge of preceding groups and not at all like the dry, powdery spore mass so frequently seen in the puffballs. The stinkhorns meet every qualification as true Gastromycetes because the spore layer is fully protected within the "egg" until it is mature. To demonstrate this one needs only to cut down through an "egg," and the fully developed head is easily seen. The spore layer is rubbery and hard to section for microscopic study, but a razor blade will overcome this obstacle so that one can observe the narrow, oblong spores (Fig. 71). They are similar throughout the entire group. Stinkhorns are commonly found in flower beds mulched with wood chips or in sawdust piles. In the woods, they occur where humus has accumulated, near decayed logs or where beetle activity in trees has left wood debris on the ground.

Edibility. The rubbery, disagreeable smelling "eggs" are apparently not poisonous but they would certainly not be a desirable food. If placed in moistened earth or leaves in a bowl in a cool place, the stalk will frequently elongate and mature and one can easily observe the process.

Key to Stinkhorns

1 a. Fruiting body club-shaped, inside at top filled with a green slimy gleba (spore mass), *339. Phallogaster saccatus*
  b. Fruiting body not club-shaped, **2**
2 a. Fruiting body without a stalk; a large-chambered head seated in the volva (Fig. 38) (see 342. *Simblum sphaerocephalum*), *Clathrus columnatus*
  b. Fruiting body with a stalk; long-cylindrical, with or without a head, **3**
3 a. Head present, smooth, wrinkled, or chambered, **5**
  b. Head absent, **4**
4 a. Tapering at apex and covered with a green slime, *340. Mutinus caninus*
  b. Upraised arms at apex with a thin strip of green slime on each arm (Figs. 36–37), *341. Lysurus borealis*
5 a. Head regularly chambered (Fig. 39); red, *342. Simblum sphaerocephalum*
  b. Head smooth, wrinkled, or chambered, green, **6**
6 a. Head chambered, green; skirtlike, netted veil hangs from lower margin of head, *343. Dictyophora duplicata*
  b. Head smooth or wrinkled, green; without a skirtlike netted veil, **7**
7 a. Head smooth, green slimy, *344. Phallus ravenelii*
  b. Head wrinkled (Fig. 99), green slimy (see 344. *P. ravenelii*), *P. impudicus*

339 PHALLOGASTER SACCATUS   Edibility unknown/Rare

Morgan ["Club-shaped Stinkhorn"]

Club-shaped, dull whitish, filled with green slime; odor disagreeable

Fruiting body 2.5–5 cm high, 1–3 cm thick, pear-shaped to club-shaped, smooth, white with a pinkish hue, upper part filled with green spores in slimy lobes and with depressions or even distinct chambers between the lobes, base narrow, attached to the substrate by many intertwined pink, rootlike cords. Odor *fetid* and disagreeable at maturity.

Spores 4–5 x 1.5–2 μ nearly cylindrical, colorless, smooth.

In small clusters on well-decayed wood in Sp. and early S. in CS, NE, SE, and So. This curious fungus has no counterpart, but it might be mistaken for the immature stage of some other stinkhorn.

340 MUTINUS CANINUS (Pers.) Fries   Nonpoisonous/Common

[Common "Stinkhorn"]

Narrow red stalk with green slime over the top; odor disagreeable

Fruiting body 6–10 cm long, 8–11 mm thick, a thin stalk which tapers to a narrow top covered with a green slime, giving way below to a chambered pinkish to red stalk, base is an oval "egg" from which the fruiting body expands, leaving a membranous volva around the stalk. Odor slightly fetid.

Spores 3.5–4.5 x 1.5–2 μ cylindrical, thin-walled, colorless.

Single to several on soil, humus, mulch, wood debris in late S. and F.; widely distributed.

The slimy spore mass covers the upper stalk without a regularly formed head, and that alone sets this genus apart from other stinkhorns.

341 LYSURUS BOREALIS (Burt.) P. Henn. Edibility unknown/Rare

Erect arms with green slime in the center; stalk white

Fruiting body 10–12 cm high with six erect hollow arms (Figs. 36–37) pale flesh color. Gleba (spore mass) olive-brown in the center of the arms. Stalk white, tapering toward the base.

Spores 3–4 x 1–2 μ narrowly cylindrical, thin-walled, olive-green.

Single or several on humus, mulch or wood debris; widely distributed fruiting during wet cool periods. This curious fungus has been found in Massachusetts, Virginia and California in S. Also known as *Anthurus borealis* Burt.

342 SIMBLUM SPHAEROCEPHALUM Edibility unknown/Rare

Schlechte ["Chambered Stinkhorn"]

Oval chambered red head on a white stalk; odor disagreeable

Fruiting body 7–9 cm high, with an oval head 10–16 mm thick; 10–15 mm high, convex, bright red, slimy spore layer with regular, chambered pits. Stalk equal, chambered, hollow, white, arises from a tough oval "egg" and leaves a regular volva. Odor strong and very disagreeable.

Spores 3.5–4.5 x 1–2 μ elliptical, smooth, yellow-brown.

Several, on ground or grass in apple orchards or in pastures in F. in So. Our deep South appears to be the northernmost distribution of this common South American stinkhorn.

The small convex pitted head is distinctive. This genus is common in South America, but in addition to this rarely encountered species, Long (1942) reports another species in North America, *Simblum texense* Long from the PSW, which is yellow and has longer spores than those described above. *Clathrus columnatus* Bosc. also has a chambered rosy red head but lacks a stalk (Fig. 38). It is found on sandy soil in the deep South and has the disagreeable odor of the stinkhorns.

343 DICTYOPHORA DUPLICATA Nonpoisonous/Infrequent

(Bosc) E. Fischer ["Netted Stinkhorn"]

Head chambered, olive, slimy; stalk surrounded by a white, long, netted veil

Fruiting body 12–17 cm high, 2–4 cm thick; head *strongly chambered* and covered with a brownish olive slime; hanging from the lower margin of the cap is a beautiful *skirtlike netted veil* up to 5 cm long and flaring at the bottom, but it is often collapsed and flattened to the stalk. Stalk 7–12 cm long, 3.5–4.5 cm thick, white, chambered, hollow, projecting from the large oval "egg" from which the fruiting body expands, leaving a membranous volva around the stalk base (much as in *Phallus*), attached to the ground by well-developed rootlike strands. Odor fetid and disagreeable.

Spores 3.5–4.5 x 1–2 μ elliptical, smooth, colorless.

Single or in groups in dense humus under hardwoods in S. and F.; in eastern North America.

The brownish olive, chambered head and the netted, skirtlike veil (indusium) form a must beautiful fungus as long as one possesses an insensitive nose!

344 PHALLUS RAVENELII Berk. and Curt. ["Stinkhorn"]   **Nonpoisonous**/Common

Head green, smooth, slimy; stalk white; odor fetid

Fruiting body 10–16 cm high, 13–15 mm thick; head green to gray-green, granulose, slimy to sticky with an open depression at the top. Stalk thick, equal, white, chambered, inserted up under the head. Base is large oval "egg" from which the fruiting body expands, leaving a membranous volva around the stalk, white to pinkish, attached to the ground by well-developed rootlike strands. Odor nauseous to fetid, sometimes detectable several yards away.

Spores 3–4 x 1–1.5 μ cylindrical, surrounded by a thin transparent envelope, colorless, often five to six arranged around top of basidium.

Usually in a dense cluster in wood debris or sawdust in S. and F.; widely distributed, but I have not seen it in RM, PNW, or PSW.

The lack of a netted long veil, the non-wrinkled green spore surface, and the whitish to pinkish "eggs" separate this species from all others. Flies are attracted by the odor and walk over the spore-bearing surface, thus disseminating the sticky spores. When the "egg" is cut in half, the immature green head and the compacted stalk are revealed. *P. impudicus* Pers. is infrequently encountered but it has a green *wrinkled* head and *no netted veil* (Fig. 99). Also known as *Ithyphallus impudicus* (Berk. and Curt.) E. Fischer.

NIDULARIALES
"Bird's Nests"

The curious resemblance to a bird's nest of the fungi in this group makes them easy to identify. They are smaller than a dime and usually many fruiting bodies are found close together. The "nest" is completely enclosed at first in most species and the spores develop inside of "eggs" (Fig. 35), called peridioles, within the enclosed nest. At maturity the top of the nest is ruptured and exposes the eggs. The eggs are ejected from the nest when raindrops literally splash them out. The two species included here have eggs attached to the nest by a thin cord; when they are splashed out, the cord wraps around nearby twigs or stems, holding the "eggs" in place until the wall of the egg disintegrates and the spores are released. These species are commonly found on sticks and wood debris.

Edibility.   The small, tough fruiting bodies are inedible. There are no reports of toxins in this group.

345 CYATHUS STRIATUS   **Edibility unknown**/Common

Pers. ["Bird's nest Fungus"]

Interior wall shiny, striate; "eggs" blackish

Fruiting body 10–15 mm high, 2–3 mm thick, vase-shaped (Fig. 98), hairy and with a thin hairy covering (epiphragm) over the top of the "nest" at first, pale cinnamon-brown to gray-brown, cover breaking away at maturity revealing a shiny, striate interior wall of the nest with several drab to nearly black eggs (peridioles) at the bottom. Each fruiting body is attached to its woody host by a cinnamon-brown pad of mycelium.

Spores 12–22 x 8–12 μ elliptical, thick-walled, colorless.

Grows in groups on bark, sticks, and other woody debris in Sp., S. and F. during wet periods; widely distributed.

There are four genera and an undetermined number of North American species. They are all close to the one above in size and general appearance; a number of them are described in Coker and Couch (1928) and A. H. Smith (1951). *Crucibulum vulgare* Tulasne (Fig. 35) is the other common species encountered in North America. Its peridioles are nearly white and the inner walls of the nest are not striate.

LYCOPERDALES
The Earthstars

This distinctive group is very closely related to the puffballs. Like a puffball (Fig. 25), the earthstars have two walls, but the outer wall splits and curves back forming the rays (Fig. 34). This exposes the inner wall which surrounds the gleba (spore mass) shown in Figure 25. An ostiole (pore, Fig. 34) develops in age and the spore sac may or may not be raised on a short stalk (Fig. 34). The spores are round (globose) or nearly so throughout the group and minutely spiny (Fig. 68) to smooth. The spores are held together by specialized hyphae called *capillitium* (Fig. 68). The various species seem to be found in specific habitats but their role in nature is not known. It is possible that they form mycorrhiza.

Edibility. I know of no reports of toxins in this group. Most of the earthstar buttons seem to be much too hard and tough to be good edibles. In addition, they are rarely found in the button stage.

Key to Earthstars

1 a. Rays open up when wet, close up around spore sac when dry; inner spore sac hairy (Fig. 33), *346. Astraeus hygrometricus*
   b. Rays remain open; inner spore sac smooth, **2**
2 a. Spore sac with one hole (ostiole, Fig. 34) at the top, **3**
   b. Spore sac with many holes (ostioles, Fig. 30) over the spore sac, *347. Myriostoma coliforme*
3 a. Inner spore sac on a short stalk (Fig. 34), *348. Geastrum coronatum*
   b. Inner spore sac not stalked (Fig. 97), **4**
4 a. Spore sac seated in a "bowl" (the surface that is exposed and visible when the ray opens up); inner surface of ray cracked, *349. G. triplex*
   b. Spore sac exposed but sessile; inner surface of ray smooth, *350. G. saccatum*

346 ASTRAEUS HYGROMETRICUS (Pers.) Morgan   Nonpoisonous/Common

Spore sac hairy; rays checkered, and open up when wet

Fruiting body 1–3 cm broad, round or somewhat depressed at first; outer skin splits into 7 to 15 pointed rays which open up when wet and flatten out (Fig. 33), exposing the inner spore sac which is covered with *dense hairs* (similar to *Geastrum*); on drying the rays again close around the spore sac; opens at top by tearing and *no regular pore* is formed. Gleba white to cocoa-brown at maturity. Sterile base absent.

Spores 7–11 μ round, thick yellow-brown wall covered with warts and spines, brown.

Scattered to numerous, in sand or sandy soil, abandoned fields, fruiting in F.; widely distributed.

The exposed inner surface of the rays is light and *conspicuously cracked* in a check-ered pattern. This combined with the hairy spore sac, dark brown gleba (spore mass), and hygroscopic rays (open when wet and close when dry), make this a distinctive fun-gus. The rays of *Geastrum* are not hygroscopic and remain open. In addition, the inner surface is not cracked.

347 MYRIOSTOMA COLIFORME (Pers.) Corda   **Edibility unknown**/Rare

Spore case with several pores

Fruiting body 1.5–8 cm broad, egg-shaped, smooth, silver-brown inner spore sac with several ostioles (pores); outer skin splitting into 5 to 7 rays; the spore sac is supported on several nearly fused short stalks. Gleba (spore mass) is brown at maturity.

Spores 4–6 μ nearly round, warty, yellow-brown.

Single to several, widely distributed throughout North America but infrequently encountered.

348 GEASTRUM CORONATUM Pers. ["Earthstar"]   **Nonpoisonous**/Common

Spore sac oval, smooth, on a short stalk; rays curved under on tips

Fruiting body 5–15 mm broad, sometimes taller than broad to egg-shaped, with rays expanded up to 4 cm broad; outer skin light yellow-brown on either side, splitting star-like into 4 to 6 rays *which recurve to support the fruiting body on their tips* (Fig. 34); inner sac (spore sac) dark purple-brown, smooth, with a silky pale area around the mouth delimited by a distinct groove or line, the mouth is lined with fine hairs, the whole *spore sac is raised 1–3 mm on a short stalk.* Gleba (spore mass) dark chocolate-brown at maturity. Below the spore sac is a short sterile stalk or base.

Spores 3.5–6 μ subglobose to globose, warted, brown.

In groups of several, usually under conifers in needle duff, in S. and F.; widely dis-tributed. I have found this several times near Kintla Lake in Glacier National Park, Montana. There are a number of closely related earthstars with stalked spore sacs. Old specimens are often found in Sp.

349 GEASTRUM TRIPLEX Jung. ["Earthstar"]   **Nonpoisonous**/Common

Spore sac round, smooth, no stalk, sitting in a "bowl"

Fruiting body 1–5 cm broad, with rays expanded up to 9 cm broad; outer skin gray-brown to wood-brown, splitting into 4 to 8 starlike rays at the top and recurving toward the base, leaving the spore sac sitting in a "bowl" (Fig. 97.), *inner surface of ray often cracked;* inner sac (spore sac) dark wood-brown to light drab, smooth with a pale circu-lar area surrounding the pore mouth which is surrounded by fine hairs. Gleba white at first to rich hazel-brown and powdery at maturity. Sterile base extends as a column into the center of the gleba.

Spores 3.5–4.5 μ round, covered with small blunt warts, brown.

Single to several in rich humus under hardwoods or mixed forests, in S. and F.; widely distributed. I have seen this in Michigan and Idaho. The important characteristics are the "bowl" which encloses the lower spore case, the large size, and the clean outer skin without much adhering debris.

350 GEASTRUM SACCATUM Fries ["Earthstar"]   **Nonpoisonous**/Common

Spore sac round, smooth, no stalk, not in a "bowl"

Fruiting body 6–25 mm broad, with rays expanded up to 5 cm broad; outer skin ochre-buff to light yellow-brown, splitting starlike at the top and recurving against the base (Fig. 97), revealing the nonstalked spore sac which has a dull brown, papery smooth wall. Gleba (spore mass) white, then wood-brown or even purplish brown, powdery when mature; pore hairs around mouth arise from a circular, *delimited disc.* Sterile base extends as a column into the center of the gleba.

Spores 3.5–4.5 μ round, covered with small blunt warts, brown.

Scattered to usually several in damp humus, during wet periods in S. but persists through F. and W.; widely distributed.

Geastrum fimbriatum Fries is similar and is not stalked, but lacks a delimited disc around the pore.

LYCOPERDALES
The True Puffballs

The true puffballs do not have a stalk (Fig. 31) and the outer peridium (skin) does not recurve back to form rays as in the earthstars (Fig. 33). In fact, the fruiting body is usually round to pear-shaped. As illustrated in Fig. 25, it commonly has a warted outer wall, smooth inner wall, and a homogeneous gleba (spore mass) with or without a sterile base. The gleba is at first either white, soft, and fleshy or purple to purple-brown and rather tough. All true puffballs have *powdery glebas* at maturity which range in color from mustard-yellow to brown or lilac. When the spore case is pressed, a cloud of spores "puff" from the mature puffball. The spores escape through a rupture in the wall or a well-formed ostiole (pore, Fig. 27) which is formed at maturity. The fruiting bodies range from the size of a marble to that of a basketball or larger. The species are either saprophytes on decayed matter or form mycorrhiza with higher plants.

Edibility.   The genus *Scleroderma* which forms a purple gleba at first should be avoided. The flesh is somewhat bitter and there are conflicting reports of sickness from those who have eaten some of the species. The rest of the true puffballs at first have a white gleba which colors only as they mature. This entire group is good to eat and forms one of the best and safest of all edible mushrooms. It is good practice to cut down through the puffball; if you find the form and gills of an embryonic mushroom this would signify that you had a mushroom button. Once puffballs are seen and collected, it is easy to make this test.

Our first meal of puffballs came from the Farmer's Market in Ann Arbor, Michigan, years ago. We bought about one-third of a basketball-sized puffball. After several trials we found that a superior dish results when the puffball is sliced and fried. During the slicing one must check for insect tunnels and discard such infected areas. For the best way to prepare puffballs, see "Breaded Puffballs" under Recipes.

Key to True Puffballs

1 a. Young puffball cut in half shows *firm purple to purple-brown fleshy spore mass (gleba)* and often a thick outer skin, **2**

    b. Young puffball cut in half shows *firm white fleshy spore mass (gleba)* and usually a thin outer skin, **5**

2 a. Fruiting body small, 3–6 cm broad; outer skin splitting into lobes but never curving back like an earthstar, **3**

    b. Fruiting body large, 4–14 cm broad; outer skin splitting and curving back like an earthstar, **4**

3 a. Outer skin covered with warts, *351. Scleroderma aurantium*

    b. Outer skin smooth (warts absent) (see 351. *S. aurantium*), *S. bovista*

4 a. Outer skin cracked and roughened, *352. S. geaster*

    b. Outer skin smooth (see 352. *S. geaster*), *S. flavidum*

5 a. Round, oval, or flattened oval shape; if pear shape it will be thicker than 8 cm, **6**

    b. Pear-shape or nearly pear-shape with a sterile base, less than 7 cm thick, **16**

6 a. Large (10–50 cm thick), larger than a softball, usually the size of a somewhat flattened *basketball*, **7**

    b. Medium to small (8 mm–15 cm thick), softball size or smaller, **9**

7 a. Very large (20–50 cm thick), **8**

    b. Usually 10–20 cm thick, cracked to form irregular scales, *353. Calvatia lepidophora*

8 a. Outer skin nearly smooth; common in the East, *354. C. gigantea*

    b. Outer skin sculptured, cracked into large flat scales, dull white; in western North America, *355. C. booniana*

9 a. Puffball grapefruit to softball size (8–16 cm thick), **10**

    b. Puffball much smaller than a softball, usually not over 6 cm broad, **12**

10 a. Outer skin covered with pointed, pyramid-shaped warts; western North America, (Fig. 24), *356. C. sculpta*

    b. Outer skin not as described above, **11**

11 a. Outer skin covered with nearly flat warts, *357. C. subsculpta*

    b. Outer skin nearly smooth, wrinkled to minutely cracked, *358. C. cyathiformis* and related species

12 a. Puffball golf ball size (3–5 cm in most cases), **13**

    b. Puffball small (8–25 mm in most cases), **15**

13 a. Round, paper thin outer skin; very light, easily detached and blown away, *359. Bovista pila* and related species

    b. Round to oval but not paper thin, remain attached, **14**

14 a. Chalky white, covered with small pointed, gray-tipped warts, odor none, *360. C. subcretacea*

    b. Smoky gray, smooth; odor disagreeable, *361. C. fumosa*

15 a. Chalky white, covered with small, pointed, pure white warts, pieces of outer skin soon cracking and falling off, *362. Lycoperdon candidum*

    b. Dull white, smooth, outer wall persistent, pore formed in old age, *363. L. pusillum*

16 a. Outer skin light brown, smooth or a few hairs at the top; in clusters on wood, *364. L. pyriforme*

    b. Outer skin covered with spines of hairs; usually not on wood, **17**

17 a. Outer skin dull whitish, covered with pointed spines which break off and leave round spots, *365. L. perlatum*

    b. Outer skin pure white to dark brown, **18**

18 a. Outer skin light to dark brown covered with long whitish, coarse hairs (3–6 mm long), *366. L. echinatum*

    b. Outer skin white to dark dingy brown with short separated hairs (up to 1 mm high) or tufts of hairs, *367. L. umbrinum* and related species

351 SCLERODERMA AURANTIUM Pers.   **Nonpoisonous**/Common

Oval, warted, yellow-brown; gleba fleshy when young, purple

Fruiting body 3–6 (up to 12) cm broad, round when young, depressed at maturity; outer skin yellow-brown, thick (2 mm), covered with cracks that set off a pattern of raised warts which may have central, somewhat darker warts, extending almost to the base, in age splitting into irregular lobes, never star-shaped like a *Geaster.* Gleba fleshy, *lilac-gray to deep violet-gray flesh,* with minute, fine, white lines, dingy to nearly black and powdery when mature, surrounded by the thick white outer skin. Sterile base a slight broadening of the outer skin, white but staining yellow to faintly pink.

Spores 8–13 μ round with fine spines and netlike ridges, brown.

Single to several or sometimes in cespitose clusters on the ground, near the base of trees or around stumps and logs, under conifers or hardwoods, in late S. and F.; widely distributed throughout eastern North America. *Scleroderma bovista* Fries is smaller (3–5 cm), usually *smooth* or sometimes cracked but without warts and with larger spores (12–18 μ in diameter). Both are bitter tasting and should be avoided.

352 SCLERODERMA GEASTER Fr.   **Nonpoisonous**/Infrequent

Oval, smooth, clay color; gleba when young is fleshy, purple with white veins

Fruiting body 4–14 cm broad, round to depressed-round; outer skin dull yellowish to clay color, very thick *(almost 5 mm),* rough, irregular cracks, in age splitting into regular lobes and recurving much like an earthstar *(Geaster),* with the *inner skin black.* Gleba fleshy, *purple when young,* compartmented by thick, white to yellowish sterile tissue; dingy dull blackish brown and powdery in age.

Spores 6.5–10 μ round, covered with spines, brown.

Single or several together, on sandy banks, hard log roads, and in drainage ditches under mixed hardwoods and conifers, in F., in So and SE, infrequent further north. *Scleroderma flavidum* Ellis and Everhart has a smooth outer skin which is roughened only in age, straw-color, spores 4–9 μ broad, and also splits like an earthstar. Both of these are bitter and not edible, but I cannot find any record of poisoning.

353 CALVATIA LEPIDOPHORA   **Edible, good**/Infrequent

(Ellis and Ever.) Coker and Couch

Oval, very large, irregular surface scales, white; gleba white, olive-brown in age

Fruiting body 5–15 cm high, 10–20 cm thick, nearly round but depressed, outer skin cracked forming irregular scales, dull white to tan or dull yellow-brown in age. Gleba white to olive-brown at maturity. Sterile base present but very small, white to yellowish in age.

Spores 4.5–6.5 x 5–6 μ nearly round, covered with spines, olive-brown.

Single to scattered in prairie or arid habitats in Sp., S., and early F.; central North America westward.

This grapefruit-size puffball is not well known, but the Ted Truebloods of Nampa, Idaho have found it not uncommon in southwestern Idaho and I have seen it there in sage-grass habitats. It is much smaller and not as sculptured as *C. booniana* but larger than most of the other western puffballs.

354 CALVATIA GIGANTEA (Batsch ex Pers.) Lloyd   Edible, choice/Common

Oval, huge, smooth surface, white; gleba white, olive-brown in age

Fruiting body large (20–50 cm broad), nearly round, outer skin *smooth like fine felt,* whitish, light gray to yellow or olive in age, base with a thick root. Gleba white to olive-brown at maturity. Sterile base lacking or very reduced.

Spores 3.5–5.5 x 3–5 μ nearly round, minute short spines or nearly smooth, olive.

Single to several together in low swales or wet areas, along edges of meadows or along small streams or drainage ditches, in S. and F.; widely distributed in central and eastern North America.

The smooth outer skin and large size distinguish this well-known *Calvatia* from all others. It is a delicious edible often collected and often sold in farmers markets. If a large *Calvatia* with a conspicuous cracked surface formed into polygons is encountered, see *C. booniana* A. H. Smith.

355 CALVATIA BOONIANA A. H. Smith   Edible, choice/Infrequent

Oval, huge, lobed and sculptured, white; gleba white to olive-brown

Fruiting body large, 10–30 cm high, 20–60 cm broad, egg-shaped but flattened; outer skin *sculptured,* soon separated into *large flat scales,* dull white to light tan or buff color, base with a thick fibrous cord. Gleba white to olive-brown mature. Sterile base lacking. Odor very disagreeable in early maturity.

Spores 4–6.5 x 3.5–5.5 μ nearly round, minute spines, olive-brown.

Single or sometimes two or three in dry open areas, in or near old corrals or arid habitats, often near or under sagebrush, in Sp. and S.; in the RM, PNW, and PSW.

The large size and distinctive configuration of the outer skin are distinctive characters of this recently recognized "western giant puffball" which A. H. Smith described in 1964 (Zeller and Smith, 1964). Without doubt this is the giant puffball that many westerners have collected and eaten since pioneer days. We find it of equal flavor with *C. gigantea.*

356 CALVATIA SCULPTA (Hark.) Lloyd   Edible, choice/Numerous

Pear-shaped with pointed, pyramid-shaped warts; small sterile base

Fruiting body 8–15 cm high, 8–10 cm thick, somewhat pear-shaped or egg-shaped, top covered with distinctive, long, pointed, pyramid-shaped warts (Fig. 24), sometimes erect, often bent over or even joined at the tip with other warts. Gleba white to deep olive-brown at maturity. Sterile base small to one-fourth of base, white to yellowish in age.

Spores 3.5–6.5 μ nearly round, minute spines, olive-brown.

Single to several in duff under or near conifers at high elevations in Sp., S., and early F., in RM, PNW, and PSW. I have encountered this species most commonly in late June and early July. The conspicuous long pointed warts are the distinctive feature of this baseball-sized puffball.

357 CALBOVISTA SUBSCULPTA Morse — Edible/Common

Flattened egg-shaped, low warts with brownish centers; sterile base over lower one-third

Fruiting body 6–9 cm high, 8–16 cm thick, nearly round, white, covered with low warts (Figs. 25–26) with brownish colored hairs at the center of each wart, often smooth to wrinkled but not warted near the base. Gleba firm, white to dark brown at maturity. Sterile base over lower one-fourth to one-third, dull white, firm.

Spores 3–5 μ round, thin-walled, nearly smooth, yellowish through microscope.

Single or several on hard, dry roadbanks, stock driveways, usually near or under conifers, in PNW, PSW, RM, and A, in late Sp., and S.

This puffball is usually of softball size, sometimes larger, but never with the pyramid-like warts of *Calvatia sculpta* (Fig. 24), which has a similar distribution and attains about the same size.

358 CALVATIA CYATHIFORMIS (Bosc) Morgan — Edible, choice/Common

Pear-shaped, large, sculptured, light pink; gleba white, yellow-brown in age

Fruiting body 9–20 cm high, 7–16 cm thick, nearly pear-shaped but with a thick base and egg-shaped top, outer skin soon minutely cracked and sculptured over the top, smooth but wrinkled over the base and sides, white to light pinkish tan, brown in age. Gleba white, yellow to purple-brown in age. Sterile base over lower one-third, chambered, white to dingy yellowish.

Spores 3.5–7.5 μ nearly round, minute spines, purple.

Scattered to numerous in grass, prairie or desert communities in late S. and F. following heavy rains; widely distributed.

Several species have a similar appearance and size. They are all edible when the gleba is young and white. The softball-size species above closely resembles *C. craniformis* (Schw.) Fries, which is easily distinguished by the bright *yellow-green* mature gleba. I have found *C. craniformis* more commonly in late F. *Calvatia bovista* (Pers.) Kambly and Lee is also similar but with an egg-shaped top and tapered narrow base; and the outer skin has flat areas separated by pronounced cracks and a dark olive-brown mature gleba. *C. rubro-flava* (Cragin) Morgan has a pinkish tan outer skin but all parts *instantly bruise bright yellow* and are orange when dry. The gleba is greenish orange when mature.

359 BOVISTA PILA Berk. and Curt. — Edible/Numerous

Round, thin and papery, bronze colored; gleba white, brown mature

Fruiting body 3–8 cm broad, round, outer skin white, fine, fuzzy, soon wears off exposing inner skin which is *thin and papery,* metallic to bronze colored but soon dingy purplish brown with well-developed round pore on top, attached to soil by a thin root. Gleba white when young to dark brown, powdery. Sterile base absent.

Spores 3.5–4.5 μ round, smooth, broken pedicel (Fig. 68), brown.

Solitary, scattered to numerous in pastures, especially where cattle have been or open woods, in S. and F.; widely distributed.

Bovista plumbea Pers. is smaller (1–3 cm) and the outer white coat shrinks to a series of white patches, bluish gray in age. The spores are large and oval (5–7 x 4.5–6 μ) and the base is attached by a series of fibers. *Bovista minor* Morg. is uncommon, small but not bluish, with a basal attachment of numerous fibers.

360 CALVATIA SUBCRETACEA Zeller   Edible, good/Common

Golf-ball size, warts with gray centers; western U.S.

Fruiting body 1.5–4 cm high, 2–5 cm thick, egg-shaped; outer skin thick, chalky white, covered with small pointed warts with *dark gray* tips. Gleba white to dark olive-brown mature. Sterile base lacking or very small.

Spores 3.5–6.5 µ nearly round, minutely spined to smooth, olive-brown.

Scattered to numerous in needles and duff in conifer forest at high elevations in Sp., S., and F. in RM, PNW, and PSW. I have very often found this fungus under Engelmann spruce in central Idaho and eastern Washington, where it appears to be most abundant.

This golf-ball size *Calvatia* looks very similar to *Lycoperdon marginatum* Vitt. and *Lycoperdon curtisii* Berk. Neither of these have the gray-tipped warts of *C. subcretacea,* and both are generally smaller and are widely distributed throughout North America. I myself have not found them at high elevations in the typical habitats of *C. subcretacea.* Both species of *Lycoperdon* have the ostiole in age but this fungus is not commonly collected or observed in age.

361 CALVATIA FUMOSA Zeller   Nonpoisonous/Common

Golf-ball size, smooth, smoky-gray; odor disagreeable

Fruiting body 3–5 (up to 10) cm in diameter, round, outer skin thick (1–2 mm) smooth, smoky gray to gray-brown. Gleba white to dark brown mature. Sterile base lacking. Odor disagreeable. Taste slightly bitter.

Spores 4.5–8 µ round to almost round, minute spines, brown.

Single to numerous under high elevation conifers in Sp. and S. in the PNW. I have usually found it under Engelmann spruce, subalpine fir, and white fir. It is commonly found fruiting in the same habitat with *C. subcretacea.* Its smooth, smoky gray outer skin is distinctive. It is the one puffball which we have found bitter and inedible but it is not poisonous.

362 LYCOPERDON CANDIDUM Pers.   Edible/Common

Small oval with sharp, white spines, flaking off in age

Fruiting body 1–3 cm high, 1–4.5 cm thick, nearly round to somewhat flattened, outer skin white, covered with white, sharp spines, sometimes several fused together, cracked and flaking off in patches at maturity. Gleba (flesh) firm, white to olive-brown or gray-brown. Sterile base white, olive-brown, chambered.

Spores 3.5–4.5 µ round, minutely spined, olive-brown.

Scattered to numerous in small clusters or even cespitose, on open ground, pastures, golf courses, in S. and F.; widely distributed.

This fungus, often called *L. marginatum* Vitt., is rather small with sharp spines which are white on the tips unlike *Calvatia subcretacea,* but the outer skin cracks and falls away unlike *L. curtisii* Berk.

363 LYCOPERDON PUSILLUM Pers.   Nonpoisonous/Infrequent

Minute, round, white, smooth; gleba brown in age

Fruiting body small, 8–20 mm thick, globose, smooth, white, with thin, long roots proliferating into the duff or ground, ostiole (Fig. 27) or pore forming in old age. Gleba white, yellow to deep coffee-brown at maturity. Sterile base none.

Spores 3.5–4.5 μ round, minutely spiny, an attached pedicel, brown.

Grows in groups on open hard ground, in Sp., S., and F. during cool wet periods; widely distributed.

I have collected this fungus under subarctic conditions in areas recently vegetated. The small size and smooth white outer skin distinguish it from other puffballs. It is probably edible but too small to be worthwhile.

364 LYCOPERDON PYRIFORME Pers.   Edible, good/Common

Pear-shaped, smooth, tan; in cespitose clusters on wood

Fruiting body 2–3 cm high, 1.5–4.5 cm thick, pear-shaped; outer skin smooth but sometimes apex has scattered hairs, soon minutely aerolate (finely cracked), pore at apex slow to form, light tan to brown in age. Gleba white to olive-brown mature. Sterile base dull white, occupies lower one-third of fruiting body.

Spores 3–3.5 μ round, smooth, olive-brown.

Single, scattered to most often dense cespitose clusters on well-rotted hardwood logs and stumps, wood debris, and sawdust in S. and F. In the western mountains look for this fungus in late July.

Old weathered spore sacs may be seen on wood at any time of the year.

365 LYCOPERDON PERLATUM Pers.   Edible, good/Common

Pear-shaped, dull whitish with deciduous cone-shaped spines; on the ground

Fruiting body 3–7 cm high, 3–6 cm thick, pear-shaped, outer skin covered with small, round, pointed spines (cone-shaped), these break off, leaving noticeable round spots, dull whitish to light tan. Gleba white to olive-brown, sometimes tinged purplish at maturity. Sterile base includes stalk composed of large chambers, white discoloring olive-brown in age.

Spores 3.5–4.5 μ round, minute spines, olive-brown.

Single, numerous to cespitose clusters in duff and humus under hardwoods and conifers, in late S. and F.; widely distributed.

The discrete, cone-shaped spines which break off and leave round scars are a unique characteristic of this species. It is very common in a wide variety of habitats. The flavor is unusual and extremely good. Also formerly known as *Lycoperdon gemmatum* Batsch.

366 LYCOPERDON ECHINATUM Pers.   Edible/Common

Oval to pear-shaped, white to brown in age with long, thin spines

Fruiting body 2–3.5 cm high, 2–4 cm broad, round to broadly pear-shaped, outer skin over the top *densely covered with long white spines (3–6 mm long)*, dark brown in age. Gleba white, yellowish, finally purplish. Sterile base small, not well developed, dingy white to purple-gray.

Spores 4–6 μ round, minute warts, purple-brown.

Single or more often numerous on moss, humus, or wood debris in late S. and F., in central and eastern North America. The long, distinctive spines separate this puffball from all others.

Pear-shaped with spines composed of fine hairs, pure white to dull brown in age

Fruiting body 1.5–8 cm high, 1.5–5 cm thick, pear-shaped, outer skin covered over the top with short, slender, separated spines (up to 1 mm high) along with some granules, honey-yellow, clay-brown to dull brown in age, pore at top (Figs. 27, 95) (3–5 mm wide), observed sooner than most. Gleba occupies rounded upper area, white but extremely variable while maturing, olive-brown to brown, usually brown with a dull purplish hue in old age. Sterile base well developed up to one half of total height, with large chambers ($\pm$ 1 mm wide), white when young to olive-brown or purplish brown.

Spores 4–5.5 μ round, minute spines or small warts, pedicel 6–10 μ long (Fig. 68), usually broken off, dull brown.

Scattered or even in loose clusters on humus, usually but not always under conifers in S. in northern latitudes but usually in F.; widely distributed.

L. pedicellatum Peck has denser, longer spines (1–2 mm), and pedicels (10–18 μ long) (Fig. 68) which remain attached to the spore. *L. umbrinum* var. *floccosum* Lloyd is a handsome shiny white variety with long, woolly soft spines. The spines are composed of a series of hairs which are bound together. This can be observed with a hand lens and should be compared with the conelike spines of *L. perlatum.*

LYCOPERDALES
The Stalked Puffballs

This group is artificially bound together on the basis of the stalk supporting the spore sac. The stalk may be thick and gelatinous as in *Calostoma* (Figs. 31–32), or somewhat tough and dry as in *Tulostoma* (Fig. 96), or woody, scaly and arising from a volva as in *Battarrea* (Fig. 28). In most of the group spores are released through an ostiole (pore) (Fig. 96), and some have well-formed teeth (Figs. 31–32) around the ostiole. In *Battarrea* the spore sac breaks open or ruptures around the middle, as shown in Fig. 28, to release the spores. In one species, *Pisolithus tinctorius,* the peridium, or outer wall, ruptures at the top and falls away exposing the spore mass. The spore mass (gleba) is powdery at maturity and not oriented into plates or penetrated by the stalk or any sterile branches as is commonly seen in the false puffball. The spores in the stalked puffballs are mostly globose (round) to almost round, with minute spines or warts, and resemble those found in the earthstars and true puffballs. *Battarrea* is a genus most commonly encountered in the desert areas of western North America. I have encountered one of the species in the Yukon in dry areas among desert shrubs. *Tulostoma* is also common in desert areas and various species of it will be found in sandy dry areas through the moist temperate zone. *Calostoma* is most prevalent in moist temperate areas, where it fruits in the winter or during cool moist periods. *Pisolithus* fruits on dry, hard soil but appears in a wide variety of habitats. Most of the species are probably mycorrhizal with higher plants.

Edibility.   This is a tough, woody, or distasteful group, and none of the species are edible.

368 PISOLITHUS TINCTORIUS Pers.   Edibility unknown/Infrequent

["Club-shaped Puffball"]

Tall, dingy brown club; gleba chambered, powdery mature

Fruiting body 5–25 cm high, 4–11 cm thick, elongate pear-shaped to club-shaped, dingy mustard-yellow to yellow-brown, tapering to an indefinite narrowed stalklike base. Gleba (spore mass) composed of numerous oval to circular chambers (peridioles) at first occupying the entire upper half of the fruiting body, gradually breaking up into a powdery mass from the top down. Base thick with coarse mustard yellowish rooting, intertwined fibers.

Spores 8–12 μ round, minutely spiny, cinnamon color.

Alone or several together, along roads, on hard ground, under conifers in the PNW, in S. and F.; widely distributed.

This ugly, stinky fungus with the very sticky, brown spores should be sectioned longitudinally. The presence of the chamber-like peridioles leaves no doubt about the identity of this most unusual fungus. It is known to form mycorrhiza and is therefore beneficial to higher plants.

369 CALOSTOMA CINNABARINA Desv.   Edibility unknown/Common

["Slimy Stalked Puffball"]

Gelatinous spore sac on gelatinous stalk (Figs. 31–32)

Fruiting body 3–5 cm high, 8–15 mm thick, outer skin a *thick, gelatinous (jelly-like) semitransparent covering* around the bright red inner spore sac, the gelatinous pieces fall away and appear like red seeds in the jelly-like covering, leaving a bright red inner spore sac; the sac has four to five raised deep red ridges at the top which break open in age to form the pore. Gleba powdery, buff color. Stalk 15–30 mm long, 5–15 mm thick, short, equal, also *covered with a thick semitransparent gelatin.*

Spores 14–20 x 6–8 μ oblong-elliptical, pitted, pale yellow.

Single to numerous or even abundant at higher elevations in the So. and SE in F., rare at low elevations.

This very unusual group of gelatinous stalked puffballs is unique and distinctive. The gelatinous outer skin over the bright red spore sac are the distinctive characteristics of *C. cinnabarina. Calostoma lutescens* (Schw.) Burnap (Fig. 31) is a long-stalked species with a thin outer gelatinous skin over a yellow spore sac capped with yellow teeth (red underneath), spores 6–8 μ round, pitted, yellow. *Calostoma ravenelii* (Berk.) Massee (Fig. 32) has no outer gelatinous layer over the spore sac which is yellow to clay-colored and has cinnamon-colored teeth around the pore. It may be found at low elevations even in winter in the SE and its spores are 10–17 x 6.5–7.5 μ, minutely pitted, white. These are unique and only the dry, thin-stalked Tulostomas resemble them to any degree. Even if edible, the small spore sacs would not be worth collecting.

370 TULOSTOMA SIMULANS Lloyd   Edibility unknown/Infrequent

["Stalked Puffball"]

Dry spore sac on narrow dry stalk

Fruiting body 2–4 cm high, consisting of a round spore case (10–15 mm thick) on a thin, *dry*, equal stalk 15–33 mm long, 3–4 mm thick (Fig. 96). Spore case dark red-brown but usually covered with dirt and debris, with a small, round elevated pore located at the top; in age the spore case disintegrates near the stalk, exposing the mature brown gleba.

Spores 4–6 μ round to nearly round, finely roughened, very pale yellowish.

Grows in groups in sand or sandy soil, often buried up to the spore sac, in S. and F.; widely distributed.

This is only one example of a genus with many species. Ten are reported in Coker and Couch (1928) from eastern North America, and White (1901) has listed fifteen species from the Southwest. They are common in arid regions or in sandy soils, beaches, or sand dunes in moist temperate to subarctic regions. We found this species near a glacial river in the Yukon Territory in sand deposited by the shifting river channel but very close to white spruce stands. One must be careful to dig down under the fruiting bodies. We have no information on their edibility.

371 BATTARREA PHALLOIDES (Dicks) Pers.   Edibility unknown/Numerous

["Desert Stalked Puffball"]

Flattened, oval spore sac; long, shaggy, woody stalk; volva present

Fruiting body consists of an ovoid spore sac on a long woody stalk which has an oval bulb at the base (Fig. 28). Spore sac 2.5–3.5 cm high, 3–5 cm thick, oval flattened, brown, splitting around the middle to reveal a dark brown sticky gleba (spore mass) which adheres to everything it touches. Stalk 20–40 cm long, 6–15 mm wide, equal, dry, woody, brown, covered with hairy scales, often appears shaggy, base is oval to nearly round with a one-layered volva surrounding the stalk.

Spores 5.5–8.5 x 4.5–6.5 μ nearly round, covered with thick, blunt warts, brown. Capillitium (hyphae) 27–50 x 5–6 μ, thick spiral bands around each cell.

Solitary or several in deserts in the PSW, PNW, RM up to A, fruiting in all seasons during wet periods. I first recorded this fungus on the Yukon-Alaskan border on dry south slopes among sage. No other genus appears even close to this strange fungus with its tall, narrow woody stem, puffball-like spore sac which splits around the center, basal bulb with a woody volva, and the sterile cells with spiral thickenings which are found among the warted spores in the gleba. *B. laciniata* Underwood has a much larger stalk (20–40 mm thick), a volva with several layers which splits apart in age, and longer elaters. It is found primarily in the PSW and not recorded from the PNW and A. Both of these are far too woody to eat.

HYMENOGASTRALES
The False Puffballs

It is important to note that the gleba (spore mass) is brittle to tough or even rocklike at maturity but *never powdery.* A powdery gleba is typically found in the true puffballs, stalked puffballs, and earthstars. The false puffballs are divided loosely into two groups. First there are those that have a homogeneous gleba but are hard as rocks at maturity. The second group has a more brittle gleba but in cross section they resemble mushrooms that never opened. Many of this latter group are closely related to various families of gilled mushrooms. They differ in that the pileus (cap) never expands and remains attached to the stalk. The basidia (Fig. 74) have lost the ability to forcibly discharge the spores. *Podaxis pistillaris* (Fig. 29) is a good example of a fungus adapted to a desert habitat. The unopened pileus protects the maturing spores from the dry, hot, desert climate. The spores in turn are thick-walled which makes them more resistant to water loss. Unlike the round to oval, spiny spores of the puffballs, most of the false puffballs have variously shaped spores, usually resembling the families of gilled fungi with which they are most closely related. Many of this group are found in harsh environments such as deserts and high mountain country, but others seem to share habitats with many other groups of fungi. Only a few of the several hundred known species in the group are covered here.

Edibility. Most of the false puffballs are not often found in large numbers. There are no edibles in this group because they are often tough, bitter, or tasteless.

Key to False Puffballs

1 a. Gleba in cross section consists only of small chambers (use hand lens); stalk absent, *372. Rhizopogon rubescens*
   b. Gleba in cross section with branched, sterile, buff flesh or fruiting body with an obvious stalk, **2**
2 a. Fruiting body pear-shaped, greenish yellow, dry; gleba in cross section with branched, sterile, buff flesh, *373. Truncocolumella citrina*
   b. Fruiting body with an obvious stalk, **3**
3 a. Fruiting body with a long stalk (15–30 cm long) and a narrow cylindric cap (6–10 cm tall); in desert regions (Fig. 29), *374. Podaxis pistillaris*
   b. Fruiting body egg-shaped; stalk barely protruding or 5–25 mm long, **4**
4 a. On conifer wood near snowbanks; cap yellow-brown, viscid, *375. Nivatogastrium nubigenum*
   b. On earth, especially in nurseries and flower beds; cap dingy white, hairy in age, *376. Endoptychum agaricoides*

372 RHIZOPOGON RUBESCENS Tulasne   Edibility unknown/Common

Rounded, rubbery with white peridium stained red; gleba chambered throughout

Fruiting body oval to round, 1–5 cm thick, white tinted yellow to yellow with red stains in age or on exposure to the air, base with white, rootlike strands which penetrate into the soil. Gleba rubbery and tough, chambered, white changing to honey-color or cinnamon-brown in age.
Spores 5–10 x 3–4.5 μ elliptical, smooth, tinged yellowish.

Several to gregarious, in duff and earth often under five-needle pines, in Sp., S., and F.; widely distributed.

There are many species of *Rhizopogon* and other genera, usually called "false truffles" or "gubers," in North America. Many of these "false truffles" form mycorrhiza with forest trees and shrubs. They are usually buried in needles and leaves or actually in the soil. Deer often feed on them and areas should be examined where they have been digging.

373 TRUNCOCOLUMELLA CITRINA Zeller   Edibility unknown/Common

Oval to pear-shaped, greenish yellow; gleba olive-gray; under western conifers

Fruiting body 15–40 mm high, 20–30 mm broad, egg-shaped with a short, narrowed base, outer skin dull greenish yellow, dry. Gleba olive-gray to hair-brown, irregular chambers, gelatinous to touch. Sterile base a short, thick stalk of firm, buff flesh which branches and proliferates up into the gleba.

Spores 6.5–10 x 3.5–4.5 µ elliptical, smooth, thin-walled, yellowish as seen with a microscope.

Solitary, scattered to numerous in needle duff under conifers, most frequently under Douglas fir, in S. and F., in PNW and RM. This fungus resembles a mushroom button but when cut in half, the gleba and sterile base are typically that of a Gasteromycete.

374 PODAXIS PISTILLARIS (L. ex Pers.) Fries   Edibility unknown/Common

Long oval head on a thin stalk; gleba black; in desert regions

Fruiting body tall and erect (Fig. 29), appears like an unopened "Shaggy Mane" *(Coprinus comatus)*, 10–30 cm high, 15–30 mm thick. Head 6–10 cm long, 10–15 mm thick, oval to elongate, dry, smooth but outer skin soon covered with large, ragged scales, appearing shaggy, pure white to tan or yellow-brown, margin appressed to stipe, occasionally breaking and opening slightly. Gleba contorted, irregular, chambered, white to yellow, often with reddish tones and finally blackish brown, occasionally slightly deliquescent (turning to black liquid like the "Inky Caps"). Stalk 15–30 cm high, 10–15 mm thick, enlarging at base to form small bulb, rest equal, dry, soon scaly, same color as head, extends through gleba to attach to the outer skin in the same way as a gilled mushroom.

Spores 10–15 x 10–13 µ round to nearly round, thick-walled with a pore at apex, yellow-brown.

Single or in closely clustered groups; in desert regions (from 65 feet below sea level to 5,000 feet above) after seasonal rain or in irrigated fields during any season but most often in Sp., in PSW and southern RM. It is found in desert regions near the equator around the world. There seems to be little doubt that this plant is related to *Coprinus,* especially *C. comatus.*

375 NIVATOGASTRIUM NUBIGENUM   Edibility unknown/Numerous
(Hark.) Smith and Singer

Head rounded, viscid, light brown, gleba like contorted gills, brown; on conifer wood

Fruiting body 15–40 mm high, 10–24 mm thick, head convex, smooth, somewhat viscid, ochre to tawny, fading to lighter or nearly white in age, sometimes the margin pulling away somewhat from the stalk, revealing the gleba, wrinkled just at the attachment to the stalk. Gleba chambered like contorted gills, cinnamon-brown. Stalk short 5–25 mm long, 5–12 mm thick, nearly equal, dry, flattened hairy fibrillose, dull white.

Spores 7–9 x 5.5–6.5 μ elliptical, smooth, honey-color.

On conifer wood at high elevations near snowbanks in Sp. or early S., PNW and RM.

This fungus looks at first like a gilled mushroom that has not yet opened. The stalk is attached in the manner of a gilled mushroom stalk and the closest relatives now known are the gilled Pholiotas. However, no spore print can be obtained because the spores are not forcibly discharged and the cap never expands.

376 ENDOPTYCHUM AGARICOIDES Czern.   Edibility unknown/Numerous

Large, oval, leather-colored; gleba like contorted gills; stalk extends into gleba

Fruiting body 1–7 cm high, 1–6 cm thick, like a young unopened cap of a gilled mushroom, oval, narrowed at the top with a short stalk, outer skin dull, hairless, dry, white at first to color of dull leather, more or less hairy or even somewhat shredded in age. Gleba chambered, white at first then yellowish brown in age. Stalk barely protruding, extends up into gleba, white at first, dull yellowish in age.

Spores 7.5–9.5 x 5–7 μ elliptical, smooth, thick inner wall, pore at apex, chocolate-brown.

Scattered to cespitose clusters in cultivated earth, nurseries, flower beds, in S. and F.; widely distributed. Also known as *Secotium agaricoides* (Czern.) Hollos.

Hetero-
basidiomycetes

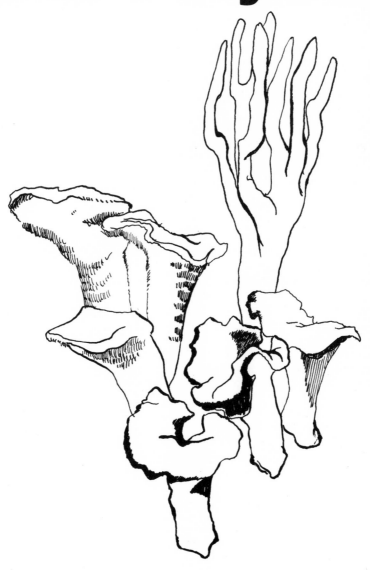

TREMELLALES
The Jelly Fungi

Certainly one of the most curious groups of fungi, these gelatinous fruiting bodies are usually found on logs, stumps, and twigs but some are on the ground. The yellow to orange species look like so many amorphous chunks of butter and have been called "Witches' Butter." All of the species of jelly fungi have a gelatinous flesh and most of them are yellow, orange, red to brown, or colorless. Almost all of the species fruit in cool, moist weather and soon disappear in dry weather. In the southern U.S., they are frequently encountered in the winter months when other fungi are not fruiting. The spores are borne over the surface of the fruiting bodies on basidia which are structurally different from those illustrated for the mushrooms (Fig. 74). Nevertheless, the jelly fungi are also Basidiomycetes, in a separate group called the Heterobasidiomycetes.

Edibility. Most species are either too small or tasteless to be worth eating, but there are no reports of toxins.

Auricularia auricula is the only species commonly eaten. In the PNW it is often so abundant near melting snowbanks that large quantities can be collected, and it could serve as a major spring edible. In China and the Far East it is grown commercially, dried, and used as a flavoring for soup. It is commonly called Cloud Ears (Yung Nge) or Wood Ears (Muk Nge) in the Orient.

Key to Jelly Fungi

1 a. Not stalked, earlike, cuplike, or irregular masses of no particular form, always on wood, **2**
　 b. Stalked clubs, sometimes with inflated heads, spoon-shaped with teeth or fan-shaped, on wood or on ground, **6**

2 a. Yellow cups or pustules (Fig. 108) grow in groups on conifer wood in PNW and RM in early spring, *377. Guepiniopsis alpinus*
　 b. Jelly-like masses of earlike lobes, yellow, orange to brown, **3**

3 a. Yellow to orange jelly-like masses, **4**
　 b. Brown to gray-brown lobes and wrinkled fruiting bodies, **5**

4 a. Yellow-orange, firm, with a white base; collapses when dried, *378. Dacrymyces palmatus*
　 b. Orange to golden-yellow, horny when dried, base not white (Fig. 107) *379. Tremella mesenterica*

5 a. Large, 2–12 cm long, ear-shaped, lobed, sometimes wrinkled; tough gelatinous; cinnamon-brown, *380. Auricularia auricula*
　 b. Smaller, 2–6 cm long, irregularly branched and lobed to wrinkled; soft, very thin; dark brown; covered with brown warts, *381. Exidia glandulosa*

6 a. Slender, coral-like, cylindrical, branched, orange-yellow, pliant, viscid stalks; on conifer wood, *382. Calocera viscosa*
　 b. Not coral-like; with an expanded, distinctive head, **7**

7 a. Gelatinous, dull whitish, translucent with small white teeth; often spoon-shaped, *383. Pseudohydnum gelatinosum*
　 b. Funnel-shaped to fan-shaped without teeth; pink to rose; firm gelatinous on ground, *384. Phlogiotis helvelloides*

377 GUEPINIOPSIS ALPINUS   **Edibility unknown**/Common

(Tracy and Earle) Bres.

Fruiting body cup-shaped, gelatinous, 4–15 mm broad, yellow to orange-red, narrowing to a small point of attachment, spore-bearing surface in the concave cup. Spores 15–18 x 5–6 μ sausage-shaped (Fig. 72) with three to four cross walls.

Abundant over logs, twigs, and stumps of conifers, usually in clusters of two to four, in early Sp.; in RM, PSW, PNW, and adjacent C.

This is certainly the most abundant single fungus during the period of snow melt in the western mountains. I have no information on its edibility.

378 DACRYMYCES PALMATUS (Schw.) Bres.   **Nonpoisonous**/Common

Fruiting body 1–6 cm broad, a lobed mass of gelatinous, tough tissue, yellow-orange to orange-red with a whitish point of attachment. Spore-bearing surface over upright colored portion of the fruiting body. Spores 17–25 x 6–8 μ oblong to sausage-shaped (Figs. 71–72), up to seven cross walls.

Often abundant on conifer logs and stumps in early Sp. to S.; widely distributed. A minute (0.5–3 mm) orange-tinted green species, *D. minor* Peck, may be seen in eastern North America in late W. to early Sp. The *Dacrymyces* species do not taste good.

379 TREMELLA MESENTERICA   **Edibility unknown**/Numerous

(S. F. Gray) Pers. ["Witches' Butter"]

Fruiting body 2–10 cm broad, 3–4 cm thick, lobed and brainlike, orange to golden yellow, tough gelatinous, horny when dried. Spore surface over the upright lobes. Spores 10–14 x 7.5–9 μ elliptical, smooth.

Single to scattered on hardwoods in late S. and F.; widely distributed. (Fig. 107). *T. foliacea* Pers. ex Fr. is a leaflike, thin-lobed, pale brown fungus usually on hardwoods but occasionally on conifer wood as well. The above species are tasteless and not edible.

380 AURICULARIA AURICULA (Hook.) Underwood   **Edible**/Common

Fruiting body large, 3–15 cm broad, thick gelatinous earlike lobes, often several from one point of attachment, brown, resembles liver, smooth on top but often ribbed beneath, dries very tough and hard. Spores over upper surfaces. Spores 12–16 x 4–6 μ sausage-shaped, smooth.

Several to abundant on logs and wood of conifers, less common on hardwoods, fruiting during cool wet periods; very common near melting snow in PNW and RM but widely distributed. This fungus is dried and eaten in China where it is commonly called Yung Nge or Muk Nge. It is one of the few jelly fungi that tastes good and is large enough to be of practical interest as an edible.

381 EXIDIA GLANDULOSA Bull. ex Fr.   **Nonpoisonous**/Common

Fruiting body in narrow rows, 5–15 cm long, leaflike to contorted, gelatinous, soft and thin, blackish olive-brown with numerous brown warts. Spores 8–16 x 3–5 μ smooth, sausage-shaped.

Several to numerous on the limbs of hardwoods, during periods of cool wet weather, especially in Sp.; widely distributed. Tasteless and not edible, much too thin to be confused with *Auricularia*.

382 CALOCERA VISCOSA (Pers. ex Fr.) Fries **Edibility unknown**/Numerous

Fruiting body 3–10 cm tall, narrow, slender, cylindric branches, viscid, pliant, orange-yellow to golden-yellow. Deeply rooted by a white base. Spores 9–14 x 3.5–5 μ slightly sausage-shaped, smooth, one cross wall at maturity.

Several to numerous on conifer logs and stumps in late S. and F.; widely distributed.

I have encountered this fungus from Maine to Alaska. It can be easily confused with a coral fungus. The pliant, viscid fruiting body is different from most coral fungi, which are brittle and moist to dry. The other genus which resembles the coral fungi is a *very tough,* dingy, white to gray fungus *(Tremellodendron)* with many stout erect branches. The most common species is *Tremellodendron pallidum* (Schw.) Burt.

383 PSEUDOHYDNUM GELATINOSUM (Fr.) Karst. **Edibility unknown**/Common

Fruiting body 1–3 cm broad, convex, gelatinous with a whitish, translucent appearance, with short teethlike spines protruding from the lower surface which is colored like the top. A short lateral stalk is most common in specimens from the SE but in the PNW the stalk is upright, longer, and supports the laterally attached head. Spores 5–7 μ nearly round, smooth, white.

Several to numerous on well-rotted wood, in early Sp. to S. in the PNW and RM, most other areas in F.; widely distributed.

There is no other jelly fungus with teeth like this; neither is there a hydnum which has a tiny translucent fruiting body. I have no reports on its edibility.

384 PHLOGIOTIS HELVELLOIDES (Fr.) Martin **Edible**/Common

Fruiting body 3–8 cm tall, fan-shaped to somewhat lobed, gelatinous, smooth, pink to rose; narrow below to a stalk. Spores 10–12 x 4–6 μ oblong (Fig. 71), smooth.

Several to numerous under conifers on hard-packed ground or in well-decayed conifer debris, Sp. and S. in the RM and PNW, and late S. and F. elsewhere, widely distributed. I have encountered large fruitings of this beautiful fungus on the Flathead Indian Reservation near St. Ignatius, Montana. I have not collected enough at one time to try the recipe for "Candied Red Jelly Fungus" recommended in *Kitchen Magic with Mushrooms.*

Ascomycetes

MORCHELLA
The True Morels

This group is probably the best known and most sought after of all the edible fungi. The various species all fruit in the spring and they are all edible. The "head" or brown top of the fungus resembles a pine cone and often is referred to by that name. The brown chambered top is actually the spore-bearing surface. Look carefully at the chambers. Unlike the false morel, the entire "head" is composed of ridges and pits (Fig. 105). The false morel on the other hand has a wrinkled, undulating "head" without distinct ridges and pits (Fig. 106). A second and important point is the attachment of the "head" to the stalk. In all the true morels, except *Morchella semilibera* (Pl. 385) the "head" is attached directly to the stalk as shown in Fig. 42. The "head" of the false morel hangs about the stalk in skirtlike fashion and is attached to the top of the stalk (Fig. 43).

Edibility. The texture and flavor of the morel ranks it as one of the best of all edible fungi. Morels fruit only in the spring and any morel-like fungi encountered in the summer and fall are probably false morels and *should be avoided.* Numerous attempts have been made to fruit the morel to be sold commercially, and although the spores germinate readily and pure cultures of the mycelium are obtainable, no one has succeeded.

Each morel should be cut lengthwise so that the hollow center can be examined for insects. The flavor is excellent so that any dish which emphasizes the mushroom is perhaps a more desirable choice. Perhaps one of the most interesting and unique recipes for the preparation of morels is "Mushroom Fondue" (see Recipes).

385 MORCHELLA SEMILIBERA   Quel.   **Edible, choice** /Common

386 MORCHELLA ANGUSTICEPS   Pk.   **Edible, choice** /Common

387 MORCHELLA ESCULENTA   Pers. ex StAmans   **Edible, choice** /Common

388 MORCHELLA DELICIOSA   Fr.   **Edible, choice** /Numerous

The characters of the true morels are described in the introductory paragraph above. Mycologists disagree concerning the number of species and many intermediate forms will be found between the four illustrated species. The characters separating these morels are found in the key. All morels are edible and choice.

Key to the Morels

1 a. Head light brown, not attached directly to the stalk but recessed somewhat; stalk long and thin, (*M. hybrida* of some) 385. *M. semilibera,* Half-free Morel or Cow's Head

   b. Head attached directly to the stalk, **2**

2 a. Ridges black, heads conical with brown pits arranged in a loose radial alignment toward the top of the head, 386. *M. angusticeps,* Black Morel

   b. Ridges brown and same color as pits, or white but not black, **3**

3 a. Ridges and pits both yellow-brown; pits not radially arranged (Fig. 105), 387. *M. esculenta.* Yellow Morel (A giant form of this species has been referred to as *M. crassipes.*)

   b. Ridges white, pits deep brown; pits not radially arranged, 388. *M. deliciosa,* White morel (A form with a narrow conical cap has been called *M. conica.*)

Because spring comes to different areas in North America at different times, I have collected morels in Maryland and Virginia most often between April 20th and May 10th. Morels are out in McCall, Idaho in late May and early June. In the Canadian Rockies in Banff Park, great numbers of morels were gathered by us during the last week in July. The season is usually about three to four weeks long.

Many attempts have been made to describe where morels grow. In the east I usually examine oak ridges, tulip-poplar stands with may apples growing beneath them, and old apple orchards. In late spring in the east *Morchella crassipes* is often found in old fields. In the west a large *Morchella* which appears to be a hybrid between *M. angusti-ceps* and *M. esculenta* is found under Douglas fir and yellow pine. In Alaska and the Yukon, *M. angusticeps* is found under spruce or in old burns. Examples of this sort add up to a wide variety of habitats where one can find the morel.

GYROMITRA, HELVELLA, AND RELATED SPECIES
The False Morels

The false morels are found throughout the growing season, but some species are more restricted in their fruiting period. Therefore, the species that may commonly be found when the true morels are out are keyed out at the beginning of the key. The false morels have a cap or head which is wrinkled (Fig. 106) to nearly smooth (Fig. 41) or saddle-shaped (Fig. 40). Not only are the typical ridges and pits of the true morel missing, but the head is joined to the stalk at the top (as shown on the right in Fig. 43) and not fused to the stalk (as shown to the left in Fig. 43). The stalk should be noted; it is usually either smooth to slightly fluted (Fig. 40) or deeply ribbed (lacunose; Pl. 392). A characteristic stalk is found in each species. The flesh of the false morels is thin and brittle.

Edibility. This is a really variable group, with toxic and edible species resembling each other confusingly. The chance for error is great. In addition, many people are adversely affected by some of the edible species. Finally, mycologists have not com-pletely explored the various species and their distribution. All of the toxins are not completely known, but the effect of one of them, monomethylhydrazine (MMH) in the bloodstream, is severe gastrointestinal disturbance and, in some cases, death. Various people have recommended parboiling a false morel (usually *Gyromitra esculenta*) and pouring off the water, but some people have suffered ill effects nonetheless. (see Group V toxins).

G. gigas in the PNW is considered one of the good edible spring mushrooms. My family and I have eaten it for many years with great enjoyment. We have also eaten *G. esculenta* but *only* in the PNW in the spring. I would recommend avoiding all large false morels and never eating any false morel collected in summer or fall. A very small portion should be consumed when first trying a species such as *G. esculenta.* Those false morels which can be eaten are usually prepared in the same manner as the true morels.

Key to False Morels

   1 a. Fruiting in the spring or under spring conditions in the mountains, **2**
      b. Fruiting in the summer and fall, **8**
   2 a. Bell-shaped, smooth or wrinkled caps; stalk equal, narrow, **3**
      b. Cuplike, lobed to saddle-shaped cap; stalk variously shaped, often ribbed, **4**

3 a. Caps smooth (Fig. 41), *389. Verpa conica*
   b. Caps wrinkled (Fig. 42), *390. Verpa bohemica*

4 a. Fruiting near or under melting snowbanks in the Rocky Mountains; yellow in KOH, *391. Gyromitra gigas*
   b. Does not fruit near melting snowbanks, **5**

5 a. Stalk deeply ribbed, white with red stains; cap large, thin, very brittle, brown, *392. G. californica*
   b. Stalk does not stain red, **6**

6 a. Cap massive, extremely wrinkled, dark brown; eastern North America, *393. G. caroliniana*
   b. Cap medium, yellow-brown to chocolate-brown, **7**

7 a. Cap saddle-shaped to irregular; stalk 2–5 cm thick, *394. G. brunnea*
   b. Cap wrinkled to irregular; stalk narrow, 1–2.5 cm thick, *395. G. esculenta*

8 a. Stalk deeply ribbed, **9**
   b. Stalk nearly smooth, **11**

9 a. Stalk white, tinted or stained red, *392. G. californica*
   b. Stalk colored, if white it does not stain red, **10**

10 a. Stalk short, ribbed, white; head cup-shaped, *396. Paxina acetabulum*
   b. Stalk long, ribbed, pale reddish to red-brown; head saddle-shaped to wrinkled, *397. Helvella lacunosa*

11 a. Stalk tall 5–10 cm, thin (3–10 mm), and white; head saddle-shaped, dark brown, *398. Helvella elastica*
   b. Stalk tall (6–8 cm), thicker (5–15 mm), white, tinted pinkish; head larger, saddle-shaped, reddish brown, *399. G. infula*

389 VERPA CONICA (Mull.) Swartz ex Pers. **Edible**/Infrequent

Cap conic, bell-shaped, 1–2 cm long, nearly smooth, brown, skirtlike and white beneath. Flesh brittle. Stalk 3–6 cm long, 8–15 mm thick, nearly equal, whitish. Spores 22–26 x 12–16 μ elliptical, smooth.

On ground in Sp. and S.; widely distributed.

390 VERPA BOHEMICA (Krombh.) Schroet. **Edibility unknown**/Common

Cap conic, bell-shaped, 2–3 cm long, distinctly wrinkled, yellow-brown, skirtlike and white beneath. Flesh brittle. Stalk 6–8 cm long, 10–25 mm thick, smooth, cylindric, sometimes furrowed, white. Spores 60–80 x 15–18 μ elliptical, smooth.

Single to several in wet areas in early Sp. (late March and April); widely distributed. It is eaten by many but it does make some people ill.

391 GYROMITRA GIGAS (Krombh.) Quel. **Edible, choice**/Common

["Snow Morel"]

Cap large, 5–20 cm broad, deeply wrinkled, yellow-brown, red-brown only in age. Flesh solid, brittle. Stalk massive, furrowed, and folded, white. Spores 24–36 x 10–15 μ elliptical, smooth to finely warted.

Several to numerous near melting snowbanks in PNW and RM in early Sp. One of the really good edibles. We have enjoyed it for years in Idaho but collect this false morel *only* near snow in the spring. Often listed as *Helvella gigas* Krombh.

392 GYROMITRA CALIFORNICA (Phill.) Raitv. Poisonous/Common

Cap large, 8–20 cm broad, spreading, lobed and folded, tan to red-brown in age. Flesh *very thin, brittle,* not solid. Stalk robust, 5–12 cm long, 2–5 cm thick, *white tinted pink,* deeply ridged, *hollow.* Spores 15–18 x 8–9 μ elliptical, smooth.

Several to numerous in RM, PNW, and PSW in early S. on ground under conifers. Also called *Helvella californica* Phill. In all likelihood MMH and not helvellic acid (Group V toxins) is the toxin present.

393 GYROMITRA CAROLINIANA (Bosc. ex Fr.) Fries Edible (?)/Infrequent

Cap very large, 5–20 cm broad, tightly wrinkled, brownish black. Flesh solid, white. Stalk robust, 8–10 cm long, 2–5 cm broad, enlarging at base, white. Spores 25–30 x 12–14 μ elliptical, minutely warted.

Single to several in Sp.; most commonly seen in the SE but widely distributed.

Easily confused with *G. brunnea,* which is not edible, therefore not recommended.

394 GYROMITRA BRUNNEA Underwood Poisonous/Numerous

Cap large, 5–15 cm broad, often saddle-shaped, lobed, often veined, chocolate-brown, white beneath. Flesh brittle, white. Stalk large, 6–15 cm long, 2–5 cm broad, slightly larger base, somewhat ribbed to smooth, white, nearly hollow to cottony inside. Spores 28–30 x 12–15 μ elliptical, finely warted.

Single or several under hardwood forests in eastern North America in early Sp. Causes severe illness from high levels of MMH, (Group V toxins). *G. fastigiata* (Krombh.) Rehm. is similar but looks as though it might be a darker brown form of southern *Helvella gigas;* however, it also contains Group V toxins. It is occasionally found at low elevations in western North America.

395 GYROMITRA ESCULENTA Fr. Edible/Common

Cap 2–8 cm broad, saddle-shaped, lobed, irregular, wrinkled, brown to red-brown in age, lighter beneath. Flesh brittle. Stalk 3–15 cm long, 10–20 mm thick, dull whitish, nearly equal enlarged somewhat just at base. Spores 20–28 x 11–16 μ elliptical, smooth.

Single to usually numerous on ground under conifers in early Sp. (May–June); widely distributed.

I will eat this only in the PNW and RM. It is easily confused with an eastern species and individuals react differently to it. Perhaps different populations of it contain different amounts of the MMH toxin. I find the stalk thinner and longer than most of its closely related species. See Group V toxins.

396 PAXINA ACETABULUM (L.) Kuntze Edibility unknown/Common

Cup large, 3–6 cm broad, deep, dark brown, light brown exterior. Flesh brittle. Stalk 1–2 cm long, 8–15 mm thick, continuous with cup, ribbed, white below. Spores 18–22 x 12–14 μ elliptical, smooth.

Single to several on ground in many habitats in Sp. and early S.; widely distributed. The large cup on a ribbed thick base set this apart from others. Often called *Helvella acetabula* (L. ex Fr.) Quél. This fungus is not eaten, but I have no information on its edibility or toxins.

397 HELVELLA LACUNOSA Fr. **Edible**/Common

Cap 2–6 cm broad, saddle-shaped, somewhat wrinkled, dark brown. Flesh brittle, gray. Stalk 4–10 cm long, 15–20 mm thick, equal, deeply ribbed, gray tinted olive or yellowish. Spores 17–20 x 11–13 μ elliptical, smooth.

Single or several on wet soil, often in burned areas, in F.; widely distributed. *H. crispa* Fries is similar but white to cream-colored overall.

I would not recommend eating it because of the closely related toxic species with which it could be confused.

398 HELVELLA ELASTICA Bull. ex Fries **Edibility unknown**/Common

Cap 2–3 cm broad, saddle-shaped (Fig. 40), nearly smooth, brown. Flesh brittle. Stalk 5–10 cm long, 2–8 mm thick, equal, smooth, white. Spores 18–22 x 11–13 μ elliptical, smooth.

Several on ground in woods, both hardwood and conifer, in S. and F.; widely distributed.

399 GYROMITRA INFULA (Schaeff. ex Fr.) Quel. **Poisonous**/Common

Cap 5–8 cm broad, saddle-shaped, finely wrinkled, red-brown, sometimes yellow-brown. Flesh brittle. Stalk 4–9 cm long, 5–15 mm thick, nearly equal, sometimes fluted but not ribbed, white to buff or tinted pink. Spores 18–24 x 8–12 μ elliptical, smooth.

Single or several on ground or in wood debris in S. and F.; widely distributed. This also contains toxins, belongs among Group V toxins and should be strictly avoided.

PEZIZA AND RELATED SPECIES
The Cup Fungi

The cup fungi usually are most abundant during cool early spring weather and again in the fall. The fruiting bodies are attached directly to the ground and only *Urnula* has a stalk. The flesh is thin and brittle and the spore-bearing surface is found within the cup. The spore sac (ascus, Fig. 77) has spores which develop and mature inside. When these are forcibly discharged the millions of spores create a small cloud of dust. The spore sacs are separated by sterile cells called paraphyses (Fig. 78). There are a number of colorful species in this large group and many small and large species too numerous to include here.

Edibility. Some of the species are disagreeable tasting and others have no taste at all. I know of no toxins in this group. *Discina* and *Peziza* are collected by many people, and I find the flavor rather good. *Sarcosphaera* is hard to clean but has a delicate flavor.

Key to Cup Fungi

1. a. Cup bright orange with or without blue stains, **2**
   b. Cup another color, **3**
2. a. Cup orange with conspicuous blue stains, *400. Caloscypha fulgens*
   b. Cup orange with no stains, *401. Aleuria aurantia*
3. a. Cup deep scarlet; lighter on outside, *402. Sarcoscypha coccinea*
   b. Cup another color, **4**

    4 a. Cup ear-shaped (Fig. 52), like a rabbit's ear, *403. Otidia smithii*
       b. Cup-shaped, flattened, or deep urn-shaped, **5**
    5 a. Deep gray, urn-shaped on a long gray stalk, *404. Urnula craterium*
       b. Cup-shaped to flattened, **6**
    6 a. Deep cup; partially buried; pink inside; margin splitting, *405. Sarcosphaera*
          coronaria
       b. Shades of brown, flattened to cup-shaped, even turned under, **7**
    7 a. Cup-shaped, rich brown, *406. Peziza badia*
       b. Flattened, disclike, even turned under on margin; tan to dark brown,
          407. Discina perlata

400 CALOSCYPHA FULGENS   **Edibility unknown**/Numerous

(Fr.) Fuckel ["Orange Peel"]

Cups 1–5 cm broad, margin wavy, pale orange with blue-green stains. Flesh brittle. Spores 10–12 x 6–8 μ elliptical, smooth.
In clusters on soil under conifers during cool Sp. or early S. weather; widely distributed; common in the RM.

401 ALEURIA AURANTIA (Fr.) Fuckel ["Orange Peel"]   **Nonpoisonous**/Common

Cups 3–8 cm broad, bright orange. Flesh brittle. Spores 18–22 x 9–10 μ elliptical, reticulate ridged.
Several to abundant on hard soil during moist weather, usually in early S.; widely distributed.

402 SARCOSCYPHA COCCINEA   **Edibility unknown**/Common

(Fr.) Lambotte ["Scarlet Cup"]

Cups 2–6 cm broad, scarlet inside, dull white outside. Flesh brittle. Spores 24–32 x 12–14 μ elliptical, smooth.
Single or several on sticks, usually in early Sp.; widely distributed.

403 OTIDIA SMITHII Kanouse ["Ear Fungus"]   **Edibility unknown**/Numerous

Cup 2–6 cm high, erect ear-shaped (Fig. 52), red-brown. Flesh brittle. Spores 18–24 x 12–14 μ elliptical, thin-walled, smooth.
Several to numerous on moist ground, in S., western United States. There are several species in North America including a yellow-brown species. All have the characteristic growth form.

404 URNULA CRATERIUM (Schw.) Fries ["Gray Urn"]   **Nonpoisonous**/Common

Cups urn-shaped, 3–4.5 cm broad, brownish black. Flesh tough and fibrous. Stalk 3–4 cm long, 5–8 mm thick, color of urn. Spores 25–35 x 12–14 μ elliptical, smooth.
Several to clustered on hardwood sticks and logs in early Sp. in So, SE, NE, and adjacent C.

405 SARCOSPHAERA CORONARIA   Edible, good/Common

(Jacq. ex Cooke) Boud. ["Pink Crown"]

Cups 4–12 cm broad, deep round cup, buried in soil at first, white, opening at surface (Fig. 48) pink inside, margin splitting crownlike (Fig. 48). Flesh brittle. Spores 15–18 x 8–9 μ elliptical, truncated ends (Fig. 65), smooth.

Several to numerous over large areas, most abundant in PNW and RM in Sp.; widely distributed. I have eaten this and like it, but it requires much cleaning. Another name for it is *Sarcosphaera eximia* (Durieu & Lev.) Maire.

406 PEZIZA BADIA Pers. ex Merat ["Pig's Ears"]   Edible/Common

Cups 3–10 cm broad, shallow cup-shape, red-brown to brown inside and outside. Flesh brittle. Spores 17–23 x 8–12 μ elliptical, broken reticulations. Paraphyses (Fig. 78) with enlarged heads.

Single to numerous on the ground in S. and F.; widely distributed. *P. repanda* Pers. is similar but often on rotten wood. *P. violacea* Pers. is violet. There are many other species of *Peziza.* No reports of toxins have come to my attention.

407 DISCINA PERLATA Fries ["Pig's Ears"]   Edible/Common

Cups 6–10 cm broad, flat or margin turned down, often veined, dark-brown, underside whitish. Flesh brittle. Spores 30–35 x 12–14 μ elliptical, pointed ends, minute warts. Paraphyses (Fig. 78) with enlarged heads.

Single to numerous in conifer woods on ground or rotten wood; widely distributed in Sp. and early S. This is edible and good. In the RM and PNW there are several edible species.

EARTH TONGUES
AND OTHER ASCOMYCETES

The other "sac fungi" (Ascomycetes) are so numerous that only a few can be included here. These fungi are either very common or so unusual that they would be noticed at once. The earth tongues are an unusual group with a variety of small fruiting bodies. They are all solitary and vary from club-shaped (Fig. 47) to spatula-like (Fig. 44) or have a wrinkled head (Fig. 46). Other Ascomycetes are found as parasites on insects (Fig. 55), while still others parasitize tubers (Fig. 56). At least two species form large, thick, gelatinous fruiting bodies on wood (Figs. 50, 51). *Hypomyces lactifluorum* (Schw.) Tul., an orange parasite, grows over the surface of members of the Russulaceae. It is not recommended because it could be parasitic on a nonedible species of mushroom.

Truffles are a rarity in North America. They occur in some areas of the deep South and on occasion elsewhere. They are most frequently encountered along the Pacific Coast but they are not found in sufficient quantity to be of interest except to the mycologist and research forester. They are usually found among leaves and needles or buried in the soil. Many of the species give off a strong odor, attracting rodents or deer. These animals will often dig up wide areas in search of truffles.

Edibility. Most of the species included here are not edible. Truffles have of course long been a prize edible in southern Europe, and although some of the good edible species can occasionally be found in America, they are never in sufficient quantity to serve as a food.

Key to Earth Tongues and Other Ascomycetes

1 a. Fruiting body with a stalk and head (For orange and green parasites, see 422. *Hypomyces*), **2**

   b. Fruiting body without a stalk, **11**

2 a. Cup on a thin stalk; very small, *408. Sclerotinia tuberosa*

   b. Head variously shaped but not cup-shaped, **3**

3 a. Fruiting body flattened at the top; yellow overall, *409. Mitrula irregularis*

   b. If flattened, not yellow, **4**

4 a. Head formed like a spatula; stalk enlarged toward base, *410. Spathularia clavata*

   b. Head round, depressed, or cylindric; sometimes branched, **5**

5 a. Head oval, wrinkled, yellow-green, viscid; stalk greenish and viscid, *411. Leotia lubrica*

   b. Head not as above, not green, **6**

6 a. Fruiting body branched, **7**

   b. Fruiting body unbranched, **9**

7 a. On wood; black, irregular finger-like clubs, *412. Xylosphaera polymorpha*

   b. On ground; not finger-like, **8**

8 a. Several fused columns, more or less erect (Fig. 58), *413. Underwoodia columnaris*

   b. Spreading ear-shaped branches from a common stalk (Fig. 53), *414. Wynnea americana*

9 a. On sticks in mountain streams; yellow, small head; brown stalk, *415. Vibrissia truncorum*

   b. Not as above, **10**

10 a. Small reddish clubs on insects (usually beetles) (Fig. 55), *416. Cordyceps militaris*

    b. Large dull brown clubs with rounded heads; parasitic on *Elaphomyces granulatus* (Fig. 56), *417. Cordyceps capitata*

11 a. Forming a hard, spindle-shaped, black growth around the branches of cherry (Fig. 60), *418. Apiosporina morbosa*

    b. Not as above or on cherry, **12**

12 a. A hard, dingy maroon to black, rounded growth on hardwood logs and sticks; concentrically zoned when cut in two (Fig. 59), 419. *Daldinia concentrica*

    b. Not hard and woody; brittle on ground or tough and jelly-like on buried wood, **13**

13 a. Flat, blackish brown, crustlike growth on needles and duff (Fig. 54), *420. Rhizinia undulata*

    b. Cup-shaped, black tough, jelly-like fruiting body (Fig. 51), on buried wood on the ground, *421. Sarcosoma globosa*

408 SCLEROTINIA TUBEROSA (Hedw.) Fuckel   **Edibility unknown**/Infrequent

Cups 10–15 mm broad, brown (Fig. 49). Flesh brittle. Stalk 3–10 cm long, 1–2 mm thick, light brown borne on a hard black fungus tissue (sclerotium). Spores 12–17 x 6–9 μ elliptical, smooth.

Several together in moist wood in Sp.; widely distributed. There are a number of species in North America that resemble it closely.

409 MITRULA IRREGULARIS (Pk.) Surand Edibility unknown/Numerous

Clubs 10–20 mm broad, flattened to contorted or club-shaped (Fig. 45), bright yellow. Stalk 10–20 mm high, 2–5 mm thick, yellowish white. Spores 6–10 x 4–5 μ elliptical, smooth.

Solitary to cespitose clusters on soil in conifer woods, in S. and F.; widely distributed.

410 SPATHULARIA CLAVATA (Schaeff.) Sacc. Edibility unknown/Infrequent

Clubs 10–25 mm broad, flattened, spatula- to spoon-shaped (Fig. 44), yellowish to brownish. Flesh brittle. Stalk 2–6 cm long, 4–10 mm thick, enlarging toward base, pallid to yellowish, hollow. Spores 35–65 x 2.5–3 μ needle-shaped, smooth.

Several on soil and litter under conifers in S. and F.; widely distributed. The earth tongues also include black club-shaped species such as *Geoglossum* (Fig. 47) and *Trichoglossum.*

411 LEOTIA LUBRICA (Scop.) Pers. Edibility unknown/Common

Club with an oval, incurved, wrinkled, viscid, yellowish head with an olive tint (Fig. 46). Flesh brittle. Stalk round to oval, viscid, slightly larger toward base. Spores 18–28 x 5–6 μ cylindric-oblong, smooth.

Several to abundant on soil or litter in hardwood and conifer forests, in S. and F.; widely distributed.

412 XYLOSPHAERA POLYMORPHA Edibility unknown/Common

(Pers. ex Merat) Dumortier ["Dead Man's Fingers"]

Club-shaped, cylindrical to finger-like, dull black, 4–8 cm long, 1–2.5 cm thick (Fig. 57). Flesh tough, white. Spores 18–32 x 5–9 μ narrow, pointed on each end, flattened on one side.

Several, usually cespitose clusters or even branched, on logs and stumps of hardwoods, in S. and F.; widely distributed.

413 UNDERWOODIA COLUMNARIS Pk. Edibility unknown/Rare

Clubs fused at base, ribbed, straight or curved (Fig. 58), brown, up to 10 cm tall, 2–3 cm thick. Spores 25–27 x 12–14 μ elliptical, warted.

On soil under hardwoods; widely distributed in northern U.S.A. and adjacent C.

414 WYNNEA AMERICANA Thaxter Edibility unknown/Rare

Several ear-shaped fruiting bodies, 6–13 cm long, arising from a central stalk (Fig. 53), blackish brown. Stalk arising from a tough, brown mass of fungal tissue (sclerotium) 4–5 cm thick. Spores 32–40 x 15–16 μ elliptical, pointed at either end and long striate (Fig. 65).

On soil under hardwoods, in S.; widely distributed in eastern North America.

415 VIBRISSEA TRUNCORUM

Edibility unknown/Infrequent

(Alb. & Schw.) Fries

Club 1–2 cm high with an oval yellow to orange head (Fig. 61), 3–5 mm thick. Stalk white to gray at apex, brown hairs below. Spores 200–250 x 1–1.5 μ fine needle-shaped, many cross walls.

On submerged sticks in mountain streams in S.; widely distributed.

416 CORDYCEPS MILITARIS

Edibility unknown/Common

(L. ex St. Amans) Link.

Club 2–5 cm tall (Fig. 55) nearly cylindric, with minute pimples over the top, red to orange-red, tapering below to a small base. Spores 3.5–6 x 1–1.5 μ, somewhat barrel-shaped.

Single or several on insect larvae and pupae (often of butterflies) in soil or on well-decayed wood, late S. and F.; widely distributed.

417 CORDYCEPS CAPITATA (Holm. ex Fr.) Link

Edibility unknown/Common

Head egg-shaped, 10–15 mm thick, brown to olive-brown (Fig. 56). Stalk narrows downward, dingy yellowish; arising from an oval false truffle. Spores 14–20 x 2–3 μ oblong (Fig. 71), smooth.

Parasitic on an oval false truffle *(Elaphomyces)* which is buried in the duff in S. and F.; widely distributed.

418 APIOSPORINA MORBOSA (Schw.) v. Arx

Edibility unknown/Common

Spindle-shaped, black knotlike growths, 3–15 cm long, on twigs (Fig. 60). Spores 16–22 x 5–6.5 μ, narrowly elliptical, two-celled, smooth.

Several to many on living cherry, the twigs affected are usually dead. Also known as *Dibotryon morbosum* (Schw.) T.&S.

419 DALDINIA CONCENTRICA

Edibility unknown/Common

(Bolt. ex Fr.) Ces. and de Not.

Convex, carbonaceous, black growths on wood, covered with minute pores. Concentric zones revealed (Fig. 59) when cut in two. Spores 12–17 x 6–9 μ elliptical, smooth dark brown.

On hardwood limbs and logs in Sp., S., and F.; widely distributed.

420 RHIZINIA UNDULATA Fr.

Edibility unknown/Common

Crustlike, flat, undulating, lobed, dark brown fruiting body, light color beneath and attached to host by numerous rhizoids (Fig. 54). Spores 22–40 x 8–11 μ smooth, pointed at each end.

On needles and conifer debris, often following fire, in S. and F.; widely distributed. An old name for it is *Rhizinia inflata* (Schaeff.) Karst.

421 SARCOSOMA GLOBOSUM   Edibility unknown/Infrequent

(Schmid. ex Fr.) Casp.

Large, tough, jelly-like, gray-black to black, 6–14 cm broad (Fig. 51). Flesh a tough, transparent jelly. Spores 25–28 x 9–10 μ elliptical, smooth.

In cespitose clusters on well-decayed conifer wood in Sp., in PNW. I have seen it in abundance in the 1971 season in Idaho, but it is usually rare. *Galiella rufa* (Schw.) Nannf. and Korf (Fig. 50) is a small, brown, gelatinous species and looks similar; it is never black, occurs more conspicuously on hardwood sticks and there are warts on the spores.

422 HYPOMYCES LACTIFLUORUM (Schw.) Tul.   Edible/Numerous

Fruiting body consists of a bright orange parasite which grows over the stalk, gills, and eventually the cap of a number of medium to large mushrooms. Most of the parasitized mushrooms are large white species of *Russula* and *Lactarius*. *Lactarius deceptivus* and *Russula brevipes* are two which I have found as hosts, but there are certainly others. The parasite will nearly obliterate the gills, leaving only blunt radial ridges. Pimple-like small swellings on the parasite contain the spores. Spores 30–40 x 6–8 μ, spindle-shaped (fusiform), roughened and slightly curved.

Single or several together in S. and F.; widely distributed.

Edible if found on an edible mushroom. There are several others, including a green species, *Hypomyces luteovirens* (Fr.) Tul., which is frequently encountered in the PNW but is widely distributed and also occurs on *Russula*.

Color Illustrations

Each color photograph bears the same number as the description of that species in the text

1

Amanita
virosa

2

Amanita
inaurata

3

Amanita
vaginata

5
Amanita
caesarea

7
Amanita
cokeri

8
Amanita
porphyria

16

Amanita
flavorubescens

17

Limacella
glischra

18

Chlorophyllum
molybdites

14a
Amanita
muscaria

14b
Amanita
muscaria

15
Amanita
flavoconia

11
Amanita
pantherina

12
Amanita
citrina

9
Amanita
rubescens

10
Amanita
brunnescens

19
Leucoagaricus
naucina

22
Lepiota
rachodes

21
Lepiota
procera

23
Lepiota
cepaestipes

25
Lepiota
clypeolaria

26
Hygrophorus
chrysodon

27
Hygrophorus
eburneus

28
Hygrophorus
subalpinus

30
Hygrophorus
psittacinus

31
Hygrophorus
conicus

32

Hygrophorus
laetus

33

Hygrophorus
nitidus

35

Hygrophorus
speciosus

38
Hygrophorus
cantharellus

39
Hygrophorus
miniatus

40
Hygrophorus
coccineus

42
Hygrophorus
russula

44
Hygrophorus
camarophyllus

45
Lactarius
piperatus

47
Lactarius
necator

48
Lactarius
affinis

50
Lactarius
torminosus

51
Lactarius
griseus

52
Lactarius
hygrophoroides

53
Lactarius
helvus

54
Lactarius
camphoratus

55
Lactarius
rufus

56
Lactarius
subdulcis

57
Lactarius
uvidus

58
Lactarius
representaneι

60
Lactarius
gerardii

61
Lactarius
corrugis

62
Lactarius
volemus

65
Lactarius
indigo

77

Russula
xerampelina

78

Russula
paludosa

79

Tricholomops
platyphylla

80
Tricholomopsis
rutilans

81
Lentinus
lepideus

85
Xeromphalina
campanella

86
Schizophyllum
commune

87
Lentinellus
ursinus

89
Rhodotus
palmatus

92
Pleurocybella
porrigens

93
Pleurotus
ostreatus

94
Panellus
serotinus

95
Phyllotopsis
nidulans

96
Pleurotus
elongatipes

99
Panus
rudis

102
Armillariella
mellea

103
Armillariella
tabescens

104
Catathelasma
imperialis

105
Catathelasma
ventricosa

106
Armillaria
ponderosa

107
Armillaria
albolanaripes

108
Armillaria
zelleri

110
Cystoderma
amianthinum

111

Cystoderma
cinnabarinum

112

Marasmius
oreades

115
Marasmius
rotula

116
Marasmius
siccus

118
Asterophora
lycoperdoides

119
Omphalotus
olearius

121
Omphalina
chrysophylla

123
Clitocybe
nebularis

124
Clitocybe
dealbata

125
Clitocybe
irina

126
Clitocybe
dilatata

127
Clitocybe
odora

128
Clitocybe
nuda

129

Clitocybe
aurantiaca

130

Clitocybe
clavipes

133

Collybia
albipilata

134
Collybia
radicata

137
Flammulina
velutipes

138
Collybia
acervata

139
Collybia
maculata

140
Collybia
exculpta

142

Collybia
dryophila

145

Mycena
haematopus

146

Mycena
leaiana

147
Mycena
pura

151
Mycena
alcalina

152
Mycena
elegantula

153
Laccaria
trullisata

154
Laccaria
laccata

155
Laccaria
ochropurpurea

156
Lyophyllum
decastes

157
Lyophyllum
infumatum

158
Leucopaxillus
amarus

159
Leucopaxillus
giganteus

160

Leucopaxillus
albissimus

163

Tricholoma
aurantium

165

Tricholoma
flavovirens

166

Tricholoma
saponaceum

167

Tricholoma
pardinum

168

Tricholoma
virgatum

169
Tricholoma
vaccinum

170
Volvariella
bombycina

175
Entoloma
abortivum

176
Entoloma
salmoneum

178
Entoloma
lividum

179

Chroogomphus
tomentosus

180

Chroogomphus
rutilus

181
Gomphidius
oregonensis

182
Gomphidius
subroseus

184
Gomphidius
glutinosus

185
Coprinus
niveus

188
Coprinus
comatus

186
Coprinus
disseminatus

189

Coprinus
atramentarius

190

Coprinus
micaceus

191

Panaeolus
separatus

192
Panaeolus
foenisecii

195
Psathyrella
candolleana

196
Psathyrella
hydrophila

197
Stropharia
aeruginosa

200
Stropharia
squamosa

201
Stropharia
ambigua

206

Naematoloma
dispersum

207

Naematoloma
capnoides

208
Naematoloma
sublateritium

209
Naematoloma
fasciculare

210
Agaricus
arvensis

212
Agaricus
placomyces

213
Agaricus
silvaticus

221

Conocybe
tenera

222

Bolbitius
vitellinus

224

Agrocybe
praecox

227

Gymnopilus
spectabilis

230

Pholiota
flammans

232

Pholiota
squarrosa

233
Pholiota
squarrosoides

236
Pholiota
albocrenulata

237
Rozites
caperata

238
Galerina
autumnalis

243
Inocybe
geophylla var.
lilacina

244
Hebeloma
mesophaeum

245
Hebeloma
crustuliniforme

246
Cortinarius
corrugatus

247

Cortinarius
collinitus

248

Cortinarius
glaucopus

249

Cortinarius
violaceus

250

Cortinarius
alboviolaceus

251

Cortinarius
traganus

253

Cortinarius
semisanguineu

255
Cortinarius
cinnamomeus

257
Cantharellus
cinnabarinus

258
Craterellus
cantharellus

259
Cantharellus
cibarius

260
Gomphus
floccosus

261

Cantharellus
tubaeformis

262

Cantharellus
umbonatus

263

Craterellus
cornucopioides

265
Gomphus
clavatus

266
Clavaria
vermicularis

267
Clavariadelphu
truncatus

268

Clavariadelphus
pistillaris

269

Clavariadelphus
ligula

270

Clavulina
cinerea

271
Clavulina
cristata

272
Clavicorona
pyxidata

273
Ramaria
formosa

279

Tylopilus
felleus

280

Fuscoboletinus
ochraceoroseus

285

Suillus
grevillei

276

Strobilomyces
floccopus

277

Gastroboletus
turbinatus

278

Gyrodon
merulioides

275a
Sparassis
radicata

275b
Sparassis
crispa

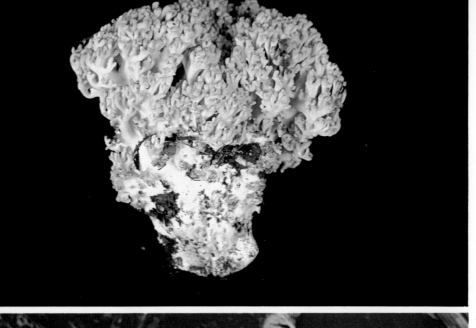

274
Ramaria
botrytis

274a
Ramaria
[spring coral]

274b
Ramaria
subbotrytis

284
Suillus
americanus

287
Suillus
luteus

288
Suillus
cavipes

289
Suillus
pictus

291
Suillus
brevipes

292
Suillus
tomentosus

294
Suillus
granulatus

296
Leccinum
scabrum

298
Leccinum
insigne

301
Boletus
parasiticus

302
Boletus
piperatus

303
Boletus
chromapes

306
Boletellus
.russellii

304
Boletus
mirabilis

307
Boletus
edulis

308
Boletus
satanus

309
Boletus
eastwoodiae

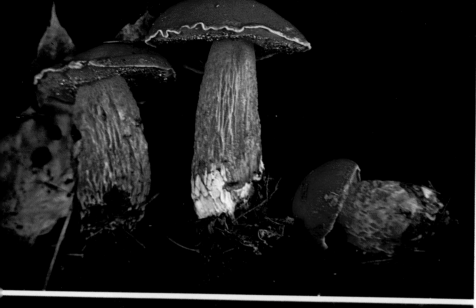

310
Boletus
frostii

313
Boletus
chrysenteron

315
Boletus
pallidus

316
Boletus
badius

317
Boletus
zelleri

318a
Fistulina
hepatica

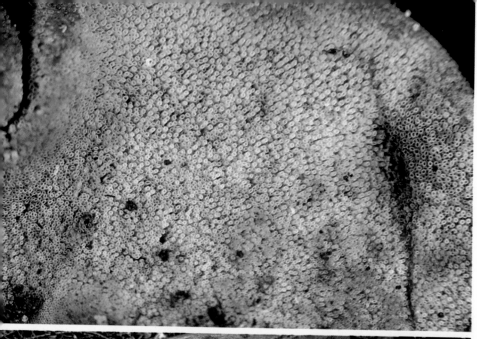

318b

Fistulina
hepatica

320

Ganoderma
tsugae

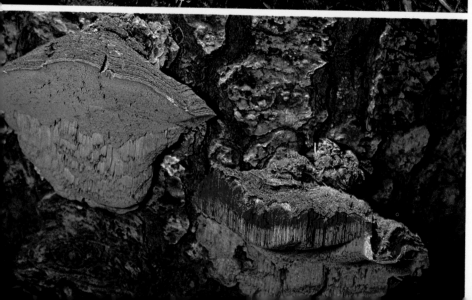

323

Fomes
pini

326a
Polyporus
sulphureus

326b
Polyporus
sulphureus

327
Polyporus
frondosus

331
Polyporus
versicolor

332
Hericium
coralloides

333
Auriscalpium
vulgare

334
Echinodontium
tinctorium

335
Steccherinum
septentrionale

340
Mutinus
caninus

336
Dentinum
repandum

337
Hydnum
imbricatum

341
Lysurus
borealis

343
Dictyophora
duplicata

344
Phallus
ravenelii

345
Crucibulum
vulgare

348
Geastrum
coronatum

349
Geastrum
triplex

350
Geastrum
saccatum

351
Scleroderma
aurantium

354
Calvatia
gigantea

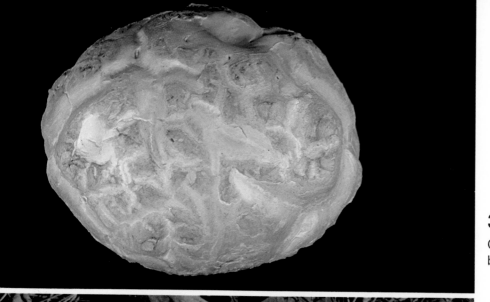

355
Calvatia
booniana

358
Calvatia
craniformis

358
Calvatia
cyathiformis

359
Bovista
pila

360
Calvatia
subcretacea

361
Calvatia
fumosa

362
Lycoperdon
candidum

364
Lycoperdon
pyriforme

365
Lycoperdon
perlatum

367
Lycoperdon
umbrinum

369
Calostoma
cinnabarina

375
Nivatogastrium
nubigenum

376
Endoptychum
agaricoides

377
Guepiniopsis
alpinus

378
Dacrymyces
palmatus

379
Tremella
mesenterica

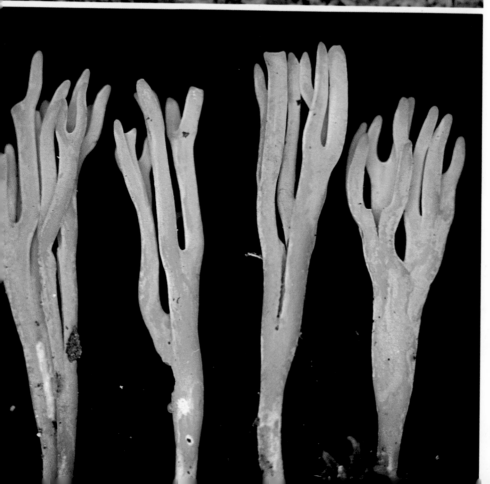

382
Calocera
viscosa

383
Pseudohydnum
gelatinosum

384
Phlogiotis
helvelloides

385
Morchella
semilibera

386
Morchella
angusticeps

387
Morchella
esculenta

388
Morchella
deliciosa

389
Verpa
conica

390

Verpa
bohemica

391

Gyromitra
gigas

392

Gyromitra
californica

393
Gyromitra
caroliniana

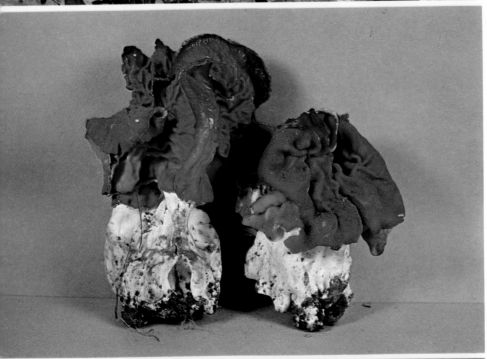

394
Gyromitra
brunnea

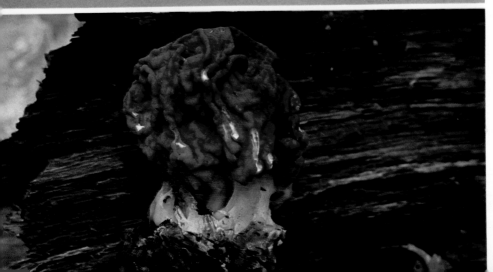

394
Gyromitra
fastigiata

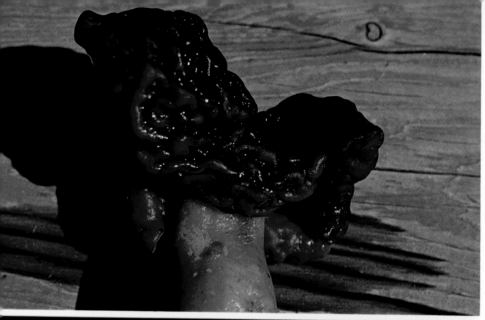

395
Gyromitra
esculenta

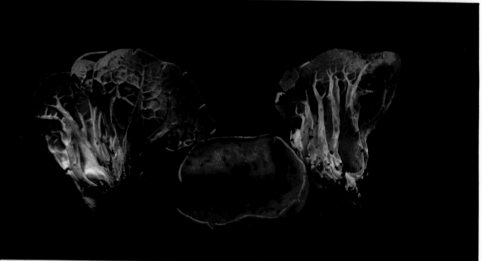

396
Paxina
acetabulum

397
Helvella
lacunosa

398
Helvella
elastica

399
Gyromitra
infula

400
Caloscypha
fulgens

401

Aleuria
aurantia

402

Sarcoscypha
coccinea

403
Otidia
smithii

404
Urnula
craterium

405
Sarcosphaera
coronaria

Chicken Elegante

4 chicken breasts
½ cup flour
1 tsp. each salt, paprika
¼ tsp. pepper
½ cup vegetable oil
2 cups cream of chicken soup
1 cup of milk
1 cup mushroom buttons or chopped mushrooms (*Agaricus* is best)
¼ cup toasted slivered almonds

Sift together flour, salt, paprika and pepper. Coat chicken in mixture. Heat oil in skillet, brown chicken on all sides. Arrange chicken in baking dish. Combine soup with milk and pour over chicken. Cover, bake in 325° oven for 45 minutes or until tender. Add mushrooms and almonds during the last 15 minutes of baking.

Mushroomers' Delight

1 cup parboiled mushrooms, chopped
2 cups boneless chicken
2 cups cream of chicken soup or chicken gravy
1 cup pineapple tidbits (drained)
½ cup slivered almonds
salt to taste

Mix all the ingredients, reserving some of the slivered almonds for garnish. Place in a well-greased casserole dish and bake for 30 minutes in a 350° oven. Serve with rice or chow mein noodles.

Rice Pilaf

4 tbl. olive oil
1 medium onion, chopped
1 cup rice
2 cups water
2 beef bouillon cubes
1 tbl. butter
½ lb. mushrooms
3 tbl. Parmesan cheese

Heat oil in skillet. Brown onion and rice. Add water and bouillon and cover. Simmer about 25 minutes until all liquid is absorbed. In a separate pan, cook mushrooms in butter; set aside. When rice is done, add mushrooms and cheese. Cook about 2 minutes.

Mushroom Gravy

Cook any edible mushroom in butter. Reserve liquid. Add to chicken broth or to the drippings of any meat. Thicken as you would any gravy and season to taste.

Mushroom and Sour Cream Casserole

4 cups chopped mushrooms (any species)
2 cups breadcrumbs
⅔ cup butter
1 cup chopped parsley
1 cup sour cream
salt and pepper to taste
(1 tbl. minced onion, optional)

In a well-buttered casserole, sprinkle breadcrumbs and parsley. Add a layer of mushrooms and dot with butter. Add a layer of crumbs and parsley. Repeat, ending with breadcrumbs dotted with butter. Bake for 30 minutes in 350° oven, adding sour cream after 20 minutes of cooking.

Recipes

Breaded Puffballs

Select only firm, white specimens. It is not necessary to peel the outer rind unless it is especially tough. Cut into ½-inch slices. Dip into beaten egg and then into cracker crumbs or flour. Then place in hot skillet with melted butter. Sauté about a minute on each side until golden brown. Salt and pepper to taste.

Mushroom and Cheese Casserole

        3 cups chopped mushrooms (any good edible species)
        1½ cups breadcrumbs
        1 cup grated cheese (cheddar, Monterey Jack, Colby, etc.)
        4 tbl. butter
        salt and pepper to taste
        (1 tbl. minced onion, optional)

In a well-buttered casserole dish place a layer of mushrooms, sprinkle it first with bread-crumbs, then with grated cheese, and dot with butter. Repeat these layers, ending with a layer of cheese. Cover and bake for approximately 20 minutes in a moderate oven (375°). To brown, remove the cover for the last few minutes.

Yorkshire Pudding with Mushrooms

        1 cup flour
        1 cup milk
        ¼ tsp. salt
        2 eggs
        ¼ cup drippings from roast (preferably beef)
        ½ cup chopped mushrooms

Preheat oven to 425°. Mix flour and salt. Add ¾ cup milk and beat until smooth. Add eggs and remaining milk and beat with a rotary beater until smooth and creamy. Pour mixture with drippings from the roast into a pan in which the mushrooms have previously been sautéed. Mix in the sautéed mushrooms gently. Bake 30 minutes. Serve while hot.

Mushroom Fondue

        1 cup mushrooms (*Agaricus, Polyporus frondosus,* etc)
        ¼ cup sherry
        1 cup grated Monterey Jack cheese

or

        1 cup morels
        ¼ cup sauterne
        1 cup Colby mild cheese

Sauté mushrooms in butter until liquid is absorbed. Blend in a blender at high speed until puréed. Place mushroom purée in a fondue pot, add sherry (or sauterne), and heat until bubbly. Add cheese by handfuls and stir until melted. Serve with cubes of French bread.

Chicken Wonder

Spread 1 can chow mein noodles over bottom of baking dish. In large bowl, combine 2 cups cooked, chopped chicken, 1 cup chopped celery, ½ cup chopped onion, 2 cups chopped mushrooms, 1 can mushroom soup, 1 cup water and 1 cup cashew nuts or almonds. Pour over noodles. Grate cheese over top. Bake in 350° oven for 15 minutes or until heated through.

416
Cordyceps
militaris

422

Hypomyces
lactifluorum

407
Discina
perlata

411
Leotia
lubrica

Coq au Vin

6 chicken breasts
2 tbl. vegetable oil
1 tbl. flour
1 tbl. minced onion
½ cup red wine
½ cup chicken broth
1 clove garlic
¼ tsp. thyme
1 bay leaf
12 small white onions
1 cup sliced mushrooms (morels, meadow mushrooms, chanterelles)

In skillet, brown chicken in oil. Add onion and cook until tender. Stir in flour, and cook until well browned. Add wine, broth, garlic, bay leaf, thyme, white onions and mushrooms. Bring to boil and then simmer 45 minutes. Serve with buttered French bread, or over rice, if desired.

Mycological Martini

Mix up a batch of your favorite martini recipe. Pour into martini glasses and place in it a small button of *Agaricus,* FRESH.

Crab Stuffed Mushrooms

12 to 18 large mushroom caps (*Agaricus* species are best)
½ cup butter, melted
1½ cups crabmeat (canned or fresh)
2 eggs, beaten
¼ cup mayonnaise
¼ cup minced onion
2 to 3 tsp. lemon juice
½ cup soft breadcrumbs (fresh bread)

Brush caps with melted butter and place the gill side up in a baking dish. Combine remaining ingredients (reserving ¼ cup of breadcrumbs) and fill caps with the mixture. Sprinkle remaining crumbs over top. Brush with melted butter. Salt and pepper to taste. Bake 15 minutes in 375° oven.

Mushroom Loaf

2 lbs. fresh mushrooms
1 large onion, thinly sliced
2 tbl. butter
½ cup dry breadcrumbs
2 eggs, lightly beaten
¼ lb. butter or margarine, melted
½ tsp. salt
dash of pepper

Sauté half of the onion in 2 tbl. butter until golden brown. Save several large mushroom caps for garnish. Chop remaining mushrooms, including stems and remaining onion, and mix with breadcrumbs, salt, pepper, and remaining butter. Stir in eggs and the sautéed onions.

Press entire mixture in a well-greased loaf pan. Arrange mushroom caps on top and press lightly. Bake in 350° oven for 1 hour. Let stand several minutes, slice, and serve with mushroom gravy.

Note: Many edible mushrooms can be used in this recipe: for example, *Agaricus, Polyporus frondosus,* puffballs, and *Armillariella mellea.* Each contributes its own special flavor.

Chicken Chipped Beef

>1 pkg. chipped beef
>6 chicken breasts, boned
>1 can mushroom soup
>8 to 9 slices of bacon
>1 cup sour cream
>1 cup mushrooms, chopped

Line bottom of greased casserole with beef. Wrap each chicken breast with bacon. Arrange chicken over beef. Mix soup, sour cream and mushrooms; pour over chicken. Bake, covered, at 275° for 3 hours. Serves 6.

Mushrooms and Scrambled Eggs

Sauté mushrooms in butter. Add well-beaten eggs to which salt, pepper and 1 tbl. milk have been added. Cook over low heat, stirring frequently until done. Serve immediately. Many mushrooms are excellent: for example, *Coprinus comatus, Agaricus,* Chanterelles and *Polyporus frondosus.*

German White Sauce for Boletes

Cook chopped boletes or other suitable mushrooms in butter and reserve the liquid. If you like mushrooms on the crunchy side, cook them for only 3 or 4 minutes. For the sauce:

>3 tbl. butter or margarine
>3 tbl. flour
>1 cup broth or water
>salt to taste
>lemon juice to taste (approx. 1–2 tsp.)
>1 egg yolk
>cream enough to make a medium cream sauce (approx. 1 cup)

Melt butter and add flour, stirring constantly to avoid lumps; add liquid from mushrooms and stir constantly. Season to taste. When thickened, stir in lemon juice. Beat egg yolk and mix into cream, then stir into cream sauce. Add mushrooms and serve over meat or on toast.

Mushroom Canape Spread

Mix two 3-oz. packages of cream cheese with 1 tbl. onion juice until light and fluffy. Add 1 cup finely chopped raw mushrooms and add ¼ tsp. lemon juice. Season to taste with celery salt, pepper, and a dash of tabasco. Spread on crackers.

Poultry Stuffing with Mushrooms

>1 cup chopped mushrooms
>1 cup chopped celery
>½ cup chopped onions
>1 stick butter or margarine (½ lb.)
>16 slices stale bread
>salt, pepper, sage, and poultry seasoning to taste
>hot water or broth.

Using a large bowl, take stale bread and moisten with warm water or broth. Add water until the bread is quite moist, squeezing out excess moisture with your hands. In large frying pan, fry onions, celery, and mushrooms. Season with salt, pepper, sage and poultry seasoning to taste. Pour mixture into bread and mix well. Stuff bird and if there is any stuffing left over put into a buttered casserole dish, cover, and bake at same time as bird. Don't forget to baste occasionally with drippings from the bird to add to the flavor. Four cups of stuffing is usually sufficient for a 4-lb. bird, 10 cups for a bird of 12 to 14 pounds. The best mushrooms to use are those with a distinctive flavor: for example, *Clitocybe nuda* or *Polyporus sulphureus.*

German Wine Sauce for Mushrooms

Cook chopped mushrooms in a small quantity of butter, reserving liquid to be used as broth. Set mushrooms aside. For the sauce:

> 3 tbl. butter or margarine
> 3 tbl. flour
> 1 cup broth or water
> salt and pepper to taste
> 2 to 3 tbl. Madeira wine (or enough to make a medium sauce)

Melt butter and stir in flour, stirring constantly to avoid lumps. Add broth, salt, pepper, and continue stirring until thickened. Add wine, stir, and then add mushrooms and serve over meat.

Hot Mushroom Dip

> 2 lbs. fresh mushrooms
> 6 tbl. butter
> 1 tbl. lemon juice
> 4 tbl. chopped parsley
> 2 tbl. minced onions
> ¾ to 1 cup sour cream
> 2 tsp. bouillon granules (or 2 bouillon cubes)

Chop the mushrooms quite fine. Place in pan with butter, add lemon juice, and simmer 10 minutes. Add the onions and sour cream. Season with salt and pepper to taste and add bouillon granules or cubes. Simmer 10 minutes more. Make a paste of 1 tbl. soft butter and 1 tbl. flour. Add to mixture and stir until thickened. Serve hot.

Mushroom Kabobs

This is a good barbecue or campers' recipe. Use any species of mushrooms which is fairly firm such as *Agaricus, Lactarius* or *Tricholoma.* Skewer them on regular skewers or green sticks. Brush with butter and salt and pepper to taste. Rotate over hot coals until slightly crisp on outside and cooked inside.

Hericium Marinade

> 1 lb. fresh, cleaned mushrooms (Hericium)
> ¼ cup vinegar (wine vinegar preferred)
> 2 tbl. olive oil
> ½ tsp. salt
> ¼ tsp. oregano
> dash of pepper
> 1 cup chopped parsley
> 1 small clove of garlic

Cook mushrooms a few minutes in boiling salted water. When slightly tender, drain. Add all the remaining ingredients. Marinate in the refrigerator for several hours or overnight. Remove garlic if it seems to be too strong, depending upon taste.

Pickled Mushrooms

> 1 cup vinegar
> ½ cup sugar
> 2 tbl. oil
> mixed pickling spices and salt to taste
> 2 cups of fresh mushrooms, approximately

Boil the above ingredients, except the mushrooms, for approximately 2 to 3 minutes. Add mushrooms, making sure the brine covers the mushrooms. Boil approximately 15–20 minutes. Pour into sterilized jars and seal. It is best to wait about 2 weeks before using. We find any species of *Morchella* to be best but any genus of mushrooms may be used.

Chanterelle Biscuits

                    2 cups flour
                    5 tsp. baking powder
                    4 tbl. shortening
                    ½ tsp. salt
                    ¼ cup milk
                    ¼ cup chopped onions or chives
                    ½ cup parmesan cheese
                    ½ cup cooked chanterelles, drained and dried with paper towels,
                        then chopped fine
                    ¼ cup butter, melted

Mix flour, baking powder, and salt. Cut in shortening. Add onion, cheese, and mushrooms, and knead lightly. Add milk and blend. Form a ball and knead lightly. Press out to a thickness of ¼ to ½ inch and cut out with biscuit cutter. Brush both sides with melted butter and bake in hot oven, 400°, for 12 to 15 minutes until golden brown.

Marasmius Cookies

Use a chocolate drop cookie mix as a foundation (any chocolate drop cookie recipe will do). To this add ½ cup fine macaroon crumbs, ½ cup chopped and drained maraschino cherries, and 1 cup of previously cooked and well-drained *Marasmius,* flavored with ½ tsp. almond extract.

Mushroom Soup

Simmer chopped mushrooms in chicken or beef stock (a bouillon cube, dissolved, may be used). In another pan add equal quantities of butter and flour and stir constantly, adding milk until thickened sufficiently. Add mushrooms. Pour soup into bowls. Grated cheese, chopped chives, parsley, or crumbled bacon pieces may be added for garnish, depending upon taste.

Freezing Mushrooms

This method may prove quite unsatisfactory but in an emergency may be used. The best way is to blanch the mushrooms or to quickly sauté them and put them into plastic containers and freeze. These may be used in casseroles or other cooked dishes. The mushrooms should not be kept frozen very long for they lose most of their taste.

Drying Mushrooms

Throughout the world mushrooms are dried and sold commercially. This is one of the least effective methods of preservation but nevertheless many people use it. After drying, it is most important that the specimens be kept dried to prevent mildew. One method of drying is to slice the cleaned mushrooms into ¼-inch slices and place slices on a wire rack or a screen. Set the racks over a hot plate or hot-air register. It usually takes two days for mushrooms to dry thoroughly. Then store in clean jars with black pepper and a few bay leaves. When reviving, soak in hot water 15–20 minutes before using.

Canning Mushrooms

Canning is considered the best method for preserving mushrooms. The following is a common method which utilizes a pressure canner. Select only fresh mushrooms. Wash, cut into small pieces. Put into vinegar water (1 tsp. vinegar to each quart of water) to prevent discoloring. Precook mushrooms in boiling water to cover for 3 minutes. Pack hot in clean, hot containers. Add 1 tsp. salt for each quart. Cover them with the water in which they were cooked. Adjust lids. Place in pressure canner and follow Processing Time Table for mushrooms for your own particular canner—usually 10 pounds pressure for 35 minutes.

Illustrated Glossary

The following drawings illustrate the various structures and parts of all kinds of fungi. Throughout the book these drawings are referred to as "figures." (A further explanation will usually be found in the regular glossary.) When the drawing shows an identifiable species, its name is given. The order of these drawings generally parallels the presentation of the various groups in the text.

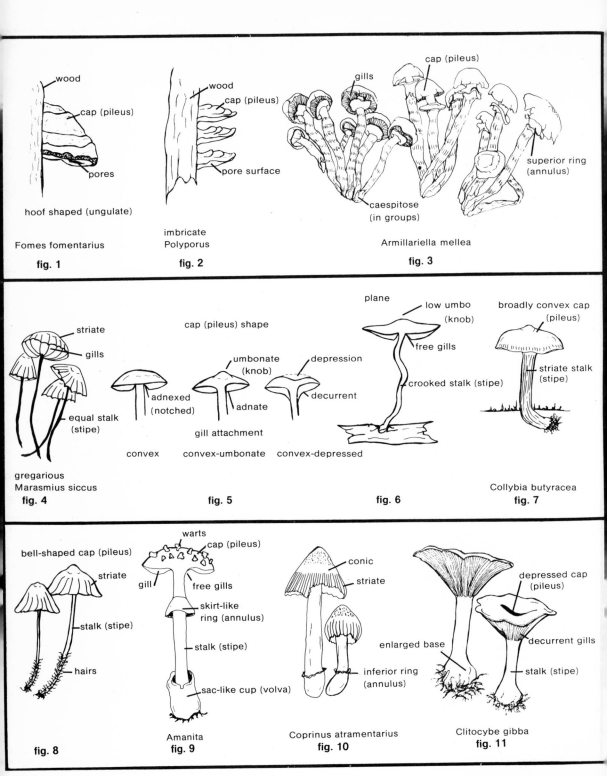

fig. 1
wood
cap (pileus)
pores
hoof shaped (ungulate)
Fomes fomentarius

fig. 2
wood
cap (pileus)
pore surface
imbricate
Polyporus

fig. 3
gills
cap (pileus)
superior ring (annulus)
caespitose (in groups)
Armillariella mellea

fig. 4
striate
gills
equal stalk (stipe)
gregarious
Marasmius siccus

fig. 5
cap (pileus) shape
adnexed (notched)
umbonate (knob)
adnate
depression
decurrent
gill attachment
convex
convex-umbonate
convex-depressed

fig. 6
plane
low umbo (knob)
free gills
crooked stalk (stipe)

fig. 7
broadly convex cap (pileus)
striate stalk (stipe)
Collybia butyracea

fig. 8
bell-shaped cap (pileus)
striate
stalk (stipe)
hairs

fig. 9
warts
cap (pileus)
gill
free gills
skirt-like ring (annulus)
stalk (stipe)
sac-like cup (volva)
Amanita

fig. 10
conic
striate
inferior ring (annulus)
Coprinus atramentarius

fig. 11
depressed cap (pileus)
decurrent gills
enlarged base
stalk (stipe)
Clitocybe gibba

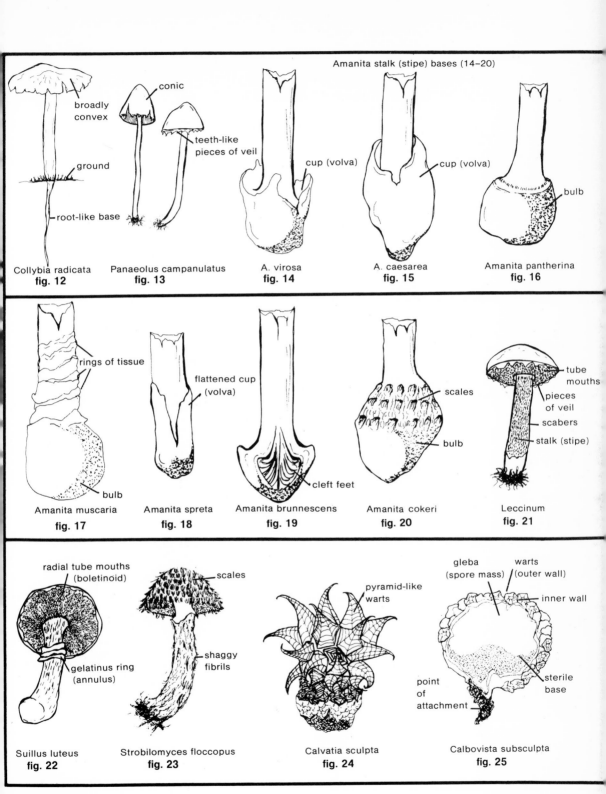

Amanita stalk (stipe) bases (14–20)

broadly convex

conic

ground

teeth-like pieces of veil

cup (volva)

cup (volva)

bulb

root-like base

Collybia radicata
fig. 12

Panaeolus campanulatus
fig. 13

A. virosa
fig. 14

A. caesarea
fig. 15

Amanita pantherina
fig. 16

rings of tissue

flattened cup (volva)

scales

tube mouths

pieces of veil

scabers

stalk (stipe)

bulb

cleft feet

bulb

Amanita muscaria
fig. 17

Amanita spreta
fig. 18

Amanita brunnescens
fig. 19

Amanita cokeri
fig. 20

Leccinum
fig. 21

radial tube mouths (boletinoid)

scales

pyramid-like warts

gleba (spore mass)

warts (outer wall)

inner wall

gelatinus ring (annulus)

shaggy fibrils

point of attachment

sterile base

Suillus luteus
fig. 22

Strobilomyces floccopus
fig. 23

Calvatia sculpta
fig. 24

Calbovista subsculpta
fig. 25

341

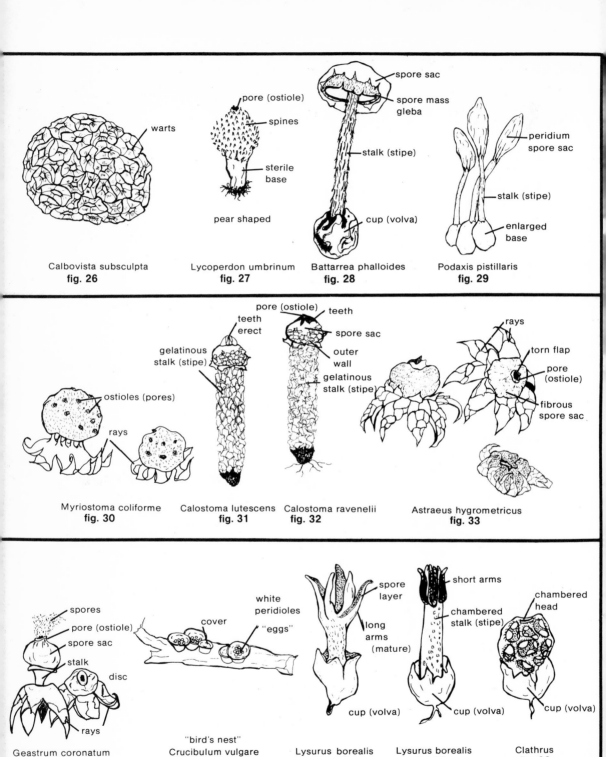

warts

Calbovista subsculpta
fig. 26

pore (ostiole)

spines

sterile base

pear shaped

Lycoperdon umbrinum
fig. 27

spore sac

spore mass gleba

stalk (stipe)

cup (volva)

Battarrea phalloides
fig. 28

peridium spore sac

stalk (stipe)

enlarged base

Podaxis pistillaris
fig. 29

ostioles (pores)

rays

Myriostoma coliforme
fig. 30

teeth erect

gelatinous stalk (stipe)

Calostoma lutescens
fig. 31

pore (ostiole)

teeth

spore sac

outer wall

gelatinous stalk (stipe)

Calostoma ravenelii
fig. 32

rays

torn flap

pore (ostiole)

fibrous spore sac

Astraeus hygrometricus
fig. 33

spores

pore (ostiole)

spore sac

stalk

disc

rays

Geastrum coronatum
fig. 34

cover

white peridioles

"eggs"

"bird's nest"
Crucibulum vulgare
fig. 35

spore layer

long arms (mature)

cup (volva)

Lysurus borealis
fig. 36

short arms

chambered stalk (stipe)

cup (volva)

Lysurus borealis
fig. 37

chambered head

cup (volva)

Clathrus
fig. 38

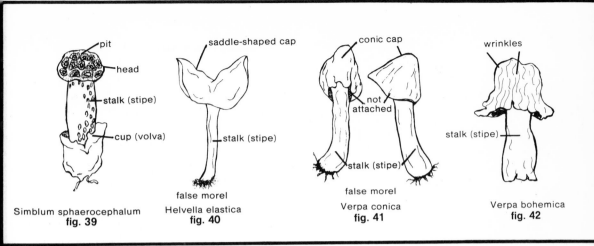

Simblum sphaerocephalum
fig. 39

pit
head
stalk (stipe)
cup (volva)

false morel
Helvella elastica
fig. 40

saddle-shaped cap
stalk (stipe)

false morel
Verpa conica
fig. 41

conic cap
not attached
stalk (stipe)

Verpa bohemica
fig. 42

wrinkles
stalk (stipe)

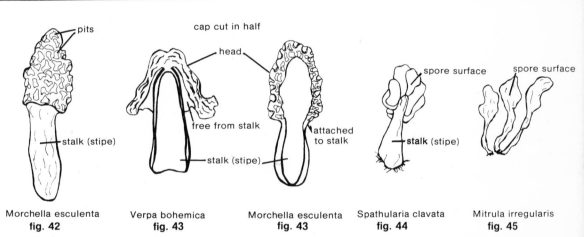

Morchella esculenta
fig. 42

pits
stalk (stipe)

Verpa bohemica
fig. 43

cap cut in half
head
free from stalk
stalk (stipe)

Morchella esculenta
fig. 43

attached to stalk

Spathularia clavata
fig. 44

spore surface
stalk (stipe)

Mitrula irregularis
fig. 45

spore surface

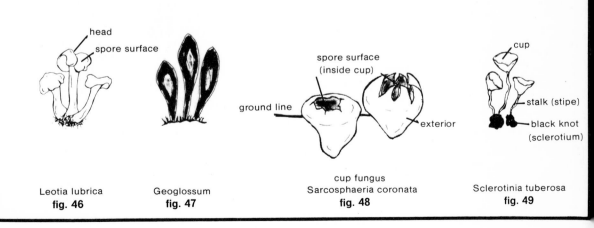

Leotia lubrica
fig. 46

head
spore surface

Geoglossum
fig. 47

cup fungus
Sarcosphaeria coronata
fig. 48

spore surface
(inside cup)
ground line
exterior

Sclerotinia tuberosa
fig. 49

cup
stalk (stipe)
black knot
(sclerotium)

343

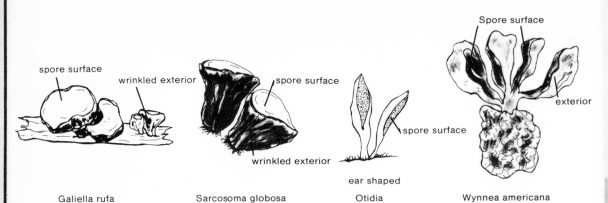

spore surface · wrinkled exterior

Galiella rufa
fig. 50

spore surface · wrinkled exterior

Sarcosoma globosa
fig. 51

ear shaped · spore surface

Otidia
fig. 52

Spore surface · exterior

Wynnea americana
fig. 53

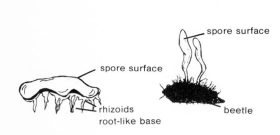

spore surface · rhizoids root-like base

Rhizinia inflata
fig. 54

spore surface · beetle

Cordyceps militaris
fig. 55

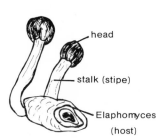

head · stalk (stipe) · Elaphomyces (host)

Cordyceps capitata
fig. 56

spore surface

dead man's fingers
Xylosphaera polymorpha
fig. 57

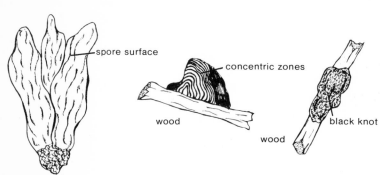

spore surface

Underwoodia columnaris
fig. 58

concentric zones · wood

Daldinia concentrica
fig. 59

black knot · wood

"black knot"
Apiosporina morbosa
fig. 60

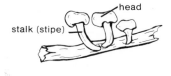

head · stalk (stipe)

Vibrissea truncorum
fig. 61

344

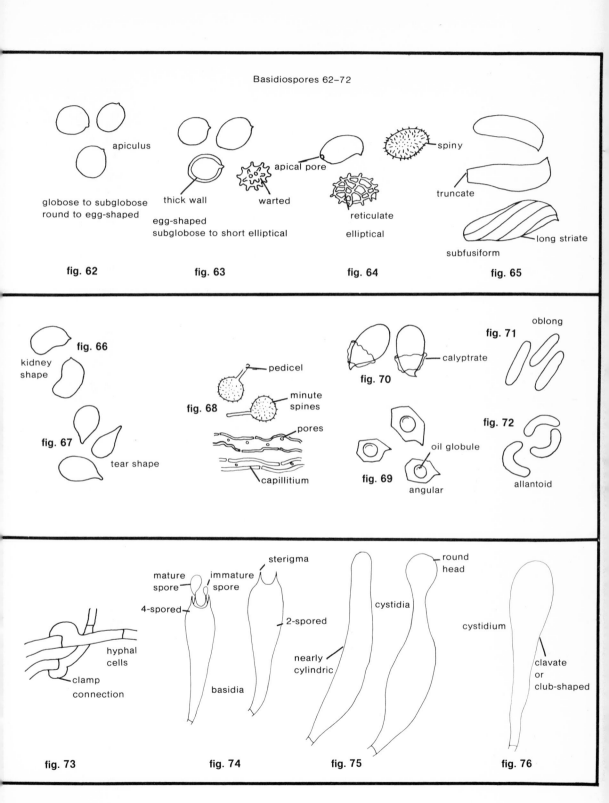

Basidiospores 62–72

apiculus

globose to subglobose
round to egg-shaped

fig. 62

thick wall

warted

egg-shaped
subglobose to short elliptical

fig. 63

apical pore

reticulate

elliptical

fig. 64

spiny

truncate

long striate

subfusiform

fig. 65

kidney
shape

fig. 66

fig. 67

tear shape

pedicel

minute
spines

pores

capillitium

fig. 68

calyptrate

fig. 70

oblong

fig. 71

oil globule

angular

fig. 69

fig. 72

allantoid

hyphal
cells

clamp
connection

fig. 73

mature
spore

immature
spore

sterigma

4-spored

basidia

fig. 74

2-spored

nearly
cylindric

cystidia

fig. 75

round
head

cystidium

clavate
or
club-shaped

fig. 76

345

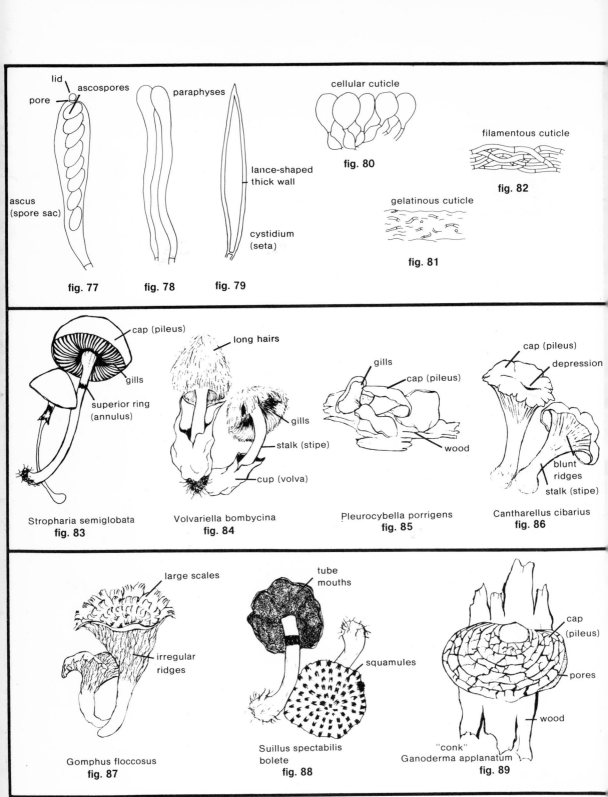

fig. 77 · fig. 78 · fig. 79

fig. 80 — cellular cuticle

fig. 81 — gelatinous cuticle

fig. 82 — filamentous cuticle

Stropharia semiglobata
fig. 83

Volvariella bombycina
fig. 84

Pleurocybella porrigens
fig. 85

Cantharellus cibarius
fig. 86

Gomphus floccosus
fig. 87

Suillus spectabilis bolete
fig. 88

Ganoderma applanatum
fig. 89

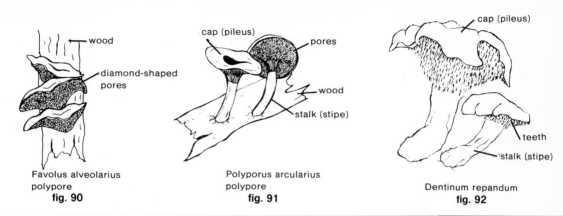

Favolus alveolarius
polypore
fig. 90

Polyporus arcularius
polypore
fig. 91

Dentinum repandum
fig. 92

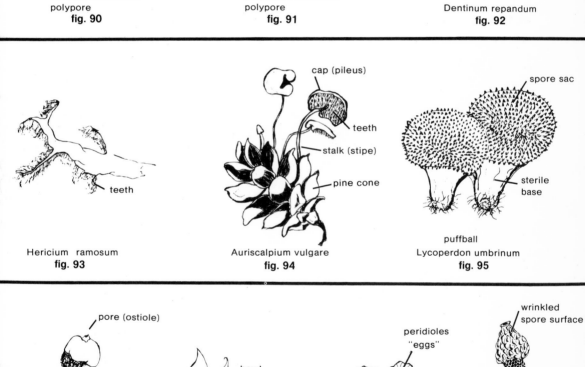

Hericium ramosum
fig. 93

Auriscalpium vulgare
fig. 94

puffball
Lycoperdon umbrinum
fig. 95

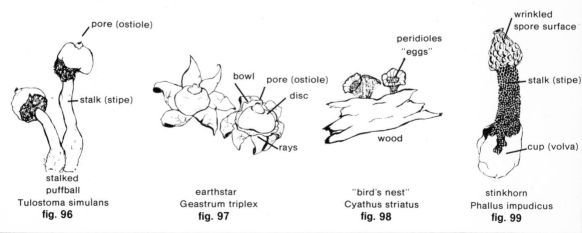

stalked
puffball
Tulostoma simulans
fig. 96

earthstar
Geastrum triplex
fig. 97

"bird's nest"
Cyathus striatus
fig. 98

stinkhorn
Phallus impudicus
fig. 99

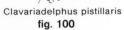

club-shaped coral

Clavariadelphus pistillaris
fig. 100

club

caespitose

coral
Clavariadelphus ligula
fig. 101

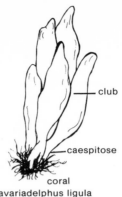

branch

fleshy base

fleshy coral

Ramaria botrytis
fig. 102

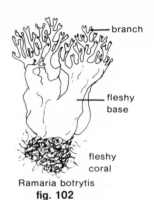

crown-like branch

coral
Clavicorona pyxidata
fig. 103

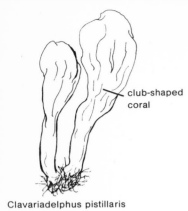

spore bearing surface

cup

cup fungus
Peziza badia
fig. 104

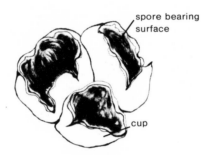

ridges and pits

attached to stalk

stalk (stipe)

Morchella esculenta
fig. 105

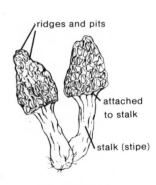

wrinkled

free from stalk

stalk (stipe)

Verpa bohemica
fig. 106

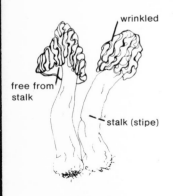

thin lobe

wood

jelly fungus
Tremella mesenterica
fig. 107

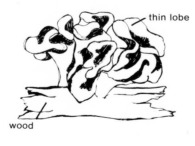

spore surface

wood

jelly fungus
Guepiniopsis alpina
fig. 108

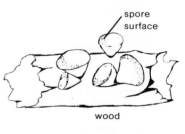

Glossary

adnate (of gills): attached directly to stalk (Fig. 5)

adnexed (of gills): notched just at stalk (Fig. 5)

aerolate: finely cracked

allantoid: sausage-shaped (Fig. 72)

amyloid (of tissue or spores): a blue reaction which takes place when an iodine solution (Melzer's reagent) is placed on the tissue or spores.

angular: the four- to six-sided spores found in the genus *Entoloma* (Fig. 69).

annulus: ring-shaped remains of the partial veil which hangs on the stalk. The superior annulus (or ring) is above the midpoint on the stalk (Fig. 3); the inferior annulus is below the midpoint on the stalk (Fig. 10).

apiculus: a small projection where the spore was attached to the basidium (Fig. 62; shown attached in Fig. 74).

appressed: lying flat or flattened against a surface, as in appressed hairs.

ascus: saclike cell in which the spores of the Ascomycetes are produced (Fig. 77).

azonate (of the cap): lacking concentric bands.

basidium: cell on which the spores of the Basidiomycetes are produced (Fig. 74)

bay: reddish brown

boletinoid: pores extending in loose radial rows from the stalk to the margin (Fig. 22).

brittle (of the stalk): rigid, and often breaking with a snap. See *pliant.*

bulb: swollen base of the mushroom stalk.

calyptrate spore: a loose covering over one end of some basidiospores (Fig. 70).

campanulate: bell-shaped cap (Fig. 8).

capillitium: coarse, thick-walled cells found among the spores in puffballs.

carbonaceous: like charcoal, leaves black stains on the fingers.

caustic potash: see *KOH.*

cellular cuticle: round or pear-shaped cells forming the outer layer of the mushroom cap (Fig. 80).

cespitose: several stalks arising close together but not joined (Fig. 3).

clamp connection: a swollen area of the fungus cell cross wall (Fig. 73).

clavate: club-shaped (Fig. 76).

cleft foot (of *Amanita brunnescens*): a deeply dissected base or bulb (Fig. 19).

conifer: a tree with needles or scales such as pine, fir, or juniper; also called an evergreen.

conk: common name of the large wood-destroying fruiting bodies of polypores and hydnums on trees.

corrugated: having long, deep wrinkles.

cortinous: weblike appearance of some partial veils, especially on *Cortinarius* and *Inocybe.*

cuticle: outer skin or layer of mushroom stalk or cap (Figs. 80–82).

cystidia: sterile cells on the spore-bearing surface of Basidiomycetes (e.g., the sides and edges of gills) among the basidia (Figs. 75 and 79).

decurrent (of gills): extending down the stalk (Figs. 5 and 11).

deliquescing (of gills): dissolving into an "inky" fluid, common only in *Coprinus.*

dextrinoid: turns red in Melzer's solution.

disc: center part of cap

distant (of gills): well separated

divergent gill tissue: in cross-section the cells of the gill appear to sweep outward from the center (as seen with a microscope).

eccentric: a stalk not attached in the center of the cap off center.

echinulate: covered with minute spines (Fig. 64).

egg: common name of the small spore sacs (peridioles) in the "nest" of the "bird's nest" fungi (Fig. 35).

family: a group of related genera, designated by the ending "aceae," such as the Amanitaceae.

ferric sulphate: a 3 percent solution in water (3 percent $FeSO_4$) yields a positive green reaction on the fresh flesh of some fungi.

fibril: minute hair.

fibrillose: composed of delicate long hairs.

filamentous cuticle: threadlike cells forming the outer layer of some mushroom caps (Fig. 82).

fleshy (of the cap and stalk): usually a soft, easily broken tissue.

floccose: cottony hairs.

four-spored (of the Basidiomycetes): having only four spores at the top of each basidium.

fruiting body: the part of the fungus plant developed for the purpose of producing and liberating spores. We have used the term "mushroom" to mean the same thing.

fungus: spore-bearing plants that do not contain chlorophyll.

fused stalks: two or more stalks arising from a common base; joined together but not cespitose.

fusiform (of spores): tapering at one or both ends; spindle-shaped.

gelatinous cuticle: slimy, gelatin-like substance forming the outer layer of mushroom cap (Fig. 81).

genus: a group of similar species. Closely related genera are placed in a family (for example, *Russula* and *Lactarius* in the Russulaceae).

glabrous (of the stalk or cap surface): lacking hairs, warts or squamules.

glandular dots: moist to viscid small dots on the stipe of some boletes.

glaucous: appearing frosted.

gleba: the spore mass inside a puffball (Fig. 25).

globose: round (Fig. 62).

glutinous (of a layer on the cap or stalk): slimy; in wet weather sometimes oozing and hanging from the margin of the cap.

gregarious: a number of fruiting bodies growing in groups or "troops" (Fig. 4).

hardwood tree: tree with broad leaves, such as elm, maple, or birch.

hyaline (of spores): transparent; lacking color.

hymenium: spore-bearing surface such as the gills of mushrooms, pores of boletes, head of morels.

hypha(ae): one or more fungal cells, collectively referred to as a mycelium (Fig. 73).

imbricate (of mushroom caps): each one growing just above another one (Fig. 2).

inferior (of annulus): located at base of stalk

interwoven gill tissue: in cross-section the cells of the gill are entwined and form no pattern (as seen with a microscope).

KOH (caustic potash): a 3 percent solution of potassium hydroxide in water in which fungal material is mounted for microscopic study.

lacerate: a torn appearance.

lamellae: gills of a mushroom.

latex: a variously colored juice; always present, for example, in **Lactarius** tissue.

leathery (of the cap and stalk): pliant without breaking easily; tough; usually difficult to tear in two.

long striate: lines or furrows extending the full length of spores (Fig. 65) or stalk (Fig. 7).

margin (of a cap): edge or outer portion.

mealy: smelling of fresh grain; or having a granular appearance.

Melzer's reagent: a solution used to test for the amyloid (blue) reaction of fungal cell walls. It contains 20 cc of water, 1.5 gm of potassium iodide 0.5 gm of iodine and 20 gm of chloral hydrate.

membranous: skinlike tissue making up the partial veil of some mushrooms.

micron (abbreviation: μ): 1,000 microns equals one millimeter; about 26,000 equal one inch; the unit used in measuring microscopic objects such as spores.

mushroom: the fruiting body of a fleshy nature characteristic of some fungi. It is applied to both edible and poisonous species.

mycelium: collective name for the filaments (cells) of the vegetative fungus plant (see *hyphae*).

mycophagist: one who enjoys eating various species of mushrooms; a mushroom fancier.

mycology: the study of fungi.

order: a group of families; designated by the ending "ales," as in the Nidulariales.

ostiole (of puffballs): a pore in the top through which the spores escape.

overlapping (of imbricate fruiting bodies):

each one growing just above the other (Fig. 2).

ovoid: egg-shaped; not quite round.

paraphyses: sterile cells among the asci on the spore-bearing surface of an Ascomycete (Fig. 78).

partial veil: a covering that extends from the unopened margin of the mushroom cap to the stalk. If part of the veil remains attached to the stalk it is called a ring or annulus (Fig. 9).

pedicle: long, tubelike extension on the spores of many puffballs (Fig. 68).

peridiole: egglike bodies in the "nest" of the "bird's nest fungi" (Fig. 35).

peridium: outer wall surrounding the spore mass in the puffballs and their allies.

pileus: mushroom cap (Fig. 9).

pip-shaped: shaped like an appleseed (Fig. 67).

plane (of the pileus): having a flat surface.

pliant (of the stalk): able to bend moderately without breaking (see **brittle**).

potassium hydroxide: see **KOH**

pruinose: as though covered with a fine powder.

pubescent (of the pileus and stalk): having minute hairs; downy.

pyriform: pear-shaped

reticulate: shallow, netlike ridges on the stalks of many species of boletes; a series of connected ridges over the surface of some spores (Fig. 64).

rimose (of a cap): cracked

ring (of the stem): remains of the partial veil clinging to the stem and forming a ring or annulus (Figs. 3 and 9).

rootlike base: a tapering extension of the stalk belowground (Fig. 12).

rugose: coarsely wrinkled

rugulose: wrinkled

scabers: small brown to blackish brown tufts of hairs on the stalks of *Leccinum* in the boletes.

scales: a piece of tissue on the cap or stalk which often looks like a shingle; it may be flat or curved backward and it is usually of a different texture and color from the surrounding tissue.

sclerotium: a hard, blackish brown knot of fungus tissue (Fig. 49).

setae: pointed, thick-walled sterile cells.

septum: cross wall in a fungus cell

(hypha) or spore.

serrate (of the gill edge): ranging from sawtooth-like to nearly ragged.

sessile: lacking a stalk; growing directly from the base.

setae: pointed, thick-walled sterile cells.

sinuate (of gills): notched at the stalk.

smooth (of a stalk, cap, or spore surface): not wrinkled, grooved, or roughened.

snag: a standing, dead, weathered tree.

species: one kind of plant or mushroom. In a species name such as *Lactarius deliciosus,* the first word is the genus name, the second is the species.

spore: reproductive body of a fungus.

squamules: small, often ill-defined, scales.

sterigma: projections from the top of the basidium (Fig. 74) where the basidiospores were formed and remain attached until they are mature.

stipe: the stalk that supports the cap of the fungus (Fig. 9).

striate (of the pileus): having minute radial line-or furrows; also longitudinal furrows or lines on the stalk. When they twist around the stalk they are said to be twisted-striate (see also *long striate).*

sub: prefix meaning nearly or almost.

subdistant (of gills): fairly well separated; between close and distant.

subfusiform (of spores): somewhat tapered but not spindle-shaped (Fig. 65).

subglobose: almost or somewhat rounded.

superior (of annulus): located near top of stalk.

terrestrial: growing on the ground.

tomentose: densely hairy, woolly.

trama: flesh inside the cap or gills.

truncate (of spores): having a blunt end

tuberculate: strongly warted.

two-spored (of the Basidiomycetes): having only two spores at the top of each basidium.

umbilicate (of the center of a cap): shaped like a navel

umbo: a knob; an abruptly raised area in the center of the cap.

umbonate (of a cap): with an umbo.

ungulate: shaped like a horse's hoof.

universal veil: a tissue surrounding the developing mushroom button.

veil: see *universal veil.*

veins: low ridges between gills.

viscid: sticky, slippery, but not a deep slimy layer of the cap and stalk (see *glutinous*).

volva: remains of the universal veil; often surrounding the stalk base like a cup (Fig. 15); or on the bulb at the stalk base as patches of tissue.

warts: unattached pieces of the universal veil (Fig. 9) which remain on the Amanita cap; a type of ornamentation of some spore walls (Fig. 63).

woody (of the cap and stalk): tough, unbending, and usually difficult to cut with a knife, as in *Fomes* and *Ganoderma* in the ploypores.

zonate (of the cap): having concentric bands or zones of color.

Bibliography

Annon. *Wild Mushroom Cookery.* Ore. Mycological Soc., Portland.

Annon 1963. *Kitchen Magic with Mushrooms.* Mycological Soc. of San Francisco, Berkeley, Calif.

Bigelow, Howard E. and Alexander H. Smith 1969. The status of *Lepista*—A new section of *Clitocybe. Brittonia,* 21:144–177.

—— 1970. *Omphalina* in North America. *Mycologia,* 62:1–32.

Coker, W. C. 1917. The Amanitas of the eastern United States. Elisha Mitchell Scientific Soc., Vol. 33:1–88.

—— 1923. *The Clavarias of the United States and Canada.* The Univ. of North Carolina Press, Chapel Hill, N.C.

—— and Couch, J. N. 1928. *The Gastromycetes of the Eastern United States and Canada.* The Univ. of North Carolina Press, Chapel Hill, N.C.

Doty, M. S. 1944. *Clavaria,* the species known from Oregon and the Pacific Northwest. Ore. State College Press.

Harrison, Kenneth A. 1961. The Stipitate Hydnums of Nova Scotia. Canadian Dept. of Agriculture, Publ. 1099, Ottawa, Canada.

Hesler, L. R. 1967. *Entoloma* in southeastern North America. Verlag von J. Cramer.

—— 1969. North American Species of *Gymnopilus.* Hafner Publ. Co., Darien, Conn.

—— and A. H. Smith 1965. North American Species of *Crepidotus.* Hafner Publ. Co., Darien, Conn.

Long, W. H. 1942. Studies in the Gastromycetes V. A white *Simblum. Mycologia,* 34:128–131.

Lowe, J. L. 1957. *Polyporaceae* of North America, the Genus Fomes. State Univ. College of Forestry at Syracuse Univ.

McKenney, M. 1962. *The Savory Wild Mushroom.* Univ. of Wash. Press.

Miller, Orson K. Jr. 1964. Monograph of *Chroogomphus* (Gomphidiaceae). *Mycologia,* 56:526–549.

—— 1968. A revision of the genus *Xeromphalina. Mycologia,* 60:156–188.

—— 1970. The genus *Panellus* in North America. *Michigan Botanist:* 9:17–30.

—— 1971. The genus *Gomphidius* Fries, with a revised description of the Gomphidiaceae and keys to the genera. *Mycologia* 53:. 1129–1163.

—— and Linnea Stewart 1971. Monograph of the genus *Lentinellus. Mycologia,* 63:333–369.

Ola'h, G. 1969. Le genre *Panaeolus,* Essai taxinomique et physiologique. Memoire Hors-Serie 10 de la Revue de Mycologie.

Overholts, L. O. 1953. *The Polyporaceae of the United States, Alaska, and Canada.* Univ. of Wash. Press.

Pilat, A. 1951. The Bohemian species of the genus *Agaricus.* Acta Musei Nationalis, Prague.

Pomerleau, R. 1966. Les Amanites du Quebec. *Naturaliste Canada,* 93:861–887.

Robbers, J. E., L. R. Brady and V. E. Tyler, Jr. 1964. A chemical and chemotaxonomic evaluation of *Inocybe* species. *Lloydia,* 27:192–202.

—— V. E. Tyler, and G. M. Ola'h 1969. Additional evidence supporting the occurrence of Psilocybin in *Panaeolus foenisecii. Lloydia,* 32:399–400.

Shaffer, R. L. 1957. *Volvariella* in North America. *Mycologia,* 49:545–579.

—— 1964. The subsection *Lactarioideae* of *Russula. Mycologia,* 56:202–231.

Simons, Donald M. 1971. The Mushroom Toxins. *Delaware Medical Journal,* 43:177–187.

Singer, R. and A. H. Smith 1943. A monograph on the genus *Leucopaxillus* Boursies. Papers of the Mich. Academy of Sciences, Arts, and Letters, 28:85–132.

Smith, Alexander H. 1957. *North American Species of* Mycena. Univ. of Mich. Press, Ann Arbor, Mich.

—— 1949. *Mushrooms in their Natural Habitats.* Sawyers Inc., Portland, Ore.

—— 1951. The North American species of *Naematiloma. Mycologia,* 43:467–521.

—— 1951. *Puffballs and their Allies in Michigan.* Univ. of Mich. Press, Ann Arbor, Mich.

—— 1960. *Tricholomopsis* (Agaricales) in the western hemisphere. *Brittonia,* 12:41–70.

—— 1963. *The Mushroom Hunter's Field Guide,* revised and enlarged. Univ. of Mich. Press, Ann Arbor, Mich.

—— and L. R. Hesler 1963. North American Species of *Hygrophorus.* Univ. of Tenn. Press, Knoxville, Tenn.

—— and L. R. Hesler 1968. The North American species of *Pholiota.* Hafner Publ. Co., Darien, Conn.

—— and R. Singer 1945. A monograph of the genus *Cystoderma.* Papers of the Mich. Academy of Science, Arts, and Letters, 30:71–124.

Smith, Alexander H. and R. Singer 1964. A Monograph on the genus *Galerina* Earle. Hafner Publ. Co., Darien, Conn.

—— and H. D. Thiers 1971. *The Boletes of Michigan.* The Univ. of Mich. Press, Ann Arbor, Mich.

—— H. D. Thiers, and R. Watling 1966. A preliminary account of the North American species of *Leccinum,* section *Leccinum. Michigan Botanist,* 5:131–179.

—— H. D. Thiers, and R. Watling 1967. A preliminary account of the North American species of *Leccinum,* sections *Luteoscabra* and *Scabra. Michigan Botanist,* 6:107–154.

Smith, Helen V. 1954. A revision of the Mich. species of *Lepiota. Lloydia,* 17:307–328.

Snell, W. H. and E. A. Dick 1970. *The Boleti of Northeastern North America.* Verlag von J. Cramer.

Thiers, H. D. and J. M. Trappe 1969. Studies in the genus *Gastroboletus. Brittonia,* 21:244–254.

Wells, V. L. and P. E. Kempton 1968. A preliminary study of *Clavariadelphus* in North America. *Michigan Botanist,* 7:35–57.

White, V. S. 1901. The *Tylostomaceae* of North America. Bulletin of the Torrey Botanical Club, 28:421–443.

Zeller, S. M. and A. H. Smith 1964. The genus *Calvatia* in North America. *Lloydia,* 27:148–186.

Index

356

List of Abbreviated Names of Authors of Fungi

Alb. and Schw.	Albertini and Schweinitz	Krombh.	J. V. von Krombholz
Atk.	F. Atkinson	Kuehn.	R. Kuehner
Berk. and Curt.	Berkeley and Curtis	Lev.	J. Leveille
Berk. and Br.	Berkeley and Broome	Lindbl.	M. A. Lindblad
Berk.	J. M. Berkeley	L.	C. von Linneaus
Bolt.	Bolton	Lund.	S. Lundell
Boud.	E. Boudier	Mass.	G. Massee
Bres.	A. J. Bresadola	Mich.	P. Micheli
Britz.	M. Britzelmayr	Morg.	A. P. Morgan
Brond.	L. Brondeau	Mull.	O. F. Muller
Bull.	P. Bulliard	Murr.	W. A. Murrill
Burl.	G. Burlingham	Nannf.	J. A. Nannfeldt
Casp.	J. Caspary	Opat.	W. Opatowski
Ces.	V. Cesati	Pat.	N. Patouillard
Cke.	M. C. Cooke	Pers.	C. H. Persoon
Curt.	M. A. Curtis	Phill.	W. Phillips
Czern.	B. Czerniaier	Pk.	C. H. Peck
D.C.	A. DeCandolle	Quel.	L. Quelet
DeNot.	G. DeNotaris	Rick.	A. Ricken
Dicks.	J. Dickson	Romag.	H. Romagnesi
Ellis and Ever.	Ellis and Everhart	Sacc.	P. A. Saccardo
Ell.	J. B. Ellis	Schaeff.	J. C. Schaeffer
Fr.	E. M. Fries	Schroet.	J. Schroeter
Gilb.	E. J. Gilbert	Schumm.	H. D. Schumacher
Gill.	C. C. Gillet	Schw.	L. D. Schweinitz
Gmel.	J. F. Gmelin	Scop.	J. A. Scopoli
Hark.	H. W. Harkness	Secr.	L. Secretan
Hedw.	J. Hedwig	Sow.	J. Sowerby
Henn.	P. Hennings	Trott.	A. Trotter
Holm.	T. Holmskjøld	Tul.	E. L. Tulasne
Hook.	W. J. Hooker	Underw.	L. M. Underwood
Huds.	W. Hudson	Vahl.	M. Vahl
Jacq.	N. Jacquin	Vitt.	C. Vittadini
Jung.	F. F. Junghuhn	Viv.	D. Viviani
Kalch.	K. Kalchbremer	Wallr.	C. F. Wallroth
Karst.	P. Karsten	Weinm.	J. A. Weinmann
Kauf.	C. Kauffman	Willd.	C. Willdenaw
Kl.	J. F. Klotzch		

Credits

All photographs are by Orson K. Miller, Jr. except where otherwise noted.
5, 10, Howard E. Bigelow; 14b, Kenneth Harrison; 25, 26, Robert T. Orr; 28, Mrs. Robert Scates; 32, Robert T. Orr; 35, Mrs. Robert Scates; 38, Howard E. Bigelow; 39, 40, Robert T. Orr; 47, 48, Howard E. Bigelow; 54, 55, Robert T. Orr; 56, Howard E. Bigelow; 57, Kenneth Harrison; 61, Howard E. Bigelow; 66, Emil Javorsky; 69, Mrs. Robert Scates; 73, Howard E. Bigelow; 77, Robert T. Orr; 80, Howard E. Bigelow; 93, Edward Degginger; 94, Mrs. Robert Scates; 95, Robert T. Orr; 102, Emil Javorsky; 110, 111, Mrs. Robert Scates; 115, Howard E. Bigelow; 116, Kenneth Harrison; 126, Mrs. Robert Scates; 127, 128, Howard E. Bigelow; 129, Emil Javorsky; 137, 138, Howard E. Bigelow; 145, Kenneth Harrison; 146, Howard E. Bigelow; 147, Robert T. Orr; 153, Howard E. Bigelow; 154, Robert T. Orr; 158, Mrs. Robert Scates; 159, Howard E. Bigelow; 160, 165, Robert T. Orr; 170, Howard E. Bigelow; 176, Kenneth Harrison; 179, 181, Mrs. Robert Scates; 185, Howard E. Bigelow; 186, Mrs. Robert Scates; 195, Kenneth Harrison; 197, Robert T. Orr; 201a, Emil Javorsky; 207, Robert T. Orr; 208, Howard E. Bigelow; 212, Mrs. Robert Scates; 219, 221, Howard E. Bigelow; 236, Kenneth Harrison; 243, 245, Robert T. Orr; 246, 248, 250, Howard E. Bigelow; 253, Kenneth Harrison; 255, Howard E. Bigelow; 257, Edward Degginger; 259, Robert T. Orr; 260, Mrs. Robert Scates; 261, Howard E. Bigelow; 262, 263, Kenneth Harrison; 265, Howard E. Bigelow; 268, Kenneth Harrison; 269, Mrs. Robert Scates; 270, Howard E. Bigelow; 273, 274b, Robert T. Orr; 275a, Mrs. Robert Scates; 276, 279, Howard E. Bigelow; 287, 289, Howard E. Bigelow; 291, 292, Robert T. Orr; 294, Mrs. Robert Scates; 306, Howard E. Bigelow; 308, Robert T. Orr; 309, Harry Thiers; 310, 316, Howard E. Bigelow; 320, Edward Degginger; 323, 326a, Mrs. Robert Scates; 326b, Kenneth Harrison; 331, Edward Degginger; 333, 334, Mrs. Robert Scates; 335, Howard E. Bigelow; 340, Kenneth Harrison; 341, 349, Howard E. Bigelow; 351a, Kenneth Harrison; 351b, Robert T. Orr; 354, Howard E. Bigelow; 365, Mrs. Robert Scates; 369, 376, Kenneth Harrison; 378, Mrs. Robert Scates; 379, Robert T. Orr; 382, Howard E. Bigelow; 383, 390, Mrs. Robert Scates; 393, Howard E. Bigelow; 397, Mrs. Robert Scates; 399, Kenneth Harrison; 401, 405, Robert T. Orr; 411, 416, Howard E. Bigelow.
 Drawings are by Robyn Morgan (illustrated glossary) and Roberta Savage (title page and section openings).

This book was planned and produced by Paul Steiner and the staff of Chanticleer Press.

Editor: Milton Rugoff
Associate Editor: Connie Sullivan
Art Director: Ulrich Ruchti, assisted by Roberta Savage
Production: Gudrun Buettner, assisted by Helga Lose
Printed by Dai Nippon Printing Co., Ltd., Tokyo, Japan